Robots and Robotics

Robots and Robotics

Principles, Systems, and Industrial Applications

Mark R. Miller
The University of Texas at Tyler

Rex Miller
State University College at Buffalo

New York Chicago San Francisco
Athens London Madrid
Mexico City Milan New Delhi
Singapore Sydney Toronto

Library of Congress Control Number: 2017941572

McGraw-Hill Education books are available at special quantity discounts to use as premiums and sales promotions or for use in corporate training programs. To contact a representative, please visit the Contact Us page at www.mhprofessional.com.

Robots and Robotics: Principles, Systems, and Industrial Applications

1 2 3 4 5 6 QVS 21 20 19 18 17

ISBN 978-1-259-85978-6
MHID 1-259-85978-9

This book is printed on acid-free paper.

Sponsoring Editor
Michael McCabe

Editing Supervisor
Stephen M. Smith

Production Supervisor
Lynn M. Messina

Acquisitions Coordinator
Lauren Rogers

Project Manager
Patricia Wallenburg, TypeWriting

Copy Editor
James K. Madru

Proofreader
Claire Splan

Art Director, Cover
Jeff Weeks

Composition
TypeWriting

To Patricia Ann Miller, wife, mother, and a great teacher . . .

About the Authors

Mark R. Miller is a Professor of Industrial Technology and Chair of the Technology Department at The University of Texas at Tyler. He has authored or co-authored more than 40 technical books and numerous technical articles. He currently serves as the Chairman of the Association of Technology, Management, and Applied Engineering (ATMAE) Board of Certification, on which he has assisted with the development of five new certification exam programs. Dr. Miller serves as the faculty advisor for the student chapter of the Society of Manufacturing Engineers and is the co-trustee for the Delta Gamma Chapter of Epsilon Pi Tau (honor society for technology professionals). He also serves as the Director of the Texas Productivity Center and is a certified Lean Six Sigma Black Belt. Dr. Miller has received numerous teaching and service awards throughout his career.

Rex Miller is Professor Emeritus of Industrial Technolgy at the State University College at Buffalo (New York), where he taught technical curriculums for more than 40 years. Dr. Miller has authored or co-authored more than 100 texts for vocational and industrial arts programs.

Contents

Preface

This book was written with a number of purposes in mind. While many people want to know more about robots and robotics, most do not have the engineering or technical background in pneumatics, hydraulics, and electronics to understand what a robot is all about internally or conceptually. Further, some people who do have the necessary background do not know where to start in looking at the future of robots in their own trade or profession. Thus *Robots and Robotics: Principles, Systems, and Industrial Applications* is intended as a comprehensive introduction to the topic.

This book is suitable for use in a first course on robotics for students in industrial electronics programs as well as in mechanical, manufacturing, or industrial technology. It is also designed to serve as a source of information on robots and robotics for robot hobbyists, professional machinists, electricians, and electronics technicians. It provides a broad view of the subject without overwhelming the reader with technical detail or jargon.

The text relies on the real world of robots to bring excitement to its pages. Up-to-date examples of industrial robots and practical applications are emphasized throughout the book. Ample illustrations are provided to clarify the discussion and to aid readers in recognizing robot parts and movements. End-of-chapter key terms sections and a comprehensive glossary at the end of the book are included to make words applicable to robots easy to understand and master. Chapter review questions and their answers are also in the book.

Whether beginners or individuals who have worked with machines for some time, readers will gain not only fundamental knowledge but also new insights into the complex field of robotics. Chapter 7 on putting a robot to work will give readers a clear idea of what these machines can and cannot do. While robots have a long way to go before they can do all the things we dream they will do, they are an exciting and dynamic force that *must* be seriously considered by everyone, no matter his or her occupation or interests. The goal of this book is to provide the necessary information in such a way that it can be used effectively by readers of all skill levels and backgrounds.

The organization of this book is flexible and allows for individual preferences in the order of study. Chapter 1 provides an overview of robotics. It includes a definition of what constitutes a robot and outlines both the positive and negative aspects of robots, as well as their interaction with human labor. Chapter 1 also contains a brief review of component programs, languages, and microprocessors.

Chapter 2 identifies various types of robots. Parts of the robot and robotic motion capabilities are also examined. Chapter 3 covers the mechanical components of robots, such as drive systems, pumps, and motors. Sensor types and sensing capabilities are discussed in Chapter 4.

Chapter 5 covers control methods for robots, including various methods of robot programming. Chapter 6 emphasizes the computer working with the robot. It also presents a better understanding of programming and control of the robot. Vision for the robot as well as object recognition is covered. Chapter 7 examines the topic of robots in industry and the future of robots and robotics. Chapter 8 provides a list of manufacturers and equipment, along with specifications, descriptive information, and illustrations from manufacturers' catalogs. Chapter 9 includes a comprehensive set of principles and practices for troubleshooting electronic controls and electric motors. Chapter 10 discusses robots of yesterday and tomorrow.

In addition to a glossary, six appendices are included to enlarge the book's usefulness. Appendix A provides a conversion chart so that international references to various values can be converted to the U.S. system with as little effort as possible. Appendix B provides a more comprehensive look at the newest robots and their controllers. Appendix C deals with the opportunities for robot repairpersons and robotics engineers. Appendix D illustrates electronics and fluid power schematic symbols. Appendix E is a cross-comparison that shows the reader the vast number of robots available; some of the information goes back to the 1970s when robots burst on the scene the world over. Appendix F provides formulas and conversion factors often needed in the work world that includes robots and their programming, control, and design. The robot operator and/or robot technician should have a scientific-type calculator handy to solve some of the problems that arise in dealing with robots and robotics. Mathematics demands a more important role in the operation and design of robots and their installation and utilization.

Mark R. Miller
Rex Miller

Acknowledgments

No book is ever completed without the energy and efforts of many people. This book is no exception, and we would like to thank the many people, both named and unnamed, whose contributions made this book a reality.

Throughout the various stages of writing this book, we received helpful comments and suggestions from a number of people, most of them professional teachers or long-time technicians who specialize in the robotics field of endeavor. Some preferred that we mention their schools and not their names:

- MacArthur State Technical College, Opp, AL
- Carroll County Area Vocational-Technical School, Waco, GA
- Amarillo College, Amarillo, TX
- Jefferson County Community College—Southwest, Louisville, KY
- Schoolcraft College, Livonia, MI
- Technical College of Alamance, Haw River, NC
- Georgia State Department of Education, Atlanta, GA
- College of DuPage, Glen Ellyn, IL
- Western Iowa Tech Community College, Sioux City, IA

Many businesses were also helpful in supplying the information and illustrations so necessary for making this book worthwhile: Automatix, Inc.; Binks Manufacturing Company; Camco/Commercial Cam Division; Cincinnati Milacron/Industrial Robot Division; Compact Air Products, Inc.; Cybotech Industrial Robots; Elicon; Emerson Electric Company; ESAB North America, Inc.; Fared Robot Systems; Feedback, Inc.; Feedmatic-Detriot, Inc.; GCA Corporation/ Industrial Manufacturing Systems Products Division; International Robomation/ Intelligence; L. S. Manufacturing, Inc.; Mack Corporation; Microbot; Microswitch, a Honeywell Division; New Jersey Zinc Company; Pick-O-Matic Systems; PRAB Robots, Inc.; RCA, Radio Corporation of America; Rhino Robots, Inc.; Schrader-

Bellows, a Division of Parker-Hannifin; Seiko Instruments USA, Inc.; Thermwood Robotics Division, Thermwood Corporation; Unimation, Inc., a Westinghouse Company; Warner Clutch and Brake Company; and Yaskawa America, Inc.

Robots and Robotics

CHAPTER 1
Introduction

Performance Objectives

After reading this chapter, you will be able to:

- Understand how robots are named.
- Know the importance of computer programs in controlling robot movement.
- Know that robots can use a number of languages.
- Know the role of microprocessors in robots.
- Know some of the positive and negative aspects of robots.
- Know how robots alleviate human drudgery in everyday work routines.
- Know the many applications of robots.
- Identify and discuss the key terms used in the chapter.
- Answer the review questions at the end of the chapter.

Robots have captured the imagination of writers and movie producers for some time. Only recently have they become useful in the production of quality products, and it is here that the greatest amount of time and effort is being spent in robot development today. The ability to produce quality products is of utmost importance because consumers benefit and manufacturers stay in business.

Definition

What is a *robot*? There are a number of definitions, but a simple definition that serves our purpose here is: A *robot* is a *reprogrammable, multifunctional manipulator* designed to move material, parts, tools, or specialized devices through variable programmed motions for the performance of a variety of tasks. A robot also can be classified as a system that simulates human activities based on computer instruction (Figure 1-1). These definitions allow us to take a closer look at the system that will produce the actions needed for a robot to perform tasks that humans can do. The computer is the secret to the system because it is, in effect, the brain of the device or system. The computer is an integral part of any robot and must be taken into consideration whenever a robot is studied as a device, a system, or a means of eliminating human effort.

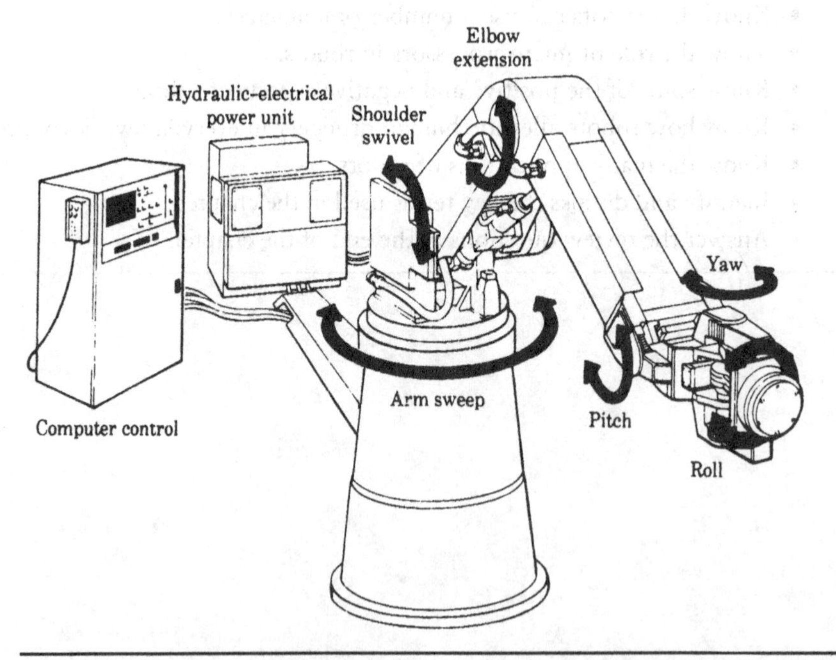

Figure 1-1 Complete industrial root system. (*Courtesy of Cincinnati Milacron.*)

Keep in mind that some nonintelligent robots do not use electronics for brains. A lot of pick-and-place robots are cam controlled (Figure 1-2). They simply pick up their load and place it elsewhere. Loading and unloading tasks are usually performed by these types of robots.

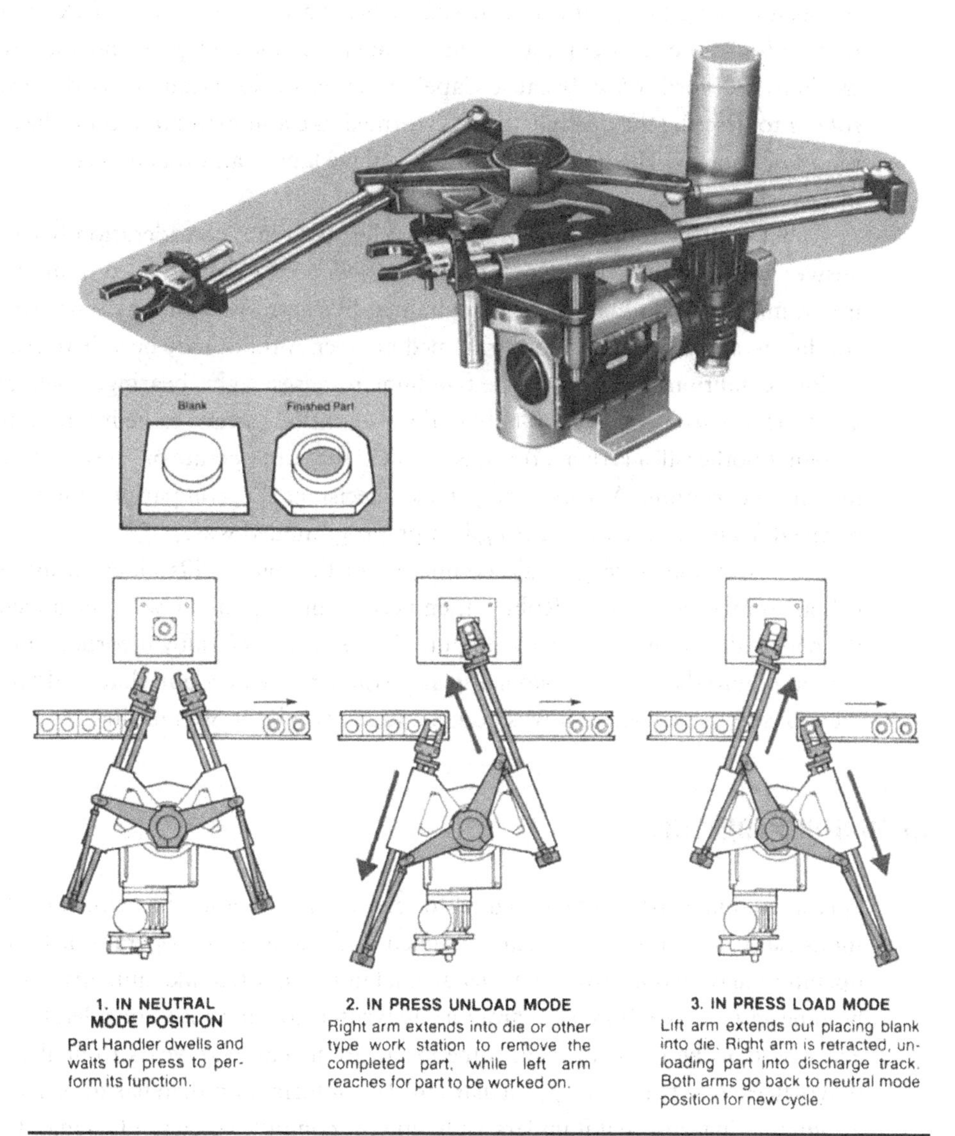

Blank
Finished Part
1. IN NEUTRAL MODE POSITION
Part Handler dwells and waits for press to perform its function.
2. IN PRESS UNLOAD MODE
Right arm extends into die or other type work station to remove the completed part, while left arm reaches for part to be worked on.
3. IN PRESS LOAD MODE
Lift arm extends out placing blank into die. Right arm is retracted, unloading part into discharge track. Both arms go back to neutral mode position for new cycle.

FIGURE 1-2 Pick-and-place robot used to load a press. (*Courtesy of Pick-O-Matic Systems.*)

Robot History

A bit of history will place robots into perspective and help to explain their popularity. Robots are a relatively recent development. Science fiction has featured robots for a long time, but it was in 1921 when Karl Capek wrote *R.U.R.* that the term *robot* came into common usage. Capek's book and play introduced the world to the word *robot.* Because Capek was Czechoslovakian, he used the word *robota* to describe the machine that performed like a human but did not have the senses of humans. The term *robota* means "slave labor" and was reduced to *robot* in English.

What is a true robot? A few ideas must be taken into consideration when you answer this question. One idea is that a *robot* is a device or system that is programmed by a human to perform human-like acts. A robot may sense various conditions and react in a preprogrammed manner. A robot may be able to react to various conditions in terms of the five human senses: sight, hearing, smell, taste, and touch. *Sensors* are available that allow all of these senses to be inserted into a system. Another idea is that a robot is a system that can operate on its own without human supervision. A robot may make decisions by comparing information received from sensors and reacting in a preprogrammed way.

The invention of large-scale computers in the mid-1950s aided in allowing robots to become a reality. Robots then became more popular with the advent of the personal computer or microcomputer in the early 1970s. By incorporating the computer into the robotic system, it was possible to create a unit that could move, talk, lift things, see where it was going, and know what it was feeling.

Computer Programs

Special computer programs designed for specific jobs are used to control robots. Industrial robots, for instance, are designed to do a particular operation. This one operation may be done over and over again, but the robot, unlike humans, does not become fatigued or bored. A *program* is written to take into consideration the exact tasks to be performed. In some instances, it may take years to analyze the moves needed to perform a particular job. This information then has to be fed into a computer in terms that it understands, and the computer then sends signals to the robot so that it will perform exactly as desired. Figure 1-3 shows two robots that are used to load and unload. By coordinating the motions of the two robots, the idle time of machine A is kept to less than 1 second. The alternate path capability

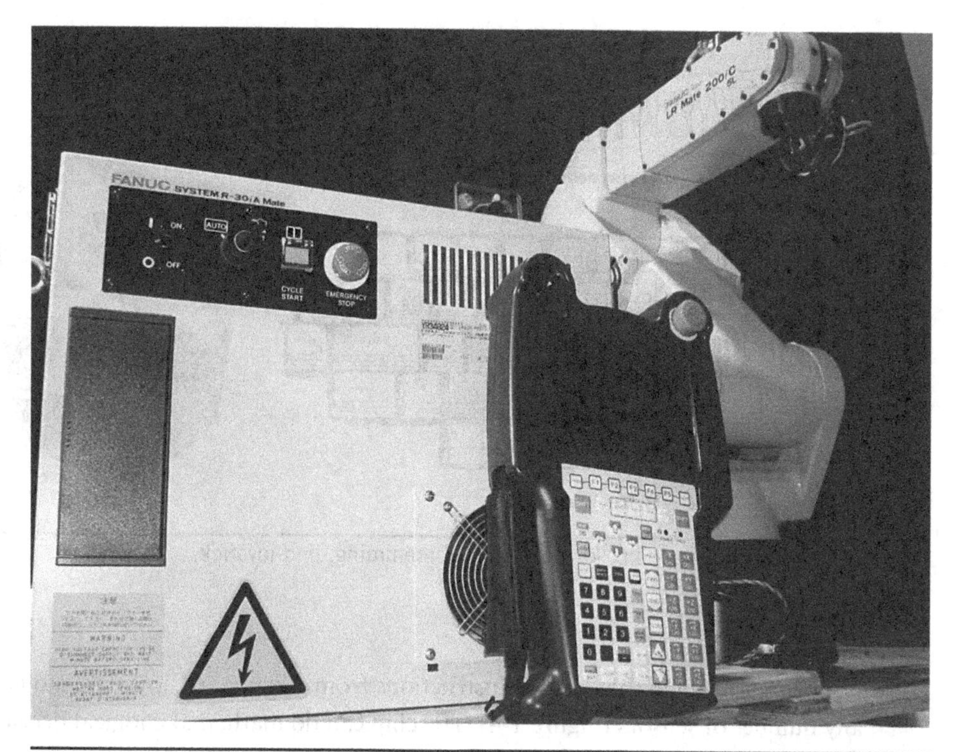

Figure 1-3 Teach pendant, controller, and LR Mate 200iC robot. (*Courtesy of FANUC.*)

built into the command module allows a number of different of sensors to detect reject parts and place them in an alternate location.

Languages

Robots use a number of computer *languages*, which are designed for specific operations. Some of the languages used with robots are AL (Stanford's Artificial Intelligence Laboratory language), VAL, AML (developed by IBM), Pascal, and ADA. These names and more on languages and programming are discussed later in this book.

Microprocessors

A *microprocessor* is just that. *Micro* means "small," and a *processor* is "a device that can process things." In this case, it is used to process information fed to it from an external source. We use the term today to describe a special-purpose chip

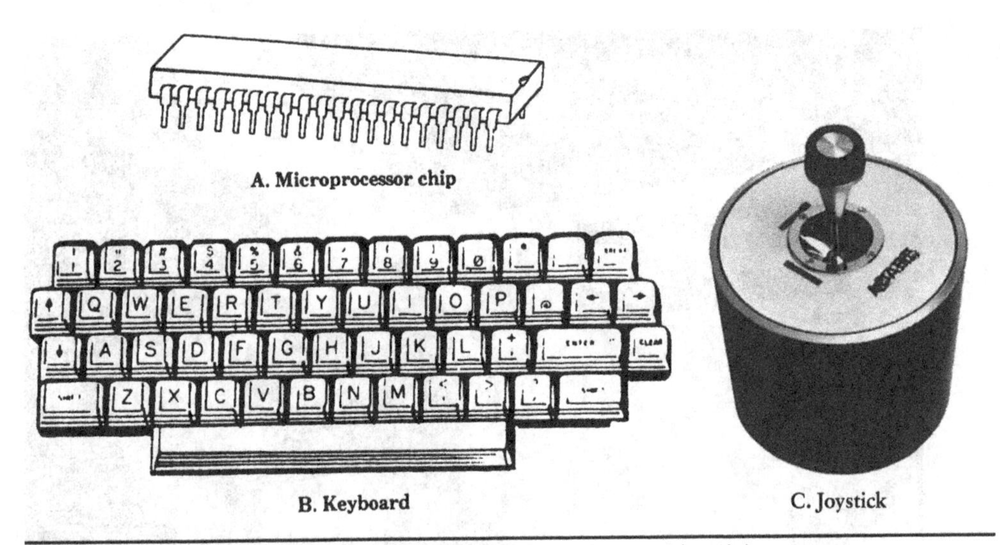

FIGURE 1-4 Microprocessor chip, keyboard for programming, and joystick.

or portion of a chip that gets its instructions from a keyboard, joystick, mouse, or any number of sensors (Figure 1-4). The chip can do math, make logical decisions, and work with words and symbols.

A robot usually has a dedicated microprocessor designed for its special job requirements. It may be a simple chip, or it may be a large mainframe computer.

Without the microprocessor, the robot is limited in its application. With the microprocessor, it is possible to have the robot operate alone without connecting wires, other than for power purposes. More on the operation of microprocessors will be found in Chapter 6.

Positive Aspects of Robots

In many instances, robots can perform work more efficiently than humans. They can work seven days a week, twenty-four hours a day, and thirty days a month without becoming bored or fatigued. The quality of their work can be checked and corrected immediately if found to be defective. Operating costs are low, and downtime is minimal. Thousands of people will be needed in the future to design, repair, and install robots. New jobs will be created, and new training programs will have to be developed to improve the use of robots. Figure 1-5 shows how a robot is used to pick up air-conditioning units from an assembly line and place them into shipping cartons, a job that would quickly tire and bore human workers.

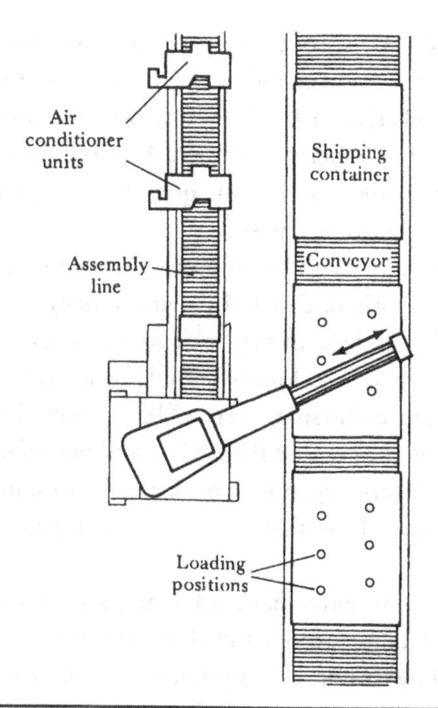

FIGURE 1-5 Robot used to pick up finished air-conditioning units and pack them in shipping containers. (*Courtesy of PWS-Kent.*)

Negative Aspects of Robots

Robots can replace humans in the labor force. They require a higher level of maintenance than do most existing jobs. Therefore, they require retraining or replacement of the humans now employed in that job. The initial cost of robots is excessive for small firms. The technology is relatively untested at this time, and downtime is expensive.

Robots, Hard Automation, and Human Labor

For many years the "American way" was the best way. This applied to manufactured goods, standard of living, and everything else that we considered a part of the American way. The United States was a large manufacturing nation that reached its level of operation during World War II. The pent-up demand for consumer goods after World War II presented a challenge to the manufacturing system as it then existed. More machines were made, and more people were employed to meet the demand for industrial products. As demand slackened, emphasis on product quality increased. People demanded quality instead of quantity. This became evident during the oil crisis in 1974. Smaller cars were imported from Japan to fill the need for more fuel-efficient vehicles. As more and more Americans began driving Japanese cars, they noticed the quality of the product and demanded the same of their American counterparts.

As the quality of American life improved, more money was needed to support it. This meant that workers demanded more money to meet their expectations. Organized labor demanded, on behalf of workers, more and more until a point was reached where it was no longer feasible to manufacture certain products in this country. Foreign companies were able to meet the demand for consumer products at lower prices because their labor and manufacturing costs were lower. American manufacturers began to look for ways to improve quality and reduce cost per item produced so that they could compete effectively with foreign manufacturers.

Inasmuch as robots have many advantages over human labor, it was only natural that manufacturers looked in that direction to satisfy their labor needs. Figure 1-6 shows how robots can do routine repetitive operations without tiring. They do, however, occasionally break down.

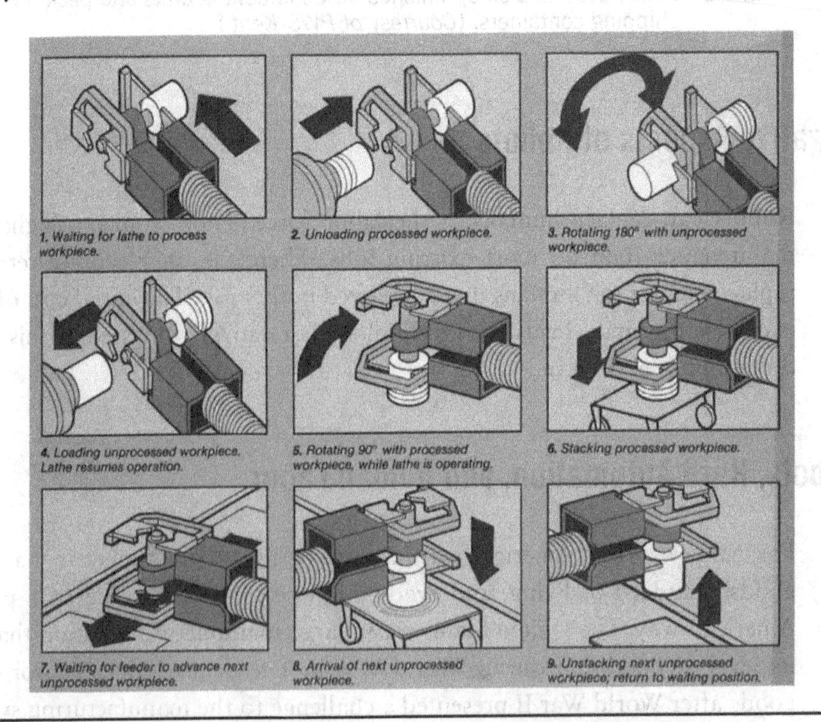

1. Waiting for lathe to process workpiece.
2. Unloading processed workpiece.
3. Rotating 180° with unprocessed workpiece.
4. Loading unprocessed workpiece. Lathe resumes operation.
5. Rotating 90° with processed workpiece, while lathe is operating.
6. Stacking processed workpiece.
7. Waiting for feeder to advance next unprocessed workpiece.
8. Arrival of next unprocessed workpiece.
9. Unstacking next unprocessed workpiece; return to waiting position.

FIGURE 1-6 Robot gripper at work. (*Courtesy of Camco, Commercial Cam Division, Emersion Electric Co.*)

Robots will work in unpleasant locations. Health hazards are not of concern to robots. Special safety equipment is not required for robots to spray paint, weld, or handle dangerous chemicals. All this adds up to reduced production costs. As the day progresses, a tired worker has a tendency to pay less attention to detail. This inattention results in a finished product of lower quality. This is especially noticeable in automobiles, where spray paint can run or sag and weld joints are not made perfectly. The panels on the car may not be aligned, and the finished product may not operate properly, resulting in a very unsatisfied customer.

Robots, however, do not tire or change their work habits unless programmed to do so. They maintain the same level of operation throughout the day. With the use of robots, it is possible for American manufacturers to compete against the lower labor costs in foreign countries. The initial investment is the only problem. After the initial investment, the overall operation of the production line is reduced or held constant. A teach pendant (Figure 1-7) can be used with small robots to train humans to operate and maintain robots.

There are some advantages to human labor. Humans can start to work immediately on learning the task. Humans can also be laid off if they are no longer needed. There are, of course, some costs, such as unemployment compensation and severance pay in terminations. If a company has very little

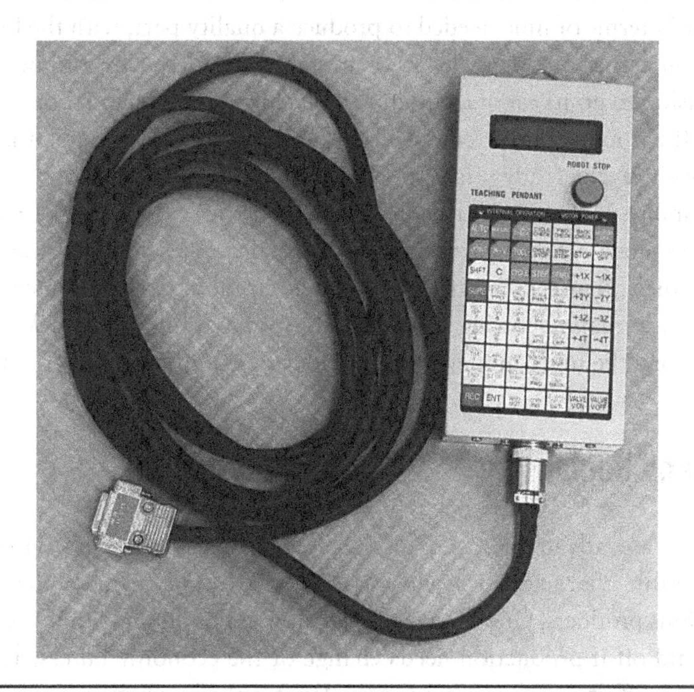

Figure 1-7 A teach pendant is used by humans to teach or train robots by programming the robot's memory. (*Courtesy of The University of Texas at Tyler.*)

money to invest in new equipment, using human labor is usually better. Robots are expensive initially.

Automated machinery will perform various operations with some degree of quality and dependability. However, if a design changes, it becomes expensive to replace the old equipment and buy new. Automated machinery is the best choice if you are going to produce a product for a long time. The investment in *hard automation* is warranted in such a case. However, if the design of the product is subject to quick change, it is best to choose robots to do the work. Robots are flexible and can be reprogrammed to do something else.

Return on investment (ROI) is the primary consideration when deciding whether to use human labor, hard automation, or robots. The robot is only part of an automated system, and probably the cheapest part of that is you. You have to have an in-feed device, an out-feed device (or parts delivery and removal system), and end-of-arm tooling, robot envelope security, and additional add-ons. All this has to be part of the equation when you ask: will the robot pay for itself in five years? Average costs can be compared here for the sake of discussion only. The cost of an industrial robot is about $70,000, whereas the cost of an automated machine designed and installed is about $225,000. Human labor, including the benefit package, will average about $30,000 a year. As far as production time is concerned, the automated machine will produce the product fastest. The robot is next in terms of time needed to produce a quality part, with the human coming in last in terms of time. The ROI must be considered by the person making the decision to go to robots, automated machines, or human labor.

If the money invested is tied up for more than five years, it is considered too long. The chief advantage of a robot in this consideration and choice is its flexibility. If the product design is changed, the robots can be reprogrammed to accommodate the design change. It usually takes humans much longer to be retrained. Figure 1-7 is an example of a retraining system. Time is money when it comes to manufacturing or production of anything. The quicker the line can be retooled, the more money that can be made and the better is the ROI.

Robots and Humans

There are advantages and disadvantages to using robots. Humans also have favorable characteristics when it comes to using them in the manufacture of various products. One of the main advantages of human labor is that a person can be laid off if production needs change or the economy falters. The robot cannot be laid off. It goes on costing every day it is there, whether it is being used or not.

It is an investment and must be considered as any other investment. The robot has to be paid for once the contract for its purchase has been issued.

Robots Versus Humans

The robot-versus-human question becomes a major consideration if robots are installed in great numbers and replace many people in a particular location. This can cause turmoil in the job market. The number of robots in use today is very small in respect to what is expected in 10 to 20 years. If 20,000 people are employed by a particular corporation, the installation of a few robots will not cause too much concern. However, if a new plant is built and robots are installed during construction, there is still very little in the way of labor problems.

However, estimates of the number of U.S. workers who will be displaced by robots range from 250,000 to 2 million within 10 years. This can accelerate or diminish according to the economy. Such displacement of workers may create problems with the American workforce. There is much speculation as to how the rapid deployment of robots will affect the social order of American industry. Many workers will have to be retrained for higher-level jobs, creating a demand for more people to be employed in retraining programs. The transition period likely will last 10 to 15 years. During this time, it will be necessary to absorb displaced workers in training programs or other types of industries. American service industries are growing rapidly, and many displaced workers will be absorbed in this type of work.

The main disadvantage of robots is downtime. When a robot breaks down, an entire plant's production schedule may be affected, causing problems with sales and distribution as well as production. One machine off-line in a production line can result in many more being made nonproductive down the line. You must maintain a proper inventory of repair parts and employ properly trained technicians to keep the equipment operating at peak efficiency. A replacement robot may be kept ready (an expensive option), or a stock of major components may be kept on hand.

Competent personnel also must be available to work quickly and effectively. In summary, robots, like other equipment, offer advantages and disadvantages. Each advantage and disadvantage must be considered with respect to the robot's intended use before a decision to acquire is made. Most industries that have investigated the use of robots have found that the advantages outweigh the disadvantages, and as a result, more robots will be making more things in the future.

Industrial Robot Applications

No doubt about it, robots are here to stay. A good example of a robot being put to use is shown in Figure 1-8, a basic robot system used for spray painting. It consists of a manipulator, control console, and hydraulic power supply. Electric-powered robots for painting can be dangerous because the fumes can easily be ignited by a spark.

Other examples of how industrial robots are used include the following applications:

- **Machine loading and unloading.** Placing parts where they are needed for machining or shipping.
- **Materials handling.** Packing parts or moving pallets.
- **Fabrication.** For making investment castings, grinding operations, and deburring; for water-jet cutting, wire harness manufacture, applying glues, sealers, putty, and caulks; for drilling, fettling, and routing.
- **Spray painting.** Painting cars, furniture, and other objects.
- **Welding.** Welding cars, furniture, and steel structures.
- **Assembly.** Electronics, automobiles, and small appliances.
- **Inspection and testing.** Quality control looking for surface and interior defects, using vision sensors and feelers.

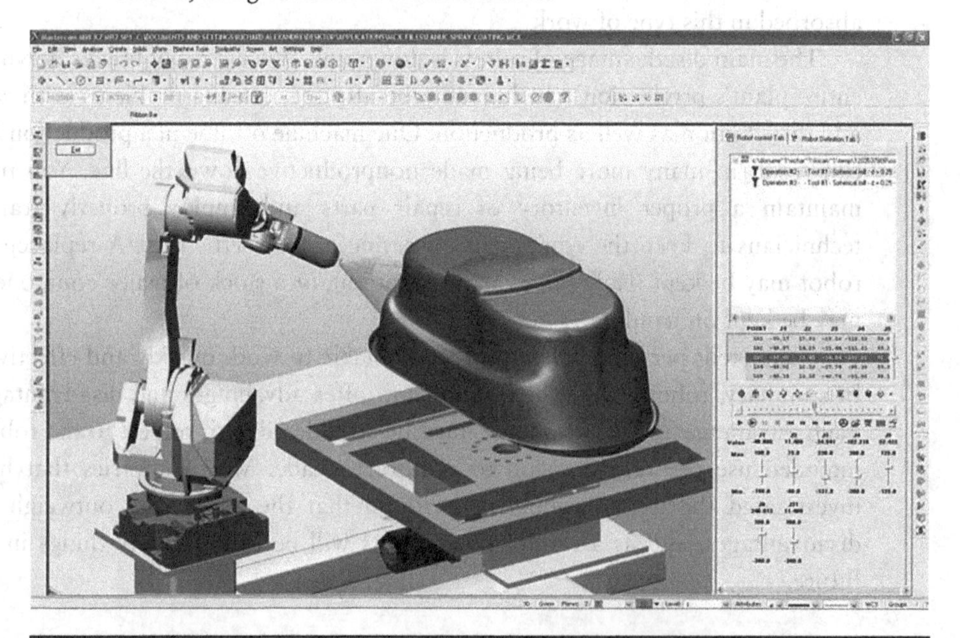

Figure 1-8 Robot spray painting a manufactured product. (*Courtesy of FANUC.*)

Summary

A *robot* is defined as a programmable multifunctional manipulator designed to move material, parts, tools, or specialized devices though variable programmed motions for the performance of a variety of tasks. The computer is the brain of the robot.

Robots is a relatively recent development. In 1921, Karl Capek used the Czech word *robota* in a book and play called *R.U.R.* The word was changed to *robot* in English.

A number of things must be considered when determining the answer to the question, What is a true robot? It is a device or system that is programmed by a human to perform human-like acts. It is a device or system that may sense various conditions and react in a preprogrammed manner. It may be able to react to various conditions in terms of the five human senses. It is a system that can operate on its own without any human supervision. It may make decisions by comparing information received from sensors and reacting in a preprogrammed way. These sensors are classified as magnetic, light activated, heat activated, and pressure activated. The robot became a reality in the mid-1950s. The development of the robot is closely tied to the development of the computer.

Special programs are used to control robots. They are designed for specific jobs and are written in special programming languages. Many languages exist for the control of robots. The microprocessor is the brain of the robot. It has the ability to take sense signals and make the robot react in a planned way. The microprocessor is an electronic device made from silicon chips.

Robots can work seven days a week without a break. They are capable of performing tasks more efficiently than humans. Robots are expensive and need highly trained technicians to keep them operational. They replace humans but create a demand for more highly skilled workers to keep them operating. Robots are tied to the improvement of quality of manufactured products. There are advantages and disadvantages to the use of robots. Each advantage must be weighed against the disadvantage before making a decision to buy robots instead of using human labor.

Robots have the advantage of being retrained rather quickly. They are flexible and can be used to do more than one thing with a minimum amount of reprogramming or retraining.

The main disadvantage of a robot is downtime. If a robot breaks down, it may hold up an entire plant's production schedule. Robots have a number of industrial applications that make them useful to larger manufacturers who can withstand the initial cost of the unit and its installation and debugging costs.

Key Terms

hard automation Use of the conventional assembly-line method of producing a manufactured product with dedicated equipment.

language A method of speaking to a robot (VAL, AL, AML, Pascal, and ADA are used; these languages are limited, precise, and rigid).

microprocessor A chip or part of a chip that has the ability to do math and react to various inputs from sensors; part of a robot's brain.

program A sequence of commands instructing a robot to perform some task.

programmable robot A robot that can be programmed or taught with a *teach box*, a keyboard, or some input device

programmer A person who teaches a robot; a person who can communicate with a robot in its language.

robot A system that simulates human activities from computerized instruction.

sensor A device used to detect changes in temperature, light, pressure, sound, and other functions needed to make a robot aware of various conditions.

teach pendant A device used to teach the robot memory a new program.

Review Questions

1. What is a robot?
2. When did robots become a reality?
3. Where does the word *robot* come from?
4. List five ways you can identify a true robot.
5. List at least five languages used by robots.
6. What is a microprocessor?
7. What are two positive aspects of robots?
8. What are two negative aspects of robots?
9. What is the difference between hard automation, robots, and human labor?
10. What is meant by return on investment?
11. What is the main advantage of a robot over humans as you see it?
12. List at least five operations that industrial robots can perform.

CHAPTER 2

The Robot

Performance Objectives

After reading this chapter, you will be able to:

- Describe the various types of robots.
- List the parts of a pick-and-place robot.
- Describe the terms used in robotics, such as wrist action, articulation, and envelope.
- Explain polar coordinates and articulate coordinates.
- List a manipulator's drives.
- Explain what a work envelope is.
- Identify robots by their drive.
- Identify and discuss the key terms used in the chapter.
- Answer the review questions at the end of the chapter.

A robot is made up of a number of subsystems that are usually standard and can be used for any number of other purposes. This makes the robot an inexpensive device to manufacture compared with what it would cost if all the parts and systems were individually designed and handmade. The designer of a robot has a purpose in mind. This purpose may be to move an object from one place to another, or it may be a complicated maneuver requiring many subsystems to get the job done.

Robots can be classified in a number of ways. The classification system we use here is the end purpose of the robot; in other words, we classify robots as industrial, laboratory, explorer, hobbyist, classroom, and entertainment devices. These are only a few of the jobs robots do. Here we will concentrate on the broad category of industrial robots, and we will introduce some of the other types as the book develops.

Industrial Robots

Industrial robots have arms with grippers attached (Figure 2-1). The grippers are finger-like and can grip or pick up various objects. They are used to pick and place. They pick up an object and place it elsewhere or move materials from one place to another. These robots can be programmed and computerized. The *teach box* is used to program the *microprocessor* that is the robot's "brain." Sensory robots, welding robots, and assembly robots usually have a self-contained microcomputer or minicomputer.

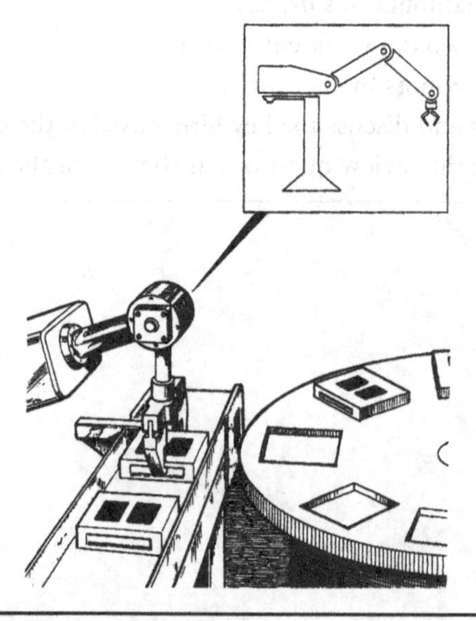

Figure 2-1 Industrial robot used to pick and place.

Laboratory Robots

Laboratory robots take many shapes and do many things. They are the beginnings of more sophisticated devices. They have microcomputer brains, multijointed arms, or advanced vision or tactile senses. Some have good hand-eye coordination and will eventually be used in industry to become more productive. Some may be mobile, and others may be stationary (Figure 2-2).

FIGURE 2-2 Laboratory robot used to handle dangerous materials. Ford Motor Company uses two of these to handle plastic lenses. (*Courtesy of Seiko Instruments USA, Inc.*)

Explorer Robots

Explorer robots are used to go where humans cannot go or fear to tread. For instance, they are used in outer space probes, to explore caves and to dive far deeper underwater than humans can, and they can be used to rescue people in sunken ships. The National Aeronautics and Space Administration (NASA) has done much to develop these robots to explore the surface of the moon and the surface of Mars. Explorer robots are sophisticated machines that have sensory systems and are remotely controlled or controlled by preprogrammed onboard computers (Figure 2-3).

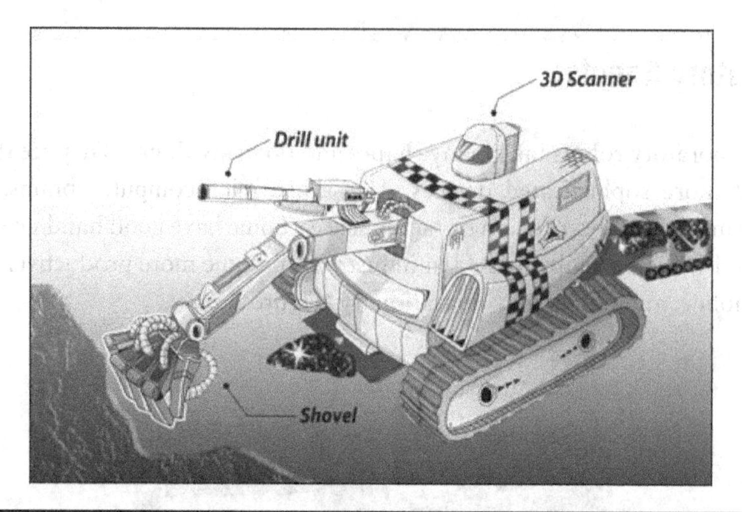

Figure 2-3 Coal mine robot. (*Courtesy of FANUC.*)

Hobbyist Robots

Most hobbyist robots are mobile. They are usually made to operate by rolling around on wheels propelled by small electric motors controlled by an onboard microprocessor (Figure 2-4). The ultimate goal of most hobbyists is to have a housekeeper robot that will do the hard work of keeping their living quarters livable. Most hobbyist robots are equipped with speech-synthesis and speech-recognition systems. Some follow a line on the floor, and others follow preprogrammed instructions. Most have an arm or arms and resemble a person in appearance.

Classroom Robots

Classroom robots are limited in application at this time. In the future, classroom robots will be able to move around the classroom and assist the teacher in various aspects of the teaching-learning process. Those now available in schools are mainly hobbyist in nature and are used to teach the fundamentals of robotics to students engaged in learning practical applications for various electronic circuits. Secondary schools, vocational institutions, junior colleges, and universities are adding robotics courses to their curricula in an effort to prepare students to enter the workforce for these advanced technologies.

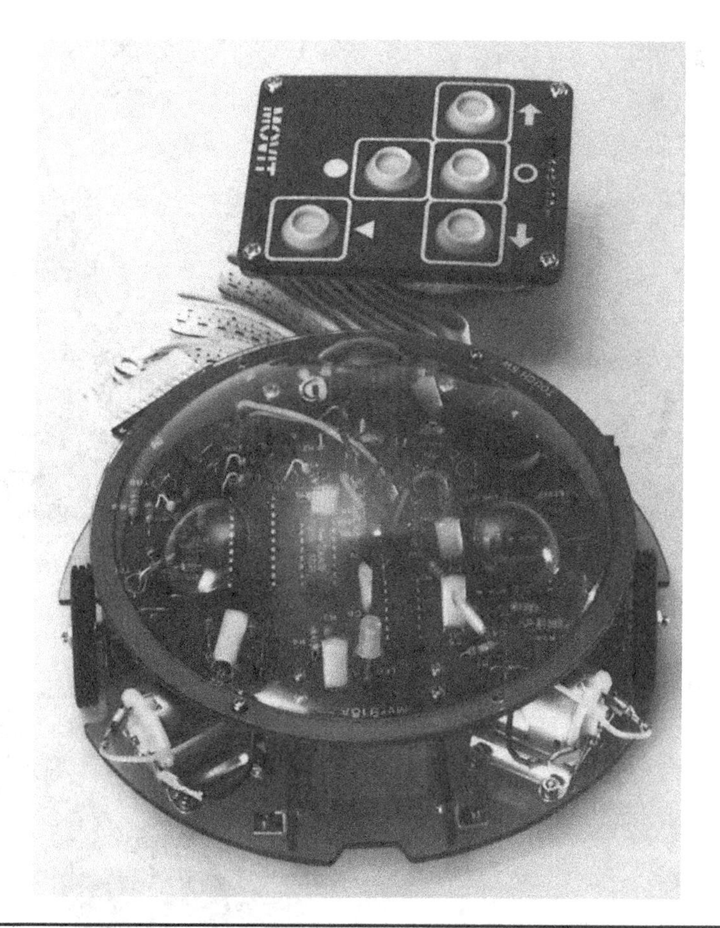

Figure 2-4 Hobbyist robot. (*Courtesy of MOVIT.*)

Because industrial robots can cost from $50,000 to $100,000, are intimidating to untrained workers, and are costly to repair, they are highly impractical for training purposes. Rhino has developed the XR Series of tabletop robotic system to meet the needs of manufacturing companies and educational institutions that require a high-quality, cost-effective means of training (Figure 2-5A). Microbot also makes an educational robot that can be used to teach the basic functions of an industrial robot (Figure 2-5B).

For about one-tenth the cost of full-scale industrial robot systems, tabletop systems can totally simulate the larger systems, providing cost-effective training that readily transfers to the industrial setting.

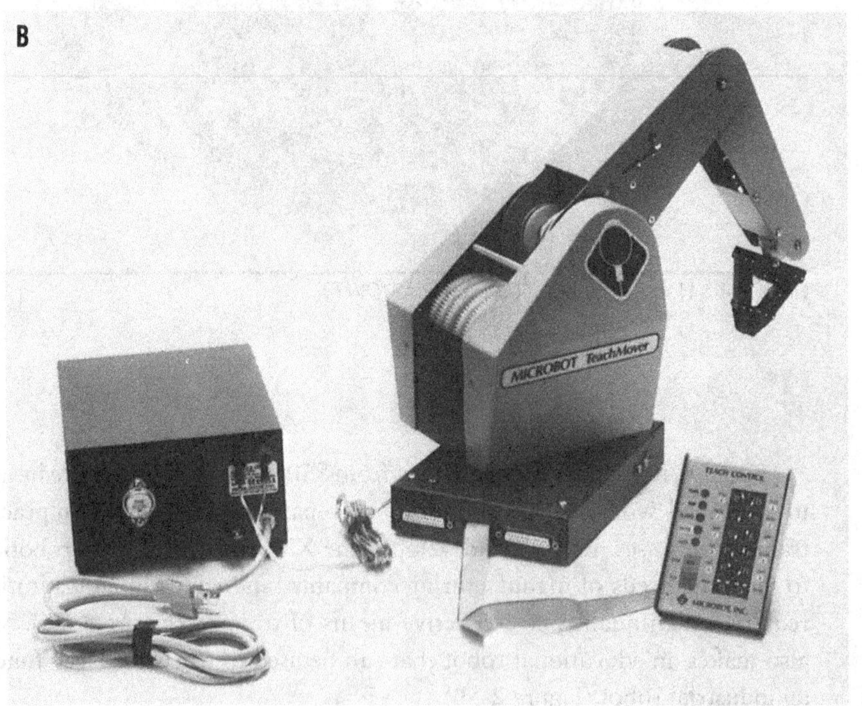

Figure 2-5 Classroom robots used to teach the fundamentals of robot operation. *((A) courtesy of Rhino Robots, Inc.)*

Entertainment Robots

Entertainment robots are just beginning to be developed and made available to the public. They have the ability to speak and respond to the spoken word. They can be used to entertain people at various events or operate as a roving advertisement. More uses for robots in the entertainment field will be forthcoming as more inexpensive, sophisticated programming becomes available. For instance, look at Figure 2-6. The entertainment industry uses robots to take pictures where it would otherwise be impossible to photograph. Special cameras readily adaptable to robot control have been developed for better photography and television coverage of events and operations of all sorts.

FIGURE 2-6 Entertainment robot with a boom arm camera control system. (*Courtesy of Elicon.*)

The Manipulator

There are about 250 manufacturers of robots in the United States, Europe, and Japan, making it very difficult to identify all the parts used in the available robots. However, there are some common components that may be examined for a better perspective on how a robot works.

The *manipulator* is one of the three basic parts of a robot. The other two are the *controller* and the *power source* (Figure 2-7). In order for a robot to do work, each of these three components must be operational.

A manipulator is classified by its specific arm movements. There are, for instance, four coordinate systems used to describe arm movement: polar coordinates, cylindrical coordinates, Cartesian coordinates, and articulate (jointed-arm, spherical) coordinates. Let's take a closer look at some of the robot components before we examine the coordinate systems. This will introduce you to the terms used in the later coordinate discussion.

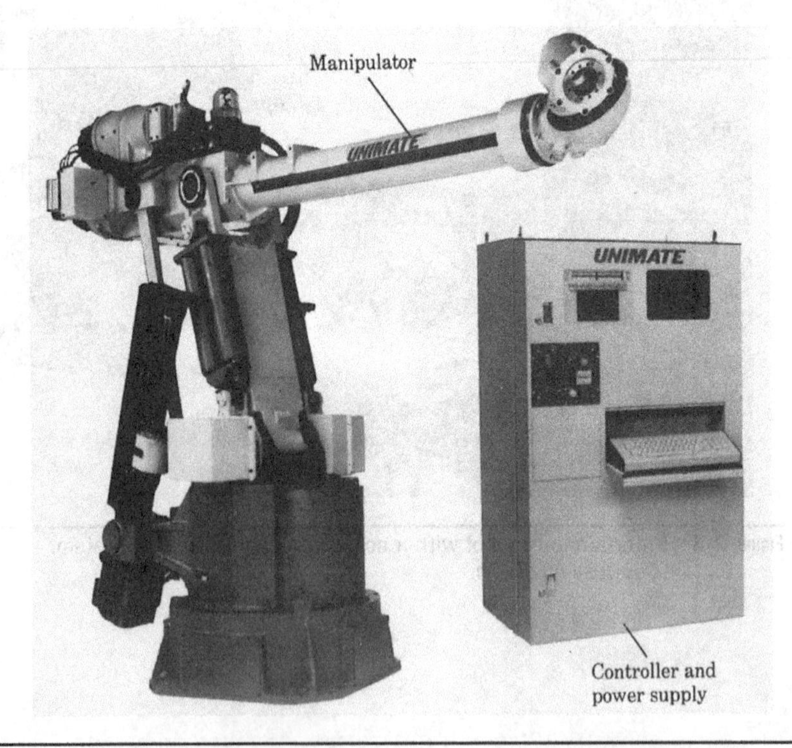

FIGURE 2-7 The three basic components of a robot. (*Courtesy of Unimation, Inc., a Westinghouse Company.*)

Base

The base of a robot is its anchor point. The base may be rigid. It is usually designed as a supporting unit for all the component parts of the robot. The base does not have to be stationary because it may become part of the operational requirements of the robot. It may be capable of any combination of motions, including rotation, extension, twisting, and linear. Most robots have the base anchored to the floor (Figure 2-8A), although, because of limited floor space, they may be anchored to the ceiling or to overhead suspended support systems (Figure 2-8B). A track or conveyor system may be used to move the robot along as needed (Figure 2-8C).

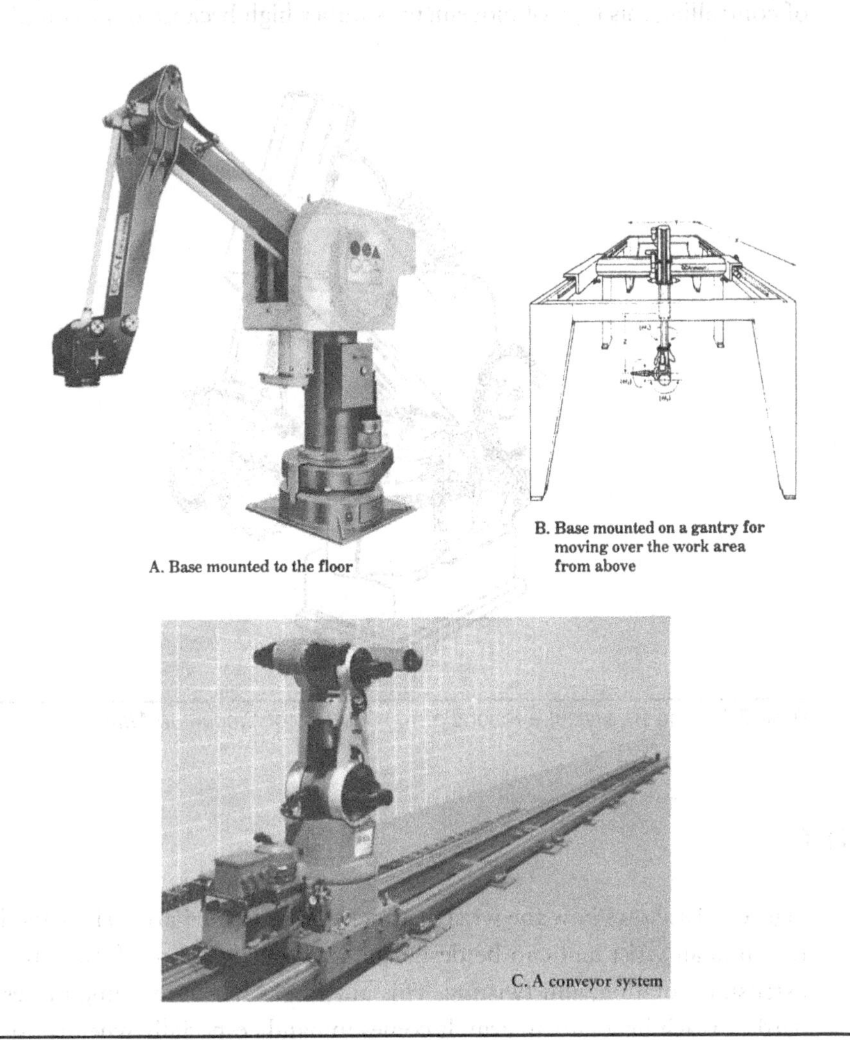

A. Base mounted to the floor
B. Base mounted on a gantry for moving over the work area from above
C. A conveyor system

FIGURE 2-8 Robot bases. *(Courtesy of GCA Corp./Industrial Systems Group.)*

Arm

Some type of arm is found on most industrial robots. It may be jointed and resemble a human arm, or it may be a slide-in/slide-out type used to grasp something and bring it back closer to the robot. A jointed arm consists of a base rotation axis, a shoulder rotation axis, and an elbow rotation axis. This type of arm provides the largest working envelope per area of floor space of any design thus far. If this is a six-axis type of arm, it requires some rather sophisticated computer control (Figure 2-9). Most arms now have some type of joint. From one to six jointed arms may be attached to a single base for special jobs. The expense of controlling this type of movement is rather high because of its complexity.

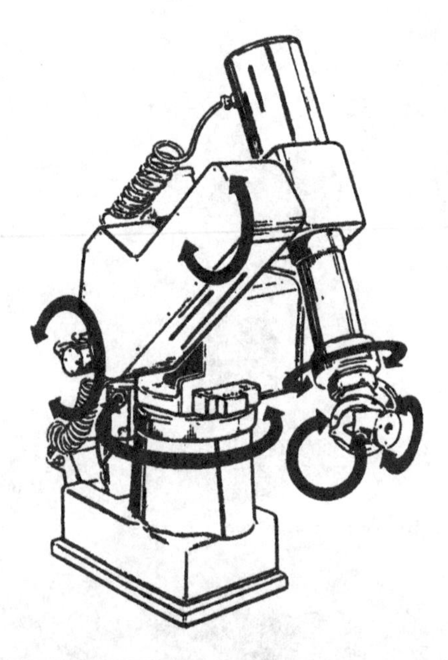

FIGURE 2-9 The six axes of a robotic arm. (*Courtesy of Cincinnati Milicron.*)

Wrist

Figure 2-10 shows how the wrist is attached to a jointed arm. The wrist is similar to a human wrist and can be designed with a wide range of motion, including extension, rotation, and twisting. This aids the robot in reaching places that are hard to reach by a human arm. It comes in handy especially when spray painting

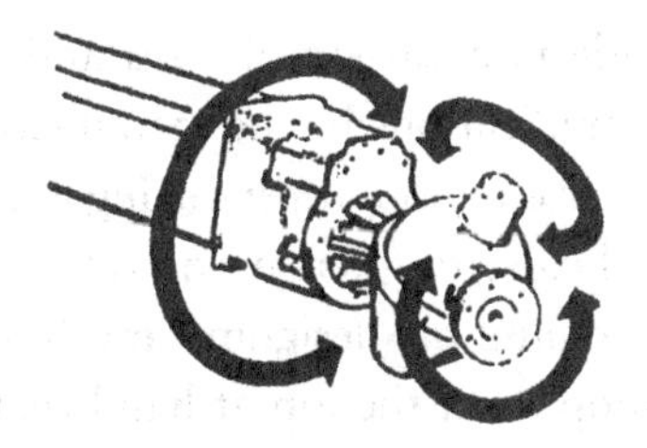

Figure 2-10 The three axes of a robotic wrist attached to a manipulator arm. (*Courtesy of Cincinnati Milicron.*)

the interior of an automobile on the assembly line. It is also helpful in welding inside a pipe. This type of flexibility will improve the manufactured products we now enjoy and add others we were unable to fabricate earlier.

Grippers

Grippers are located at the end of a wrist. They are used to hold whatever the robot is to manipulate (Figure 2-11). Pick-and-place robots have grippers to move objects from one place to another.

Some robots have end-of-arm tooling instead of grippers. In such cases, the robot is used primarily for one type of operation, such as spray painting or welding. If a tool is attached, it is unnecessary to have a gripper on the end of the arm. A pneumatic impact wrench can be fitted at the end of the arm just as easily as a set of grippers.

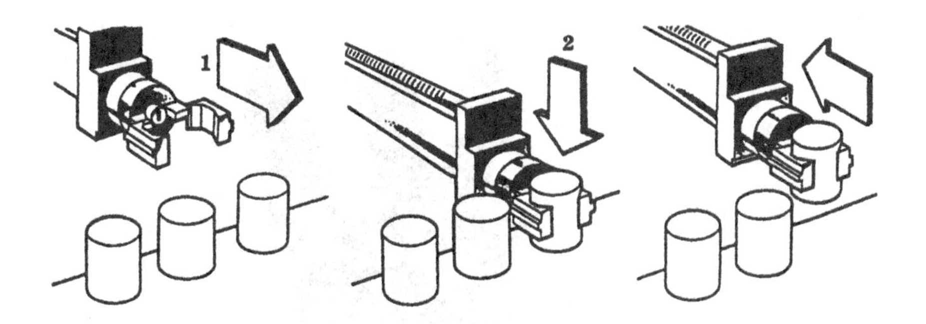

Figure 2-11 Robot arm (manipulator) with grippers for reaching and picking up an object on a line and bringing it back to place it elsewhere. (*Courtesy of Cincinnati Milicron.*)

Grippers are made in a variety of sizes and shapes (Figure 2-12). They are designed for special applications by the manufacturers of robotic equipment. The simplest type of gripper is a motion-producing device that is joined by two fingers. The fingers open and close to grasp an object. In most instances, only the finger mounts are purchased, and the designing and building of the fingers are done in the local machine shop to fit the job at hand (Figure 2-13). Variations in design account for much of the design time spent on robots.

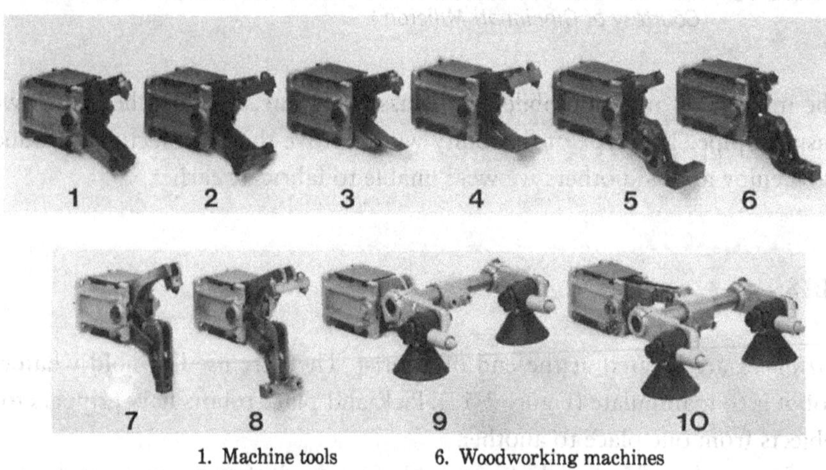

1. Machine tools
2. Stamping presses
3. Moulding machines
4. Die cast machines
5. Forging presses
6. Woodworking machines
7. Welding fixtures
8. Extrusion billets
9. Transfer equipment
10. Glass handling

FIGURE 2-12 Different types of grippers made for the end of a manipulator arm. (*Courtesy of I.S.I. Manufacturing, Inc.*)

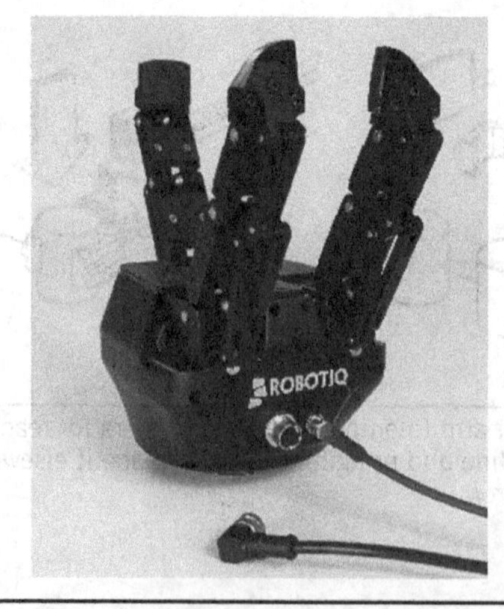

FIGURE 2-13 Three-fingered gripper. (*Courtesy of Robotiq.*)

Various types of grippers are made to fit the job being done by the robot. For instance, it is possible to have a donut-shaped inflatable piece of rubber that slips over the neck of a bottle, is inflated, and the bottle then moved. When it is time to release the bottle, the donut is deflated, the bottle slides out, and the robot arm moves upward and onto the next bottle that needs to be lifted onto the line. In some cases, rubber suction cups are used to pick up materials (see Figures 2-9, 2-10, and 2-12). A vacuum is applied to pick up the object, and a blast of air is added to cause the object to drop.

All Together It Becomes a Manipulator

The manipulator is really a combination shoulder, arm, wrist, and hand (Figure 2-14). The grippers are the hand. This combination makes it possible for a robot to reach for an object, pick it up, carry it, and put it down where desired.

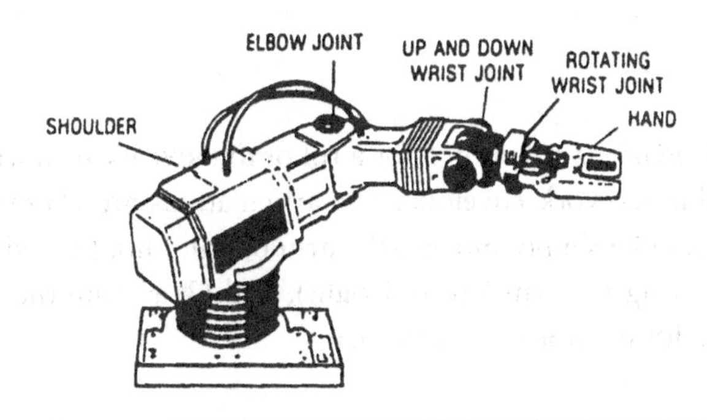

FIGURE 2-14 The parts of an industrial robot used to pick and place. (*Courtesy of Radio Shack.*)

Work Envelope

The *work envelope* is also referred to as the *sphere of influence* or *work area* because it is not always spherical in shape. This is the space a robot occupies when it swings around and up and down to do the work for which it was designed (Figure 2-15).

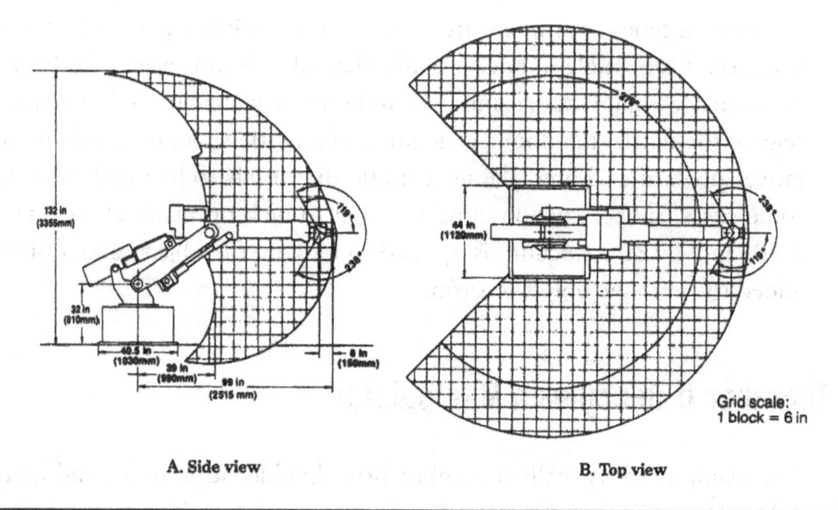

Figure 2-15 Teardrop-shaped work envelope.

Articulation

Three *articulations* are needed for a robot to move its arm, wrist, and hand to any place within its work envelope. The articulations are (1) extend and retract the arm (this can be simply moving the arm out and in), (2) swing or rotate the arm (this is moving the arm left and right), and (3) elevate the arm (this is nothing more than lifting or lowering the arm).

Wrist Motion

The three types of motion for a robot's wrist are (1) bending or rolling forward and backward, (2) yawing (spinning) from right to left or left to right, and (3) swiveling, which is nothing more than rolling down to the right or left (Figure 2-16).

Figure 2-16 Wrist action known as yaw. (*Courtesy of Cincinnati Milicron.*)

Degrees of Freedom

A robot has six *degrees of freedom* if it can move its wrist three ways and its arm three ways (see Figure 2-9). This is very limited when you compare it with human shoulder, arm, and wrist motions. Humans have forty-two degrees of freedom. As you can see, the robot arm needs improvement if it is to be as versatile as the human arm. There is some question as to whether the human arm and wrist should be emulated. However, such emulation would call for some rather involved engineering to get the job done right. At present, it is possible for a robotic hand to do what the human hand can do: grip, push, pull, grasp, and release.

Robot Motion Capabilities

Robots have four basic motion capabilities: (1) linear motion, (2) extension motion, (3) rotating motion, and (4) twisting motion. These four motion capabilities are referred to as the LERT classification system (*L* stands for "linear motion," *E* stands for "extensional motion," *R* stands for "rotational motion," and *T* stands for "twisting motion"). A superscript is used with the classification system to indicate the number of times the robot is capable of a particular motion. Keep in mind that most robots are mounted on the floor on a base. However, it is possible for them to be mounted to the ceiling or onto a mobile platform. Each axis is listed in the order it is mounted to the first component or base. For instance, L^3 indicates there are three linear motions. If you see R^2L^3, you have two rotational motions and three linear motions.

Coordinates

The manipulator *arm geometry* refers to the movement of the robotic arm. There are four systems of classification for robot axis movement: articulate, Cartesian, cylindrical, and polar. Each of these *coordinate systems* describes the arm movement through space. Various axis movements are the result of design characteristics.

Cartesian Coordinates

The simplest and most easily understood system is the Cartesian coordinate system. *Cartesian coordinates* are used to describe the *X*, *Y*, and *Z* axes or planes.

The reference point for these planes is the intersection of all three (Figure 2-17). The centerline of the robot is the reference point for all three axes. Note the location of the center point or origin point.

The manipulator uses the X, Y, and Z planes to reach its target, so these planes become the operational axes of the manipulator arm. Note how the movement along the Z axis is in a linear motion with an up-and-down movement from point *a* to point *b* (Figure 2-18). The Y axis is also a linear motion but with an in-and-out movement. This Y axis provides a reaching type of movement of the arm from point *b* to point *c*. The X axis is the side-to-side motion of the manipulator. This means that the side-to-side motion produces a movement that causes the whole manipulator to rotate about its base.

A rectangular-shaped work envelope is produced by a Cartesian coordinate robot. This rectangular working area is caused by the limitations that the X, Y, Z axes produce. It is not possible to reach areas outside the rectangular work envelope with a Cartesian coordinate system. It is impossible for the arm to move from point *a* to point *c* directly. The movement is limited to the X, Y, and Z directions.

The Cartesian coordinate system robot is one of the simplest in operation. It can be used to load and unload and do point-to-point operations. This type of robot can be further classified as a low-technology type. Cartesian coordinates in and of themselves do not make a robot low technology. Most air-operated pick-and-place robots, being low technology, are of the Cartesian design, however (Figure 2-19).

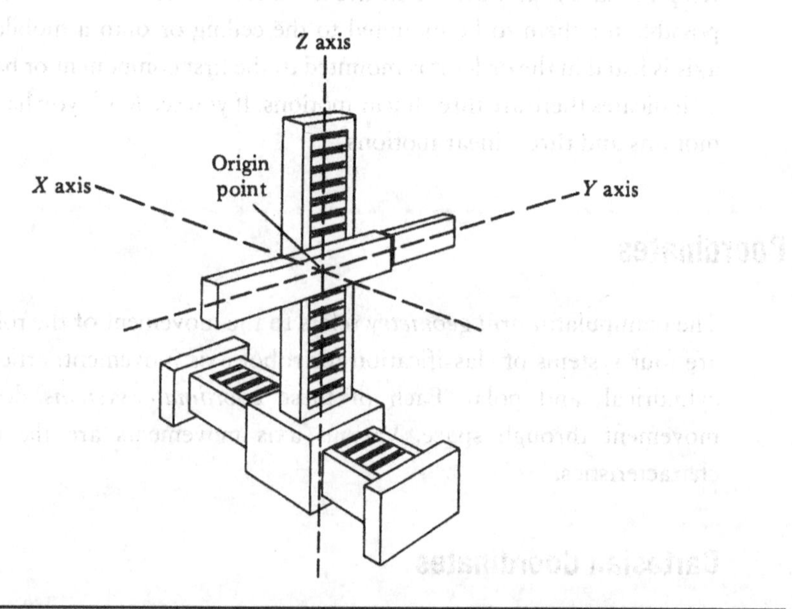

FIGURE 2-17 Cartesian coordinates or X, Y, and Z axes. (*Courtesy of PWS-Kent.*)

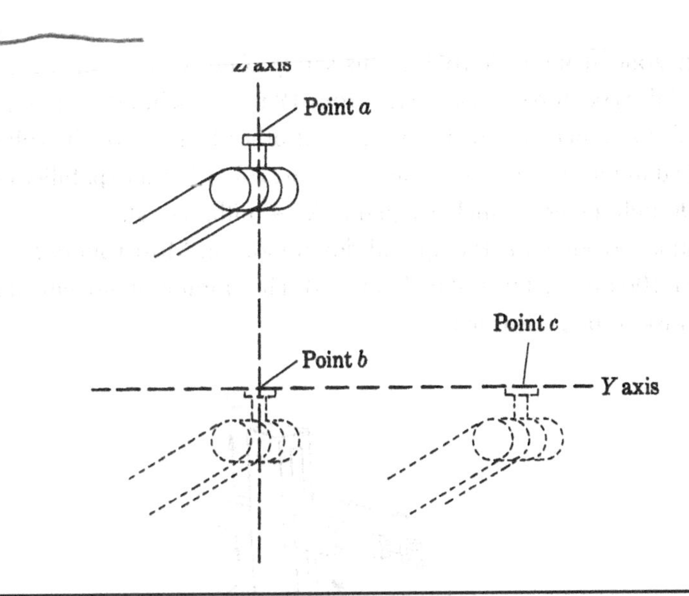

FIGURE 2-18 Manipulator arm movements for Cartesian coordinate system. (*Courtesy of PWS-Kent.*)

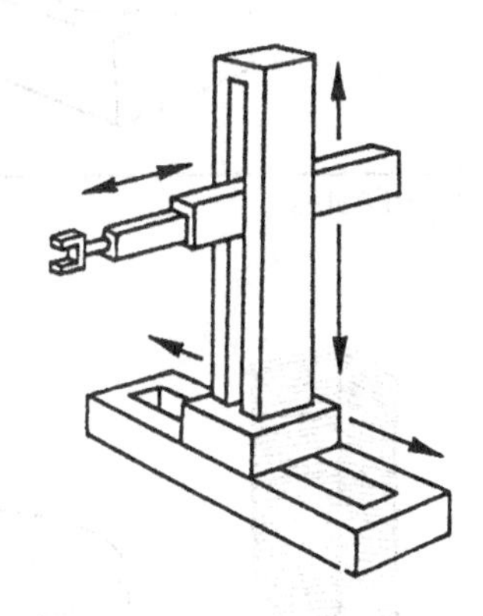

FIGURE 2-19 Cartesian coordinates.

Cylindrical Coordinates

The cylindrical coordinate system also uses three axes; however, the names for the axes are different (Figure 2-20). Cylindrical coordinates are labeled *theta* (a Greek letter that looks like a zero with a horizontal line through the middle of it, i.e., θ) and describe the rotational axes (Figure 2-21). The *R* axis is the reach or in-and-out axis. The *Z* axis indicates the up-and-down movement. This type of movement results in tracing out a cylindrical shape when its work envelope is examined. Keep in mind that the rotation of the base is described by the Greek letter θ. The cylindrical robot can rotate up to 300 degrees. The extra 60 degrees are used for a safety zone around the robot. This safety zone is called the *dead zone* (Figure 2-22). This type of robot can reach from 19 to 59 inches depending on the design and task to be performed. The reach and other features of the robot are usually described in metric measurements. Thus the robot has a capability of reaching 50 to 1,500 millimeters (1 inch is equal to 25.4 millimeters).

Z-axis movement is the up-and-down motion. Most robots can travel within 100 to 1,100 millimeters (~4 to 43 inches). The amount of movement is determined by the task to be performed.

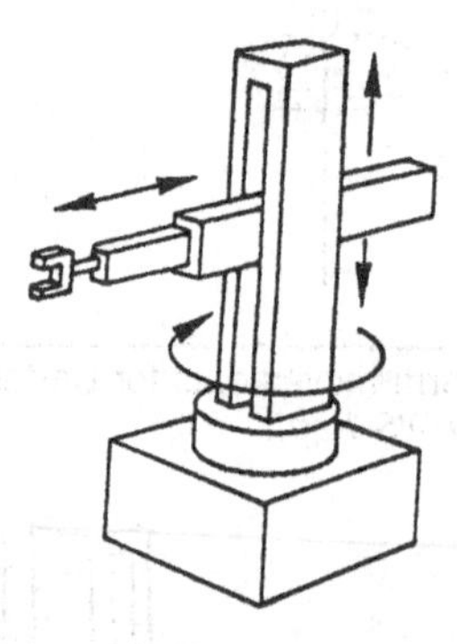

FIGURE 2-20 Cylindrical coordinates.

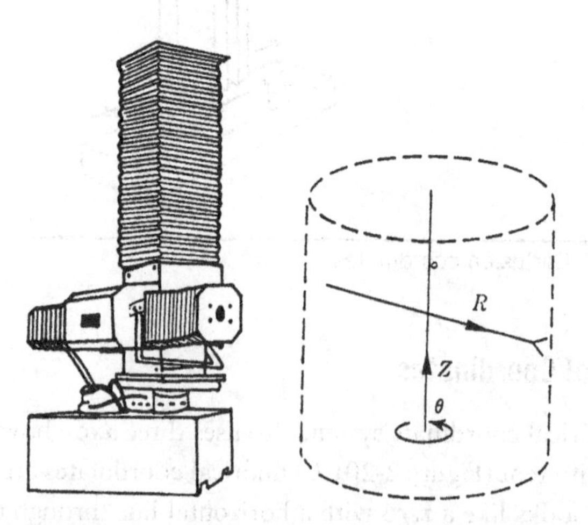

FIGURE 2-21 Cylindrical robots and resulting cylinder traced by the arm's movements. (*Courtesy of PRAB Robots, Inc.*)

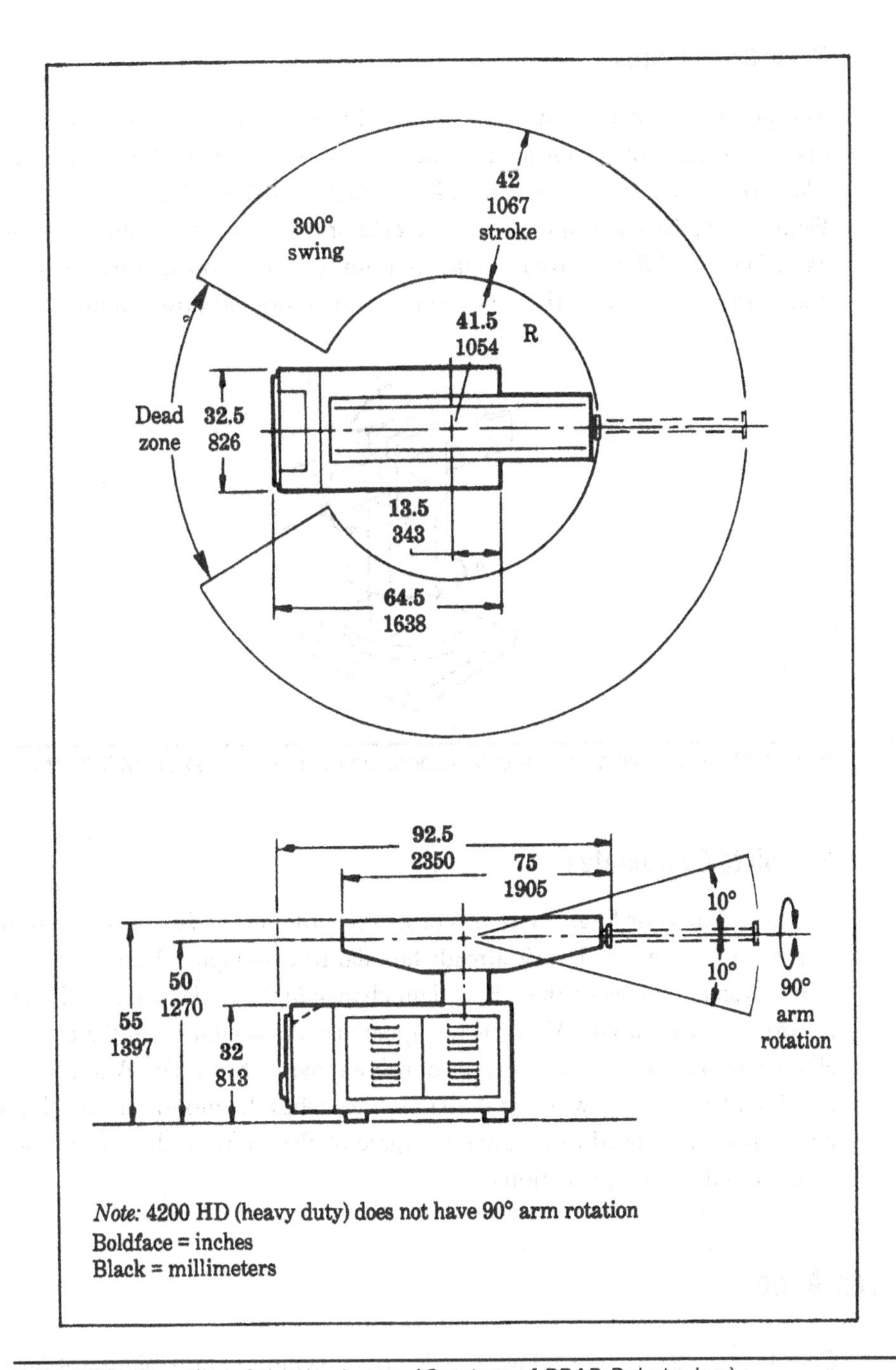

FIGURE 2-22 Location of the dead zone. (*Courtesy of PRAB Robots. Inc.*)

Polar Coordinates

The polar coordinate system is slightly different, but it too has three axes of operation. The polar coordinate system describes a *spherical* movement pattern. The three axes of this system are θ, *R* (or reach), and beta (β). As you can see from Figure 2-23, the polar system and the cylindrical system basically have the same axes. The θ and *R* axes are the same in both types of robots. However, the β axis allows the entire arm of the robot to move in an up-and-down motion.

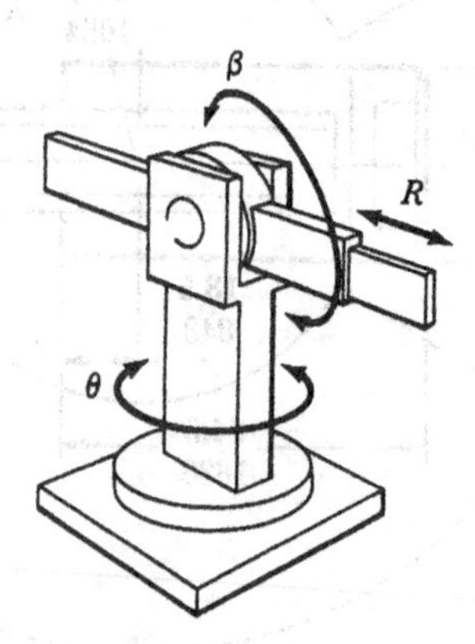

FIGURE 2-23 The three axes of a polar coordinate robot. (*Courtesy of PWS-Kent.*)

Articulate Coordinates

The articulate coordinate system is descriptive of a *jointed* system. The θ motion or rotation around the base is already familiar to you (Figure 2-24). It, too, is used in this system. However, there is a slight change *in* the designation and motion of the other two axes. The *W* axis is the upper arm or shoulder, and the *U* axis is the elbow motion. Two axes in this type of robot move or bend. The *W* axis represents the shoulder rotation, whereas the *U* axis describes the motion of the elbow and its rotation. This produces a greater degree of flexibility in the robot, making it ideal for industrial applications.

Wrist Action

Coordinate systems are also used to describe the motion possibilities of the wrist. The manipulator arm is limited in its ability to do work without some type of *end effector* to act as a hand. To make this hand work, it is necessary to have a wrist action to cause it to approach the abilities of the human hand to do work.

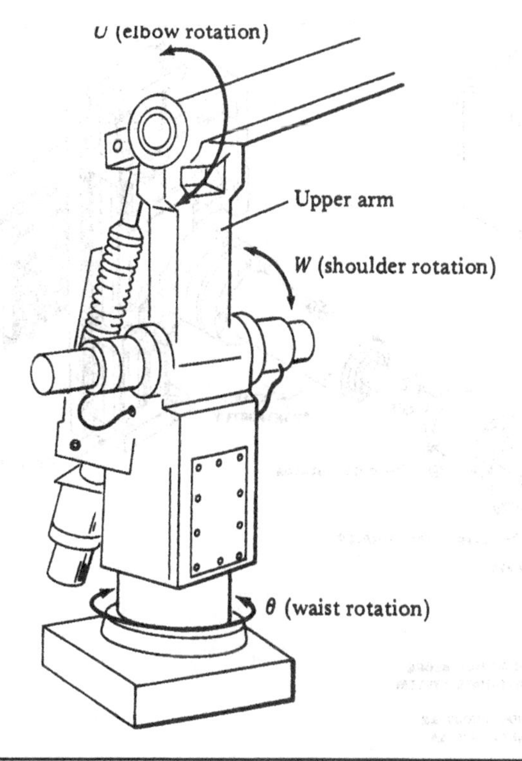

Figure 2-24 The three axes of an articulate coordinate robot. (*Courtesy of PWS-Kent.*)

Four coordinate systems are used to describe this wrist action. You can see the possibilities of the wrist in Figure 2-25. By adding the wrist, it is possible to have a robot with six axes of motion. This added dimension increases the flexibility of the robot and its possible application list. The axes that the wrist adds to the robot are identified as the pitch axis, the yaw axis, and the roll axis.

The pitch axis is usually limited to 270 degrees of movement. It moves up and down. This 270 degrees can be moved gradually or degree by degree depending on the task to be performed (Figure 2-26). The yaw axis describes the side-to-side movement of the wrist.

The movement range is 90 to 270 degrees. The roll axis refers to the movement of the end of the wrist. It is possible to obtain 360 degrees of rotation with an end effector.

With these three additional axes of rotation, it is possible to have the robot do things that humans cannot. This is especially true in painting automobiles on an assembly line. It makes possible the painting of the inside of the car body at the same time as the outside is sprayed, which results in better protection of the body metal with decreased possibility of rust later.

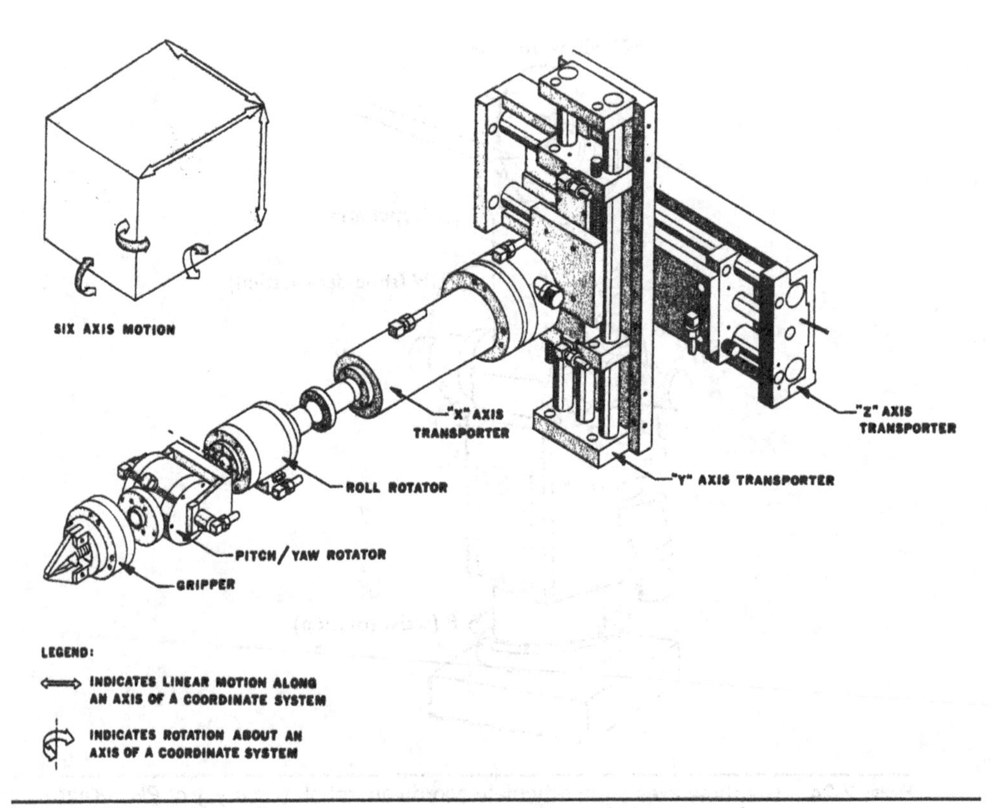

Figure 2-25 Wrist action (note how the movements are made by rotators). (*Courtesy of Mack Corporation.*)

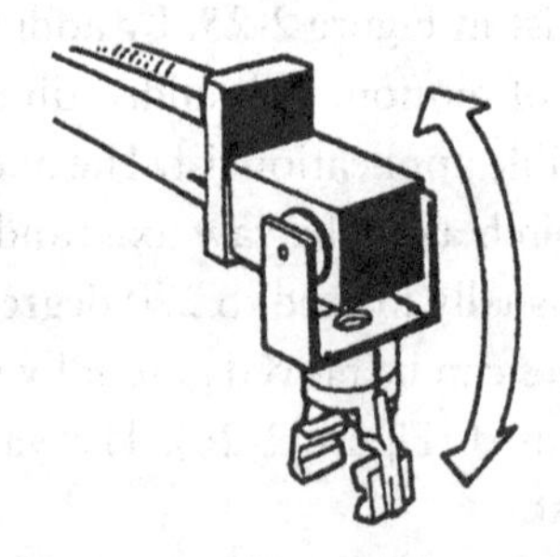

Figure 2-26 The wrist action known as pitch. (*Courtesy of Cincinnati Milicron.*)

Work Envelopes

The Cartesian coordinate robot has a rectangular-shaped work envelope. This envelope is very important to anyone who works on or near robots. This is the area in which the robot can move. Since the robot cannot see, it is possible for its arm to hit a person or equipment that is left in this area. The cylindrical coordinate robot has a cylindrical work envelope, which means that you also have to be careful of what is above the robot as well as around it.

The polar coordinate robot (see Figure 2-27) has a spherical work envelope, and the articulate coordinate work envelope is teardrop shaped. Keep in mind that all points that can be programmed within the reach of the robot are part of the work envelope. Figure 2-28 provides a better idea as to the space occupied by various types of robots in their normal operational modes.

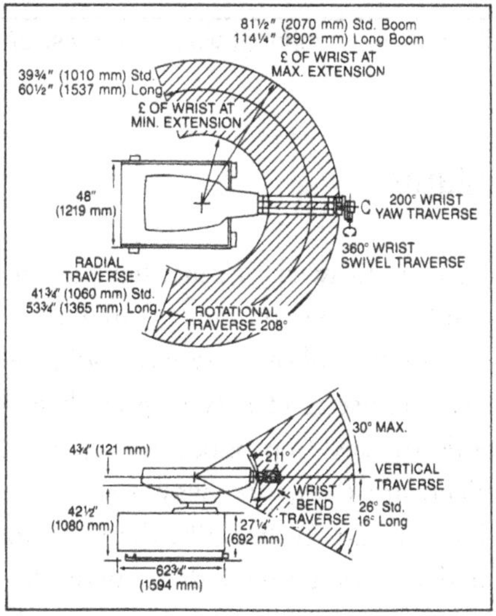

Figure 2-27 Spherical robot and its work envelope. (*Courtesy of Unimation, Inc., a Westinghouse Company.*)

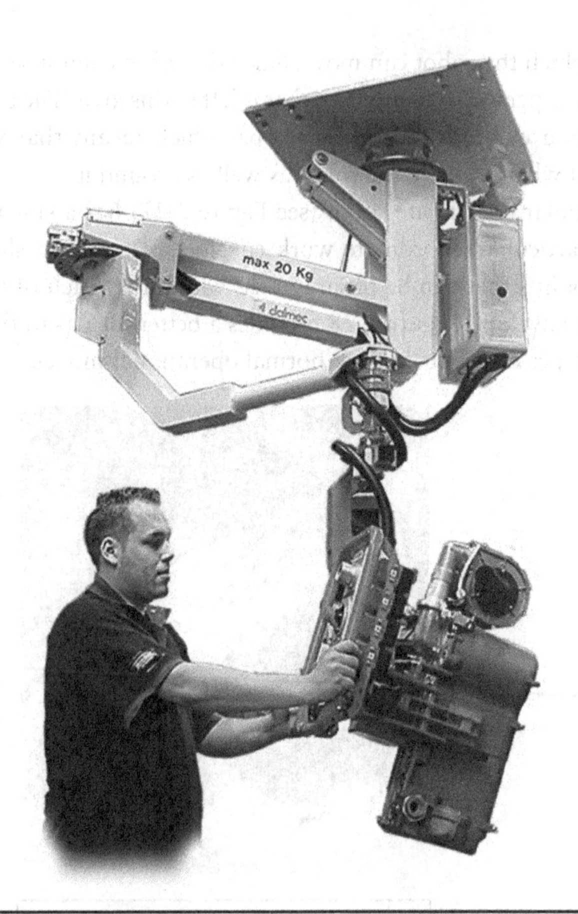

FIGURE 2-28 A ceiling-mounted robot at work. (*Courtesy of FANUC.*)

Moving the Manipulator

For the manipulator to do work, it must have a *power source*. Industrial robots work on 220- and 440-volt systems available within the plant or factory. Entertainment and smaller robots may work with batteries that need recharging occasionally. All robots use electricity to give them their basic power. However, other drive methods are used to produce motion and do work. They are hydraulic and pneumatic. There are, then, three ways to drive a manipulator and do the task a robot was designed to do. Technically, there are four. Mechanical drive robots can also be counted as representing a drive method.

Pneumatic Drive

About one-third of industrial robots use pneumatic power to drive the axes. If some of the pneumatic-powered pick-and-place units are not considered robots, then the percentage may be less. *Pneumatic power* is generated by compressed air. Air is pressurized or compressed within the plant and fed through pipes to different locations. When the pressurized air is fed to a cylinder attached to some part of a robot, it is converted into motion. The air pressure causes the desired movement of the arm or the whole robot (Figure 2-29).

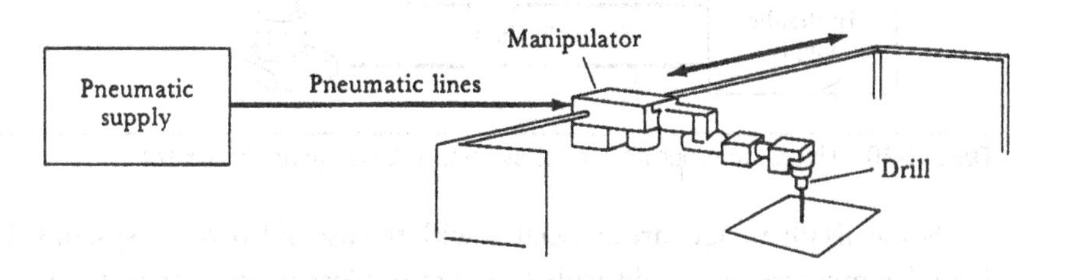

Figure 2-29 Pneumatic-operated robot system. (*Courtesy of PWS-Kent.*)

Pneumatic drive systems have some advantages, among them cost. They are the least costly of the three drive sources. However, pneumatic drive systems do have some limitations. They can develop only enough torque to pick up about 6.5 to 10 pounds (3 to 4.5 kilograms). Pneumatic drive systems are used on robots that do simple assembly operations, die-casting operations, materials-handling jobs, and machine loading and unloading. Their greatest limitation is their inability to move large payloads. Larger payloads require electric or hydraulic drives (see Appendices).

Hydraulic Drive

About 45 percent of the industrial robots in use today are driven by hydraulic systems. Hydraulic fluid is placed under pressure by a pump operated by an electric motor. The pressurized fluid is used to move the axes of the robot by applying this pressure to a cylinder, hydraulic motor, or rotary actuator. The cylinder is extended by the pressure of the fluid. The direction of fluid flow in the cylinder determines if it extends or retracts. By moving the cylinder in and out, it is possible to move the axes of the robot.

Hydraulic drive systems are capable of lifting 50 to 300 pounds (22.7 to 135.9 kilograms). They are usually referred to as medium-technology robot systems, but some hydraulic systems can be considered high technology. Low-, medium-, and high-technology classifications depend on the type of control, not the power source.

Hydraulic drive systems can be used to load and unload up to their lifting ability. They are also used in arc welding, spot welding, die casting, spray painting, and press loading. One of their major uses is in spray painting because they are safer to operate in a volatile atmosphere than are electric-powered robots. Figure 2-30 shows the simplicity of the hookup for a hydraulic drive system.

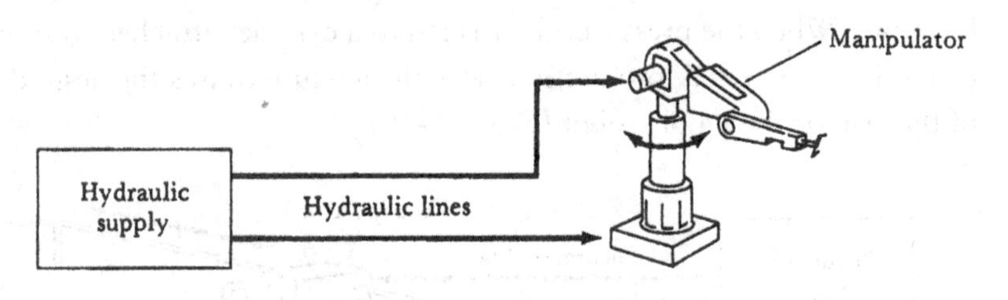

FIGURE 2-30 Hydraulic-operated robot system. (*Courtesy of PWS-Kent.*)

Some disadvantages are associated with the use of hydraulic systems. The fluid is under pressure and is difficult to contain. Cost is another factor to consider. Depending on the size of the unit being considered, cost can be of importance. Generally speaking, hydraulic systems are the most expensive of the three drive systems mentioned here. However, they are employed where a lot of torque is required to lift heavy loads.

Electric Drive

Electric drive systems use electric motors for their power source. The motors may be operated by direct current (DC) or alternating current (AC). Torque is developed by gearing the speed of the motor down to the required motion or movement of the arm (Figure 2-31).

Some advantages of electric drive systems are their ability to allow smooth startup of payloads and their smooth deceleration and stopping. Operating costs

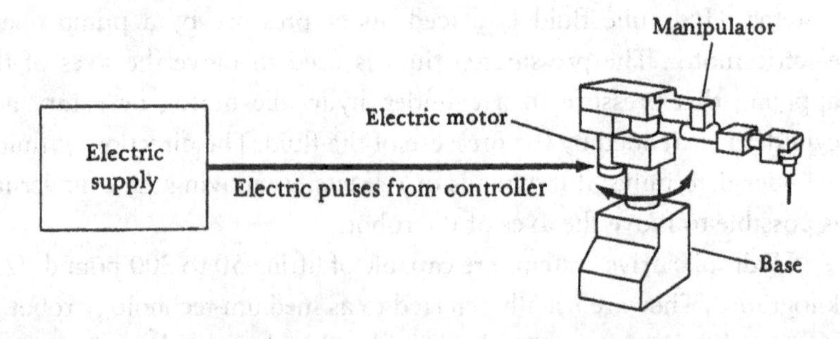

FIGURE 2-31 Electric-operated robot system. (*Courtesy of PWS-Kent.*)

are minimal in terms of electrical power used and maintenance of system motors. Electric robots have a higher repeatable accuracy than do hydraulic robots.

Electric drive robots cannot handle as heavy a payload as hydraulic drive systems. The payload for an electric drive is between 6.6 and 176 pounds (3 and 80 kilograms). However, electric drive systems are very versatile in operation.

Electric drive systems are used in arc welding, spot welding, machine loading and unloading, materials handling, and deburring as well as in assembly operations. Development of explosion-proof motors will make it possible to use them safely in spray-painting atmospheres. Electric drive systems are used in high-technology robots.

Summary

Robots are made up of a number of subsystems. A number of methods are used to classify robots. The classification system used in this book is based on the end purpose of the robot.

Industrial robots have arms with grippers attached. The grippers are finger-like and can grip or pick up various objects. They are used to pick and place. Robots can be computerized and operate without human supervision.

Laboratory robots take many shapes and do many things. They have microcomputer brains, new multijointed arms, and advanced vision or tactile senses. Some have hand-eye coordination.

Explorer robots are used to probe outer space and to explore caves, dive underwater, and explore areas where no human can exist.

Most hobbyist robots are mobile. They are still experimental and have been part of an effort to develop a housekeeper robot that resembles the human form.

Classroom robots have limited application today. However, remarkable things are planned for them in the near future. Entertainment robots are just beginning to be developed and made available to entertain people and act as roving advertisements.

The manipulator is one of three basic components of a robot. The other two are the controller and the power supply. The manipulator can be classified by four coordinate systems used to describe arm movement.

The base of the robot is its anchor point. The base may be either rigid or mobile. Some type of arm is found on most industrial robots. It may be jointed and resemble a human arm, or it may be a slide-in/slide-out type used to grasp something and bring it back closer to the robot.

The wrist is attached to the jointed arm and can be designed with a wide range of motions. And the gripper is located at the end of the wrist and is used to hold whatever the robot is to manipulate.

The manipulator is really a combination shoulder, arm, wrist, and hand. The gripper represents the hand, and the work envelope is also referred to as the sphere of influence or work area for the robot. Articulations are (1) extend and retract the arm, (2) swing or rotate the arm, and (3) elevate the arm.

The robot has six degrees of freedom if it can move the wrist three ways and the arm three ways. Human workers have forty-two degrees of freedom.

The four basic motion capabilities of robots are linear motion, rotating motion, twisting motion, and extensional motion. These four motions are the basis of the LERT classification system. The manipulator *arm geometry* refers to the movement of the robot arm. There are four systems of classification for robot axis movement: articulate, Cartesian, cylindrical, and polar. Each describes the movement of the arm through space and within its work envelope. The manipulator uses the X, Y, and Z planes to reach its target. However, there are also the theta (θ), beta (β), W, and U axes to be considered in the operation of a robot. Coordinate systems are also used to describe the motion possibilities of the wrist. The manipulator arm is limited in its ability to do work without some type of end effector to act as a hand. The axes that the wrists add to the robot are identified as the pitch axis, yaw axis, and roll axis.

Cartesian coordinate robots have a rectangular-shaped work envelope. The polar coordinate robot has a spherical-shaped work envelope, and the articulate coordinate type is teardrop shaped. The work envelope for the cylindrical coordinate robot is cylindrical in shape.

Drive systems for robots are classified as pneumatic, hydraulic, and electric. Each type has its applications due to physical limitations of the method used to do the work. Each type of drive has its advantages and disadvantages.

Key Terms

articulations The ability of a robot to extend and retract, swing or rotate, and elevate its arm.

Cartesian coordinates Simplest of the coordinate systems because it refers to up-and-down, back-and-forth, and in-and-out movements of a robot arm.

controller Supplies the direction to the robot.

coordinates, points, and planes In reference to the movement capability of a robot arm.

dead zone Safety zone where a robotic arm does not move during normal operation.

degrees of freedom The number of axes found on a root.

end effector A device mounted on the end of the manipulator or robot arm to physically do the work.

grippers Located on the end of the manipulator arm and used to pick up things.

LERT A classification system for robots based on four basic motion capabilities: linear, extensional, rotational, and twisting.

manipulator One of three basic parts of a robot.

microprocessor A device, a semiconductor chip used to provide the brain for robot control.

pneumatic power Using air pressure to drive the manipulator.

power source Supplies power to a robot.

target The point at which a robot's arm is expected to reach for picking up an object.

work envelope The space in which a robot's arm moves during its normal work cycle.

Review Questions

1. List six types of robots based on end purpose.
2. What is a manipulator?
3. What is the anchor point of a robot called?
4. How many axes of motion does a robot's wrist have?
5. What is a gripper?
6. What is an end effector?
7. What are the four basic motion capabilities of a robot?
8. What does arm geometry mean?
9. What is the target in robot terminology?
10. What is a Cartesian coordinate?
11. List the four classifications of robots based on the coordinate system used.
12. What is a work envelope?
13. What type of coordinate system produces a teardrop-shaped work envelope?
14. What does yaw mean?
15. What is a roll axis?
16. What are the three uses for the electric drive type of robot?
17. What is the advantage of the pneumatic method of moving a robot?
18. What is the disadvantage of the electric drive type of robot?
19. How much can a hydraulic drive robot lift?
20. How much can an electric drive robot lift?

CHAPTER 3

Drive Systems

Performance Objectives

After reading this chapter, you will be able to:

- Understand the role of hydraulics in the operation of robots.
- Know how pressure is created, pumps are used, and the word *pneumatics* got its name.
- Know the role of electric motors in robots.
- Know the types of electric motors used in robots.
- Know what slip is as well as what end effectors, grippers, end-of-arm tooling, and positioning are.
- Understand what is meant by the terms *repeatability* and *accuracy*.
- Know how drives work and how gears are used.
- Identify ball screws and bevel gears and know what harmonic drives can do.
- Identify the types of belts used in robots.
- Identify the types of chains used in robots.
- Identify and discuss the key terms used in the chapter.
- Answer the review questions at the end of the chapter.

Robots need some type of power to allow them to function. In order to make the arm or any other part of the system move, it is necessary to develop the power in a usable form and in some type of readily available unit. Three ways used to power a robot and its manipulator have already been mentioned. They are hydraulic systems, pneumatic systems, and electricity. Of course, electricity is also used to make the pneumatic and hydraulic systems operate. However, there is also an all-electric type of drive available in the form of electric motors that are attached directly to the manipulator or moving portion of the robot.

Hydraulics

Of the three types of drive systems for robots, the hydraulic system is capable of picking up or moving the heaviest loads, which is why it is so popular (Figure 3-1). It is also ideally suited for spray painting, where electric systems can be hazardous. Automobiles use a hydraulic system for braking. As the brake pedal is depressed, it puts pressure on a reservoir of liquid that is moved under pressure to the brake cylinders. The brake cylinders then apply pressure to the pads that make contact with the rotor that is attached to the wheel. By applying the pressure to the pads in varying amounts, forward motion of the car is either stopped or slowed as desired. The hydraulic system used for robots is similar to the braking system of an automobile.

Hydraulics is derived from the Greek word for water. However, oil, not water, is used in robot drive systems. (Some water-based hydraulic fluids are used in foundry and forging operations, however.) It is this oil that applies the pressure to the correct place at the right time in order to move the manipulator or grippers. Hydraulic pressure in a hydraulic system applies force to a confined liquid. The more force applied to the oil, the greater is the pressure on the liquid in the container.

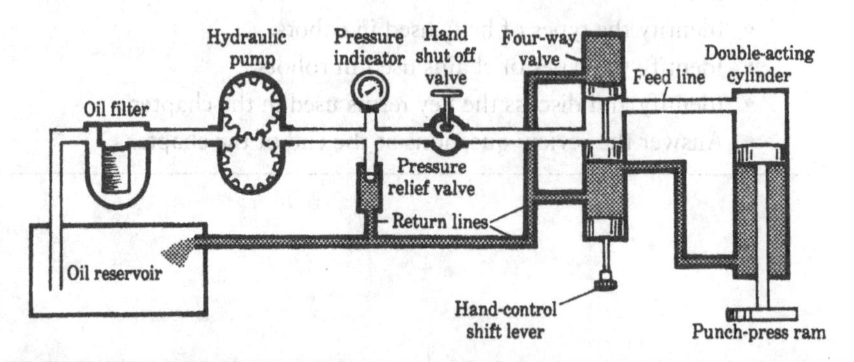

Figure 3-1 Hydraulic system.

Pressure

The force applied to a given area is called *pressure*. Pressure is measured in pounds per square inch, which can be abbreviated as lb/in^2 or psi. The metric unit for pressure is the pascal. The pascal is one newton per square meter (newton is the unit of force). Because the pascal is such a small unit, kilopascals and megapascals are used, For example, a typical automobile tire pressure is 190 kilopascals. Typical industrial air pressure lines use a pressure of 700 kilopascals (1 pound per square inch is equal to 6.896 kilopascals, and 1 kilopascal is equal to 1.45 pounds per square inch).

The pressure applied to a hydraulic system is in terms of how much push or force is applied to a container of oil (Figure 3-2). There are several ways to develop the pressure needed in a robot. The pressure developed in a hydraulic circuit is in terms of how much push or force is applied per unit area. Pressure results because of a load on the output of a hydraulic circuit or because of some type of resistance to flow. A pump is used to create fluid flow through the piping to the point where it causes movement of the gripper or manipulator of the robot. A weight can also be used to generate pressure.

One of the disadvantages of using a hydraulic system is its inherent problem with leaks. While fluid is under pressure, it has a tendency to seek the weakest point in the system and then run out. This means that each joint along the piping must be able to withstand high pressure without leaking. It also means that the moving part of the robot, the manipulator, must be able to contain the fluid while

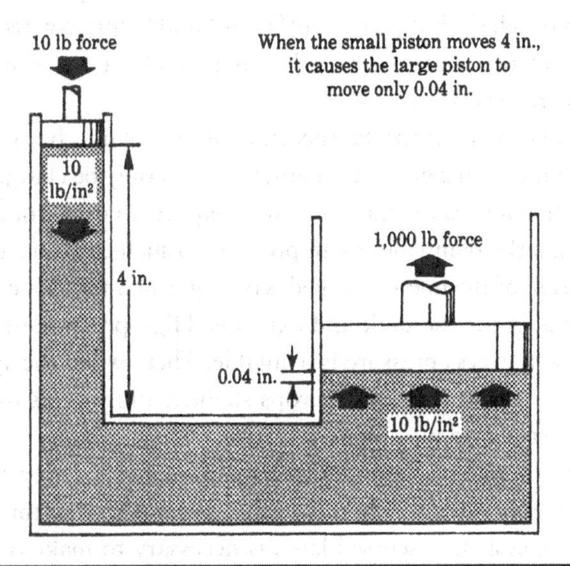

FIGURE 3-2 Hydraulic pressure changes (the 10-pound force can cause 1,000 pounds to be lifted, but notice the distance of each).

it is under pressure. In most cases, the O-rings used to seal joints are not able to contain the fluid, and leaks occur. O-rings are very important in sealing the moving end of the manipulator or terminating end of the fluid line. This is one of the maintenance problems associated with robots used for picking up heavy objects. The hydraulic system is used to pick up the weight, and then it has to be able to maintain the integrity of the system to keep from losing too much fluid. The fluid is slippery when spilled on the floor and must be contained to prevent accidents by humans working in the area.

Because of this pressure buildup in a hydraulic system, it is necessary to prevent its continuing to increase beyond the capacity of the pipes and containers. This is where a relief valve becomes important; a relief valve is used as an outlet when the pressure in the system rises beyond the point where the system can handle it safely. The valve is closed as long as the pressure is below its design value. Once the pressure reaches the point of design for the relief valve, the valve opens. Then the excess fluid must be returned to the reservoir or tank. This means that a relief valve and the return of the fluid to the tank must be part of the system.

Hydraulic systems need filters to keep the fluid clean; even small contaminants can cause wear quickly. The filter has to take out contaminants that measure only microns in size. The filters have to be changed regularly to keep the fluid and system in good operating condition.

Pressures of several thousands of pounds per square inch are not uncommon in the operation of a robot. This means that a leak as small as a pinhole can be dangerous. If you put your hand near a leak, it is possible that the fluid being emitted through the hole can cut off your hand before you feel it happen. However, this will not occur at pressures less than 2,000 pounds per square inch. High pressures are very dangerous.

Additional volume for the operation of a robot may be obtained by pressurizing the surge tank with a gas. An accumulator can only be charged to the main system pressure. In some cases, the robot will require a larger volume of fluid for rapid motion than the pump alone can provide. In such cases, a surge tank on the high-pressure side of the line is charged with a gas in a flexible container. This flexible container is inside the tank and expands. High pressure of the fluid compresses the gas when excess pressure is available. Then, when the system needs the fluid for rapid motion, the pressure drops slightly, and the gas in the tank expands to force the extra fluid into the system.

Hydraulic motors mounted on a manipulator are operated via signals from a transducer either to open the valve for a given period of time or to keep it closed. The control system, discussed later, is necessary to make the robot do what it is supposed to do when it is supposed to do it. Hydraulic motors and cylinders are

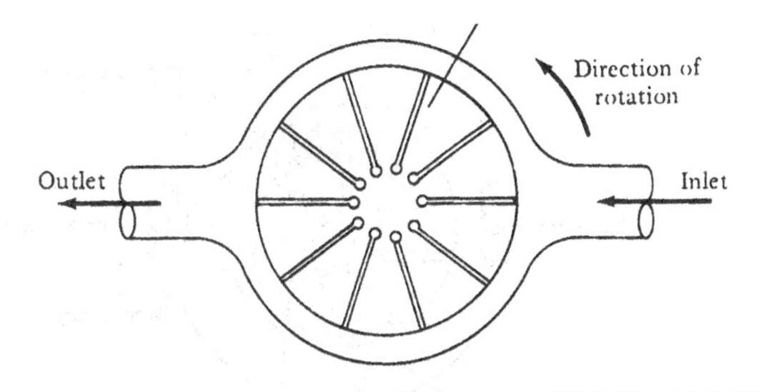

FIGURE 3-3 Hydraulic motor (pressure on hydraulic fluid causes the motor to turn in a counterclockwise direction; reversing the direction of fluid flow reverses the direction of motor motion). (*Courtesy of PWS-Kent.*)

compact and generate high levels of force and power. They make it possible to obtain exact movements very quickly (Figure 3-3).

Pumps

A hydraulic system must have a pump to operate. The pump converts mechanical energy usually supplied by an electric motor to hydraulic energy. Pumps are used to push the hydraulic fluid through the system.

There are two basic types of hydraulic pumps: hydrodynamic and hydrostatic. A *hydrodynamic* pump is a low-resistance pump and is not found in robot systems. Therefore, a closer look at the hydrostatic pump is in order.

Hydrostatic pumps are further classified into gear pumps and vane pumps. These pumps are called on to deliver a constant flow of fluid to the manipulator. A gear pump produces its pumping action by creating a partial vacuum when the gear teeth near the inlet unmesh, causing fluid to flow into the pump to fill the vacuum. As the gears move, they cause the fluid or oil to be moved around the outside of the gears to the outlet hole at the top. The meshing of the gears at the outlet end creates a pushing action that forces the oil out the hole and into the system.

A vane pump gets its name from its design, as does the gear pump. The vanes cause the fluid to move from a large-volume area to a small-volume area by pushing the fluid into a smaller area; as this occurs, the pressure on the fluid increases because the fluid itself is not compressible (Figure 3-4). A vane pump can be used to supply low to medium pressure capacity and speed ranges. They are used to supply the manipulator with the energy to lift large loads. Pressures can

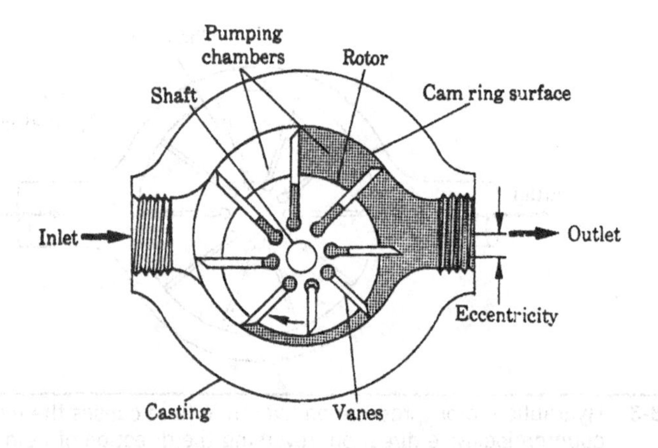

FIGURE 3-4 Vane-type pump/motor (note that this hydraulic pump can also be used as a motor when driven by the hydraulic fluid under pressure).

reach as high as 2,000 pounds per square inch with about 25 gallons of fluid per minute moving in the system.

Piston-type rotary devices are also used to increase the pressure in a hydraulic system. The pistons retract to take in a large volume of fluid and then extend to push the fluid into the high-pressure output port. There are usually seven or nine cylinders. These vary with the different configurations of machines. In-line piston pumps are good for robot applications. They have a very high capacity, and their speed ranges are from medium to very high. Pressures can get up to over 5,000 pounds per square inch (Figure 3-5).

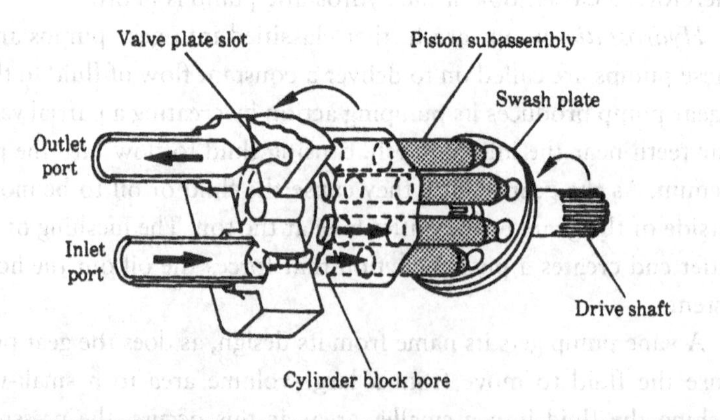

FIGURE 3-5 Piston-type rotary pump.

Pneumatics

Pneu simply means "air" in Latin. Pneumatics is that part of physics that works with air and gases. Robots use pneumatics to control their grippers and, in some cases, the manipulator arm. Therefore, in the field of robotics, compressed air is the medium used.

Pneumatic systems are used in industry to power hand tools and to lift and clamp products during machining operations. They use a compressor with a tank to store the compressed air. The compressor is driven by an internal combustion engine or an electric motor. A filter is used to remove contaminants, and a condensation trap is added with a drain to remove moisture. In some cases, a mist of oil is added to lubricate the parts being serviced by the air supply (Figure 3-6).

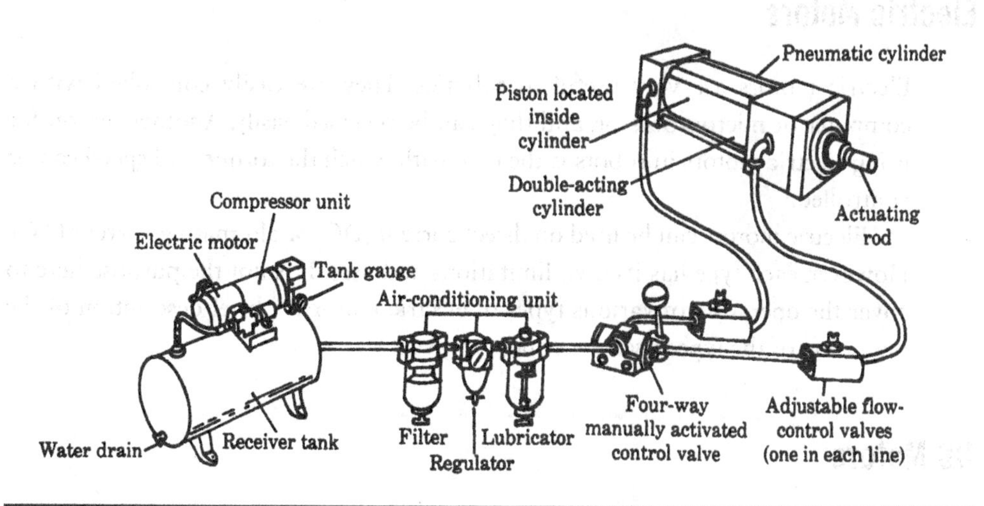

FIGURE 3-6 Pneumatic system and pneumatic cylinder (the piston and actuating rod are pushed forward by air pressure).

In order to maintain a constant pressure, a motor turns on when the pressure in the storage tank drops to a predetermined level. The motor and compressor build up the pressure and then turn off until needed again. An advantage that pneumatic systems have over hydraulic systems is that they can exhaust compressed air into the atmosphere, whereas hydraulic systems need a return system to contain the fluid for reuse.

Pneumatic systems contain a motor-driven compressor, a storage tank, and lines to carry the air from the compressor or storage tank to the using device. Along the way, the system has several methods of control. There is the hand shutoff valve and pressure-relief valve to control the air. Airflow or pressure can be regulated by a regulator and a three-way valve. Notice the similarity of the pneumatic and hydraulic systems. However, the pneumatic system simply exhausts the air into the atmosphere when it is finished using the pressure to operate the device. Silencers keep down the noise of the exhausted air. The hydraulic system has to have a return line from the user to the storage tank. This makes the hydraulic system more expensive to install and operate.

The pneumatic cylinder is the load device designed to use air pressure to do work. This part changes the mechanical energy of air into linear motion that drives a press ram. Pneumatic load devices can also be used to produce rotary motion. About the only problem in maintaining a pneumatic system is keeping the air supply free of moisture. The lines have to be kept clean and dry.

Electric Motors

Electric motors are very useful in robotics. They are easily controlled with a computer or microprocessor, and they can be reversed easily. Another reason for using electric motors in robots is the ease with which the torque and speed can be controlled.

Electric motors can be used on direct current (DC) or alternating current (AC). However, each type has its own limitations and uses. It is not the purpose here to cover the operation of various types of electric motors. A brief description of the operation of the types used for robots is given.

DC Motors

DC motors have series, shunt, and compound configurations (Figure 3-7). Each type has advantages and disadvantages. The series configuration, for instance, can be used where there is a need for lots of torque to lift or move something. However, it cannot be used for devices that are belt driven or where constant speed is needed. The shunt configuration does not have the starting torque but does have the constant speed characteristic. The compound configuration has some of the qualities of both the series and shunt but has other limitations.

Most DC motors have brushes. They also have commutators that consist of small pieces of copper separated by a mica insulator. A maintenance problem can result because of brush wear and the arc that occurs between the brushes and the commutator as the motor is loaded.

Speed control of DC motors can be accomplished by regulating their voltage or current or both. Electronic controls have been designed to aid in the speed

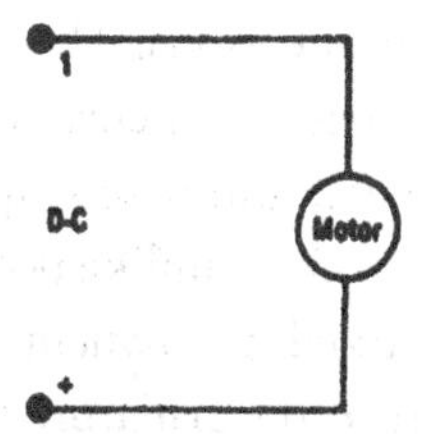

PERMANENT-MAGNET, BRUSHLESS DC, PRINTED CIRCUIT, SHELL-TYPE ARMATURE. To reverse: transpose motor leads.

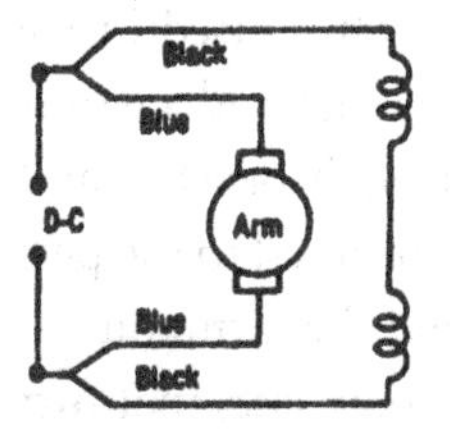

SHUNT WOUND. To reverse: transpose blue or black leads.

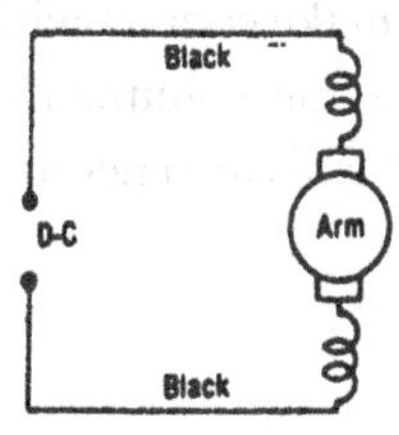

SERIES WOUND (2 LEAD). Non-reversible.

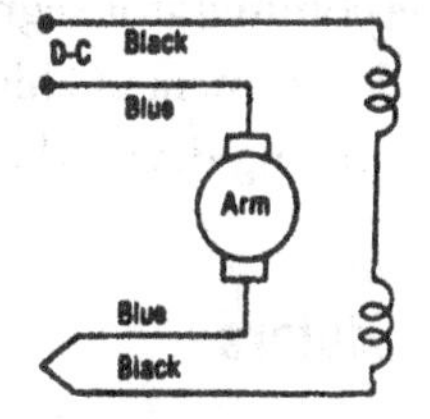

SERIES WOUND (4 LEAD). To reverse: transpose blue leads.

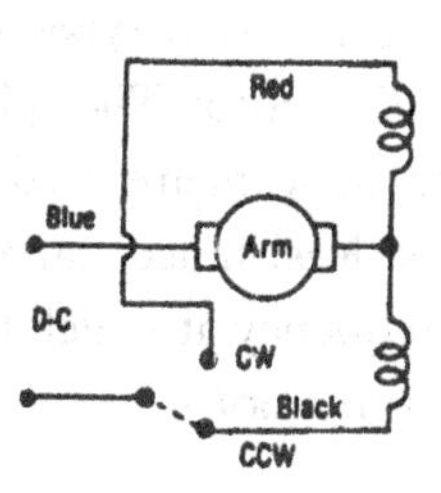

SERIES WOUND (SPLIT FIELD). To reverse: connect other field lead to line.

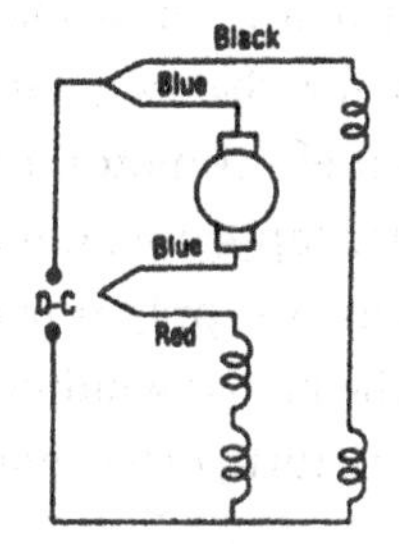

COMPOUND WOUND (5-WIRE REVERSIBLE). To reverse: transpose blue leads.

FIGURE 3-7 Wiring diagram for DC motors (the loops indicate field coils, and "arm" indicates the armature or rotor).

control of DC motors. They are easily reversed by reversing the polarity of the electric current supplied to their windings. Variable resistors can also be used to control the speed of DC motors. A resistor is inserted in series with the field windings and adjusted to increase or decrease the voltage available to the motor.

Permanent-Magnet Motors

Permanent-magnet (PM) motors are used to drive small toys and such things as electric seats and electric windows in automobiles. They have other uses as well.

They are not often used in robots except in toy applications. They consist of a permanent magnet and a wound rotor or armature. The direct current is fed to the armature through brushes and commutator segments, and it sets up a magnetic field that is attracted to the permanent magnet's field. Once unlike poles have been attracted and the armature rotates toward the fixed permanent magnet, the commutator segments that furnished the DC power to the armature move over to two other segments and cause a different pole to be energized and magnetized. This pole will then be attracted to the permanent magnet. But, as it approaches the permanent magnet, the energy in the coil of wire making up this segment will be removed by the commutator segment moving on to make contact with the brushes on the next two segments. This continues because the commutator is nothing more than a switching device that determines which coils are energized and when.

DC Brushless Motors

Newer DC brushless motors use transistors or *Hall-effect* magnetic devices to reverse the current through the field coils. Whereas the commutator does the switching of current in the armature of regular DC motors, DC brushless motors use electronics for the switching operation. There are two types of brushless motors: the Hall-effect motor and the split-phase PM motor. The split-phase PM motor has center-tapped windings (Figure 3-8). Transistors are used to set up an oscillator circuit to supply power to the windings. Resistance, capacitance, and inductance of the motor windings determine the frequency at which the oscillator operates. This, in turn, determines the operation of the motor.

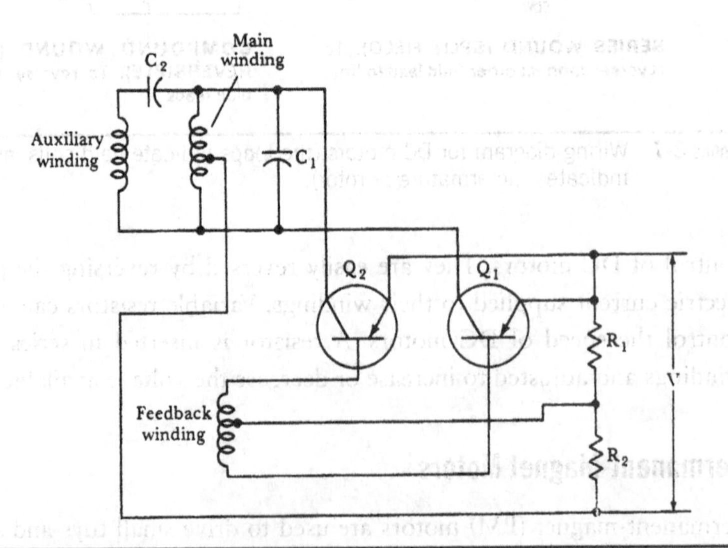

FIGURE 3-8 Split-phase permanent-magnet DC brushless motor. (*Courtesy of PWS-Kent.*)

The Hall-effect motor uses that quality of a transistor to react to the presence of a magnetic field to do its switching. When a magnetic field passes near a transistor, the resistance of the semiconductor in the transistor decreases. This decrease in resistance causes the current in the circuit to increase. The increased current flow is fed to the electronic switch that controls the current being fed to the field winding of the motor (Figure 3-9).

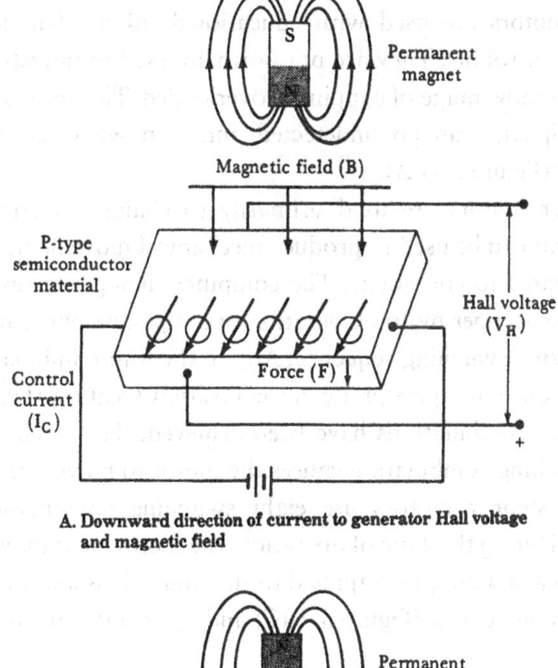

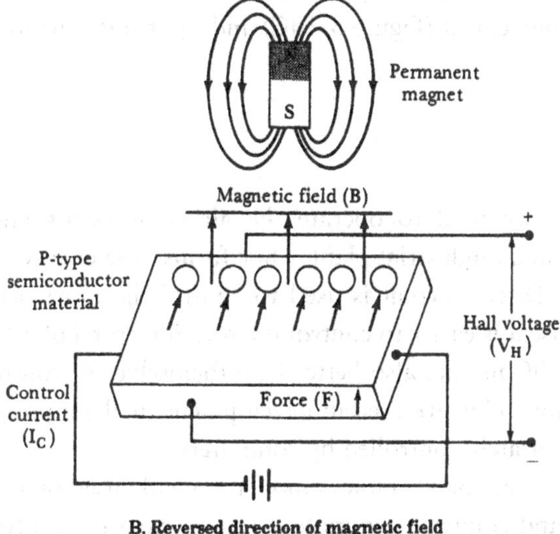

FIGURE 3-9 Hall effect used for controlling motor speed. (*Courtesy of PWS-Kent.*)

nature
shaft rather than being mounted as field magnets. The field coil is wound. The advantage of using a brushless motor is its long life. Elimination of the brushes and commutator makes it practically a maintenance-free motor. The main disadvantage of the brushless motor is the low torque developed by the split-phase PM type.

Stepper Motors

Stepper motors are used with educational robots that demonstrate the basic operation of robots. They are not normally used in industrial applications. They have the disadvantage of slipping if overloaded. This means that the error created by the slippage can go undetected and ruin whatever is being machined or processed (Figure 3-10A).

Stepper motors are used primarily to change electrical pulses into rotary motion that can be used to produce mechanical movement, which is why they are so well mated to computers. The computer then generates the pulses needed to operate the stepper motor. Operation of a bipolar stepper motor is accomplished in a four-step switching sequence. Any of the four combinations of switches 1 or 2 will produce an appropriate rotor position location (Figure 3-10B). After the four switch combinations have been achieved, the switching cycle repeats itself. Each switching combination causes the motor to move one step.

Some stepper motors use eight switching combinations to achieve *half-stepping*. During this type of operation, the motor shaft moves half its normal step angle for each input pulse applied to the stator. This allows for a very precise and controlled movement (Figure 3-10C) and eight stator windings.

AC Motors

AC motors are used to operate the air compressors and materials handling equipment in an industrial plant. They furnish the energy to move materials and equipment. Direct current is used to control the movement of robots, in some cases because it is easier to control the rotation speed of a DC motor than it is an AC motor. DC motors also better lend themselves to control by computers than do AC motors. However, recent developments make it possible for AC motors to be more accurately controlled by computers.

The two types of AC motors most frequently used on robots are wound-rotor induction and squirrel-cage motors. They operate on 120, 240, and 440 volts based on the voltage available in the plant where the robot is located.

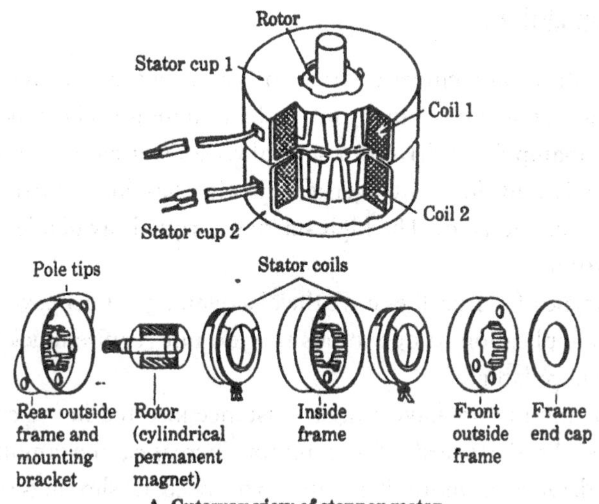

A. Cutaway view of stepper motor

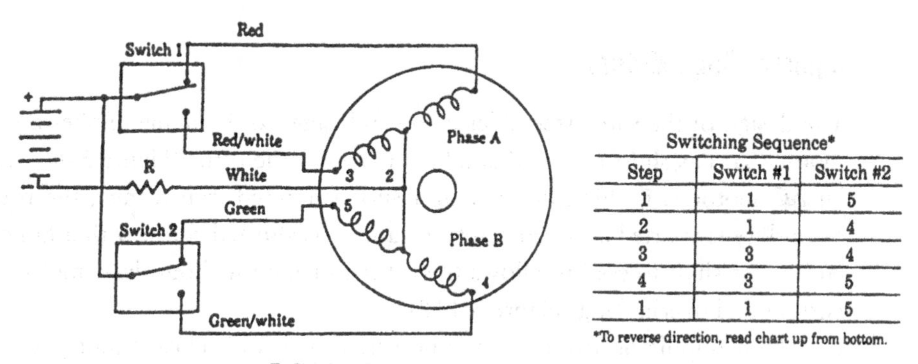

Switching Sequence*

Step	Switch #1	Switch #2
1	1	5
2	1	4
3	3	4
4	3	5
1	1	5

*To reverse direction, read chart up from bottom.

B. Wiring diagram and switch combinations

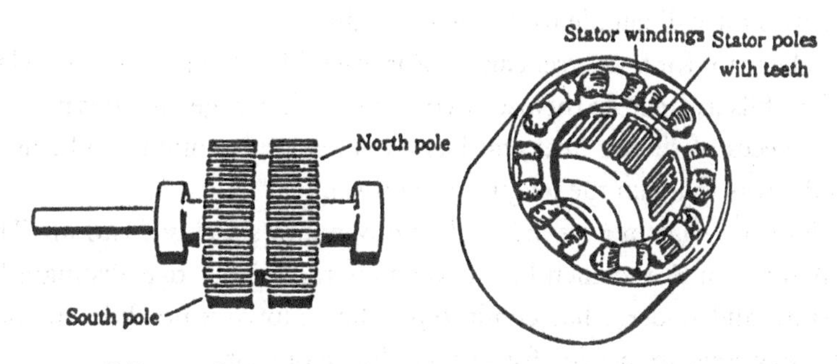

C. Rotor and stator of a bipolar, permanent magnet stepper motor

FIGURE 3-10 DC stepper motor. (*Courtesy of Superior Electric Company.*)

Induction Motors

This classification includes a large variety of motors. However, we will limit our discussion to the wire-wound rotor induction motor. This type of motor is located on many manipulator drives for several reasons. It has smooth acceleration under heavy loads and does not overheat. It also has high starting torque and good running characteristics. The high starting torque is available because of the wire-wound rotor.

This type of motor does have the disadvantage of being very slow to start. This limits its application to operations that do not require quick motion on the part of the manipulator.

Induction motors have a high resistance inserted in the circuit with the rotor windings. As the speed of the motor increases, the amount of resistance is decreased. Once the motor has come up to speed, its slip rings remove the resistance box, and the motor operates as an ordinary squirrel-cage motor.

Squirrel-Cage Motors

The design of the rotor resembles a squirrel cage, so the name implies a type of motor that has short-circuited conductor bars in the rotor (Figure 3-11A). Since an AC motor is nothing more than a short-circuited transformer, the rotor or secondary is shorted and allowed to rotate. This shorted rotor, with a fan on the end of the shaft to cool it, allows the rotor to rotate without drawing too much current and overheating (Figure 3-11B).

The manipulator arm is usually powered by a squirrel-cage motor. Different design characteristics meet the particular needs of a manipulator. Most of these design characteristics concern the way the rotor is made, its speed, and the amount of current it will handle to do the job right.

The rotor for a squirrel-cage motor resembles that part of a wheel that squirrels and gerbils use to keep themselves occupied when caged up. It is made of laminated core pieces of silicon steel and has a copper or aluminum end ring and bars for conductors that run the length of the rotor.

Squirrel-cage motors are broken down into six classifications. They are listed as A through F, and each has its own characteristics that distinguish how it will operate and under what conditions. The following breakdown shows why the classifications are needed for proper choice of motor.

- **Class A:** Normal torque, normal starting currents, most popular type of squirrel-cage motor
- **Class B:** Normal torque, low starting currents

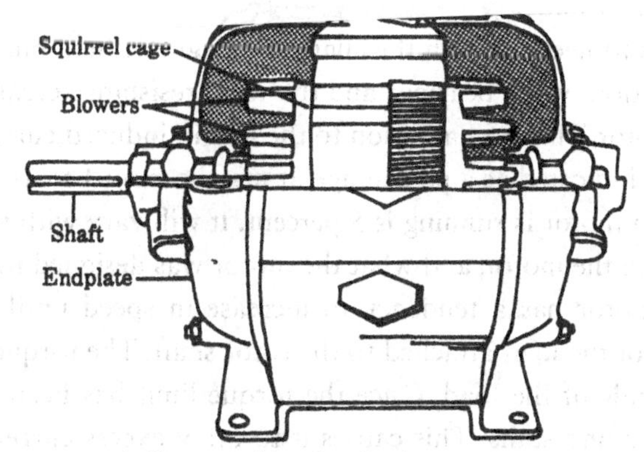

A. Cutaway view of squirrel-cage motor

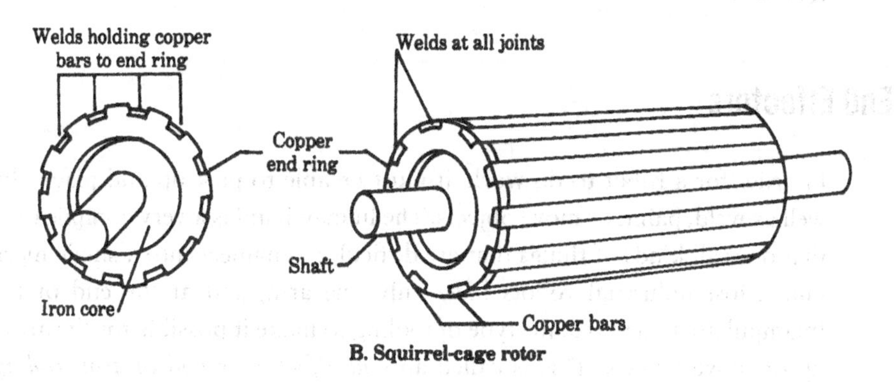

B. Squirrel-cage rotor

Figure 3-11 AC squirrel-cage motor.

- Class C: High torque, low starting currents
- Class D: Very high slip percentages, most often used for robot applications
- Class E: Low starting torque, normal starting currents
- Class F: Low torque, low starting currents.

Slip

The rotor of an induction motor cannot keep up with the changing magnetic field in the pole pieces or stator. The difference between the speed of the rotor and the changing magnetic field generated by the 60-hertz alternating current is known as *slip*. Slip results when there is a loss of induced current between the stator poles and the rotor conductors. Load torque is what causes the motor to develop slip. The difference in speed of the rotor and the changing magnetic field is what makes the motor turn. If the rotor and magnetic field were rotating at the same speed, the difference in the induced magnetic field and the one in the stator would inhibit the rotor from producing the torque to maintain its speed; even with no load, the rotor tries to keep up with the changing magnetic field but cannot do so because of the friction of the bearings and the wind resistance created by the fan and the moving rotor. This is in addition to the loss in induced current in the rotor.

Slip is indicated as a percentage. It may be from 1 to 100 percent. The normal slip after a motor is running is 5 percent. It will vary with the design of the rotor, the load on the motor, and what the motor was designed to do.

The motor has a tendency to increase in speed until it reaches the torque demands of the load attached to the rotor shaft. The torque will increase to meet the demands of the load. Once the torque limit has been reached, the motor is overloaded and stalls. This causes it to draw excess current and smoke and get very hot.

End Effectors

In order for a robot to do work, it must be able to pick up and place objects as well as weld, paint, or move objects. The human hand is a very complicated device that does all kinds of things that are difficult to engineer into a single mechanical unit. Most industrial robots have only one arm, and at the end of that arm (manipulator) there is some type of tooling to make it possible for the arm to pick up or move objects. This is called an *end effector* or *end-of-arm tooling*. Both terms are found in the literature, but *end effector* seems to be gaining acceptance as the term to be used.

The manipulator is used to move the end effector to where it is supposed to do its job. Once the end effector is at the point where it is supposed to do its work, there must be some way for it to grasp or pick up the material. And, in the case of a vacuum-operated end effector, there must be some way of turning off the vacuum to release the part once the robot has reached the programmed place. Many different designs of end effectors are available from manufacturers. They operate by using a vacuum, a magnet, or a mechanical gripper.

End effectors are divided into two groups: grippers (similar to the human hand) and end-of-arm tooling.

Grippers

Grippers perform no operations on the parts they handle. They are used to pick up a part and place it somewhere else (Figure 3-12). A gripper may be used to place a part in a hot furnace and remove it once it is properly heated. Then it can move the piece to a bath that will treat the metal properly for a given period of time. Once the piece has been treated properly, it can be placed on a conveyor belt or a pallet for further work at another station. A gripper is like a hand; it takes an object and holds it. The gripper may hold the object or move it while an operation is being performed on it. In either case, the gripper is designed by the manufacturer or in some cases the owner of the robot to perform a specific function.

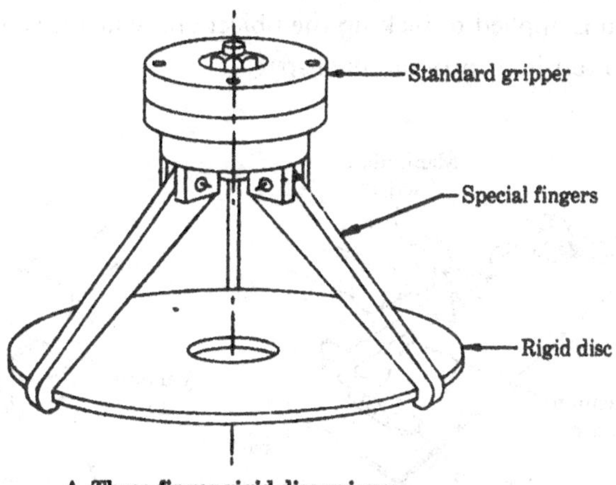

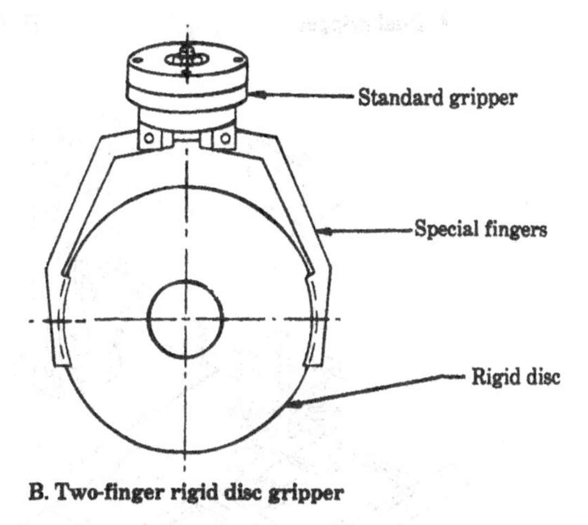

Figure 3-12 Two- and three-finger grippers. (*Courtesy of Mack Corporation.*)

Vacuum Grippers

Finger-grasping grippers are not the only grippers available. Vacuum cups are also available as standard items (Figure 3-13). They are chosen to do a job of lifting based on the weight of the object to be moved. Vacuum cups are attached to the end of the manipulator, and a vacuum hose is attached to the cups. A good vacuum system will be able to cause the suction to hold a flat part until it can be moved and deposited in its desired location. The vacuum must be turned off once the part has reached its programmed place. The controller then is used to determine when the suction is applied to pick up the object and when it is to be released to allow the part to rest in its programmed space.

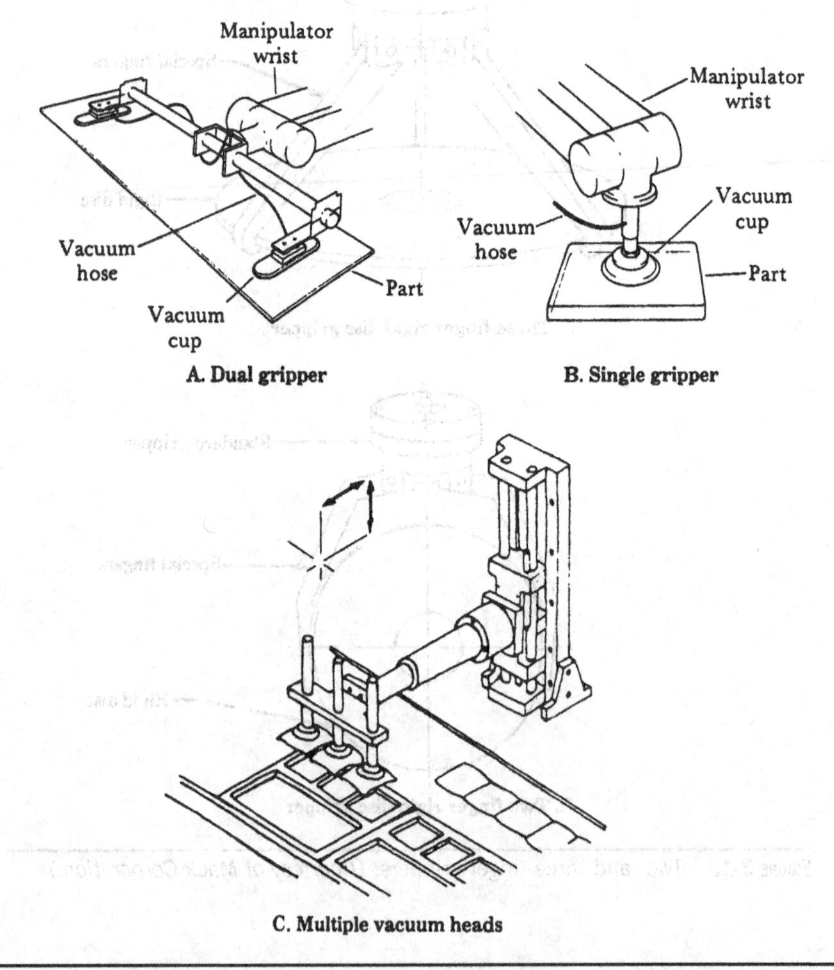

FIGURE 3-13 Vacuum grippers. (*(A) and (B) courtesy of PWS-Kent; (C) courtesy of Mack Corporation.*)

Magnetic Grippers

An electromagnet may be used as an end effector. The electromagnet is designed so that it can fit the object to be picked up. In this instance, the object must have a flat surface and must be ferromagnetic. This means that it has to be easily picked up by a magnet (Figure 3-14).

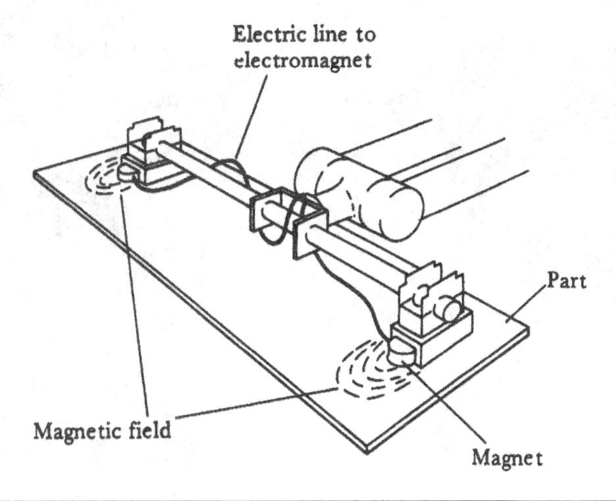

FIGURE 3-14 Dual magnetic grippers. (*Courtesy of PWS-Kent.*)

End-of-Arm Tooling

If a robot is to be used for welding or spray painting, another type of end effector can be used. A cutting torch can be fitted on the end of the manipulator or a spray hose can be connected to a nozzle to allow painting to a precise degree. The end-effector mounting flange holds the spray nozzle or welding or cutting torch so it can be manipulated easily and precisely.

The end of the manipulator usually has holes that accommodate attachment of the different types of end-of-arm tooling (Figure 3-15). The ability of the manipulator to lift certain loads is critical to what the robot will be able to do. This payload or lifting capacity is adjusted in relation to the weight of the end-of-arm tooling. The tooling is also part of the weight that has to be lifted by the manipulator. Payload is usually given in kilograms. So, if you have an end-of-arm tooling that weighs 4 kilograms and the limit of the manipulator is 10 kilograms, this means that the load to be picked up cannot weigh more than 6 kilograms. The end-of-arm tooling is sometimes connected to the manipulator by a safety joint. The safety joint is designed to protect the tool in case the robot crashes.

FIGURE 3-15 A Kuka welder robot. (*Courtesy of Kuka.*)

The path for the tool is determined by the controller. The controller is very important when it comes to using a robot for welding. The path of travel has to be precise. This type of welding is done in large manufacturing plants, such as in the manufacture of automobiles and appliances. Spot welding, arc welding, and gas welding are done by robots in large numbers. These welding tools are mounted on the end of the manipulator. Look at Figure 3-15 again.

Positioning

One of the most important and demanding aspects of designing robots is the ability of the robot to position the manipulator. The robot must keep the manipulator there as long as it takes to perform the task called for by the controller.

Robot controllers are classified as either low, medium, or high technology. These classifications indicate the level of operation the robot is able to accomplish (Figure 3-16). (Note that not everyone in the robotics field agrees with this low-, medium-, and high-technology classification. Therefore, the terms are used here to provide continuity of thought for this book.)

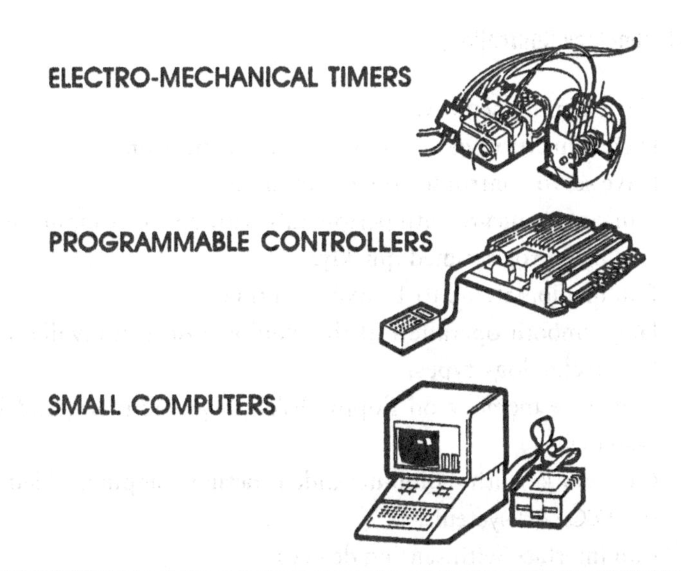

FIGURE 3-16 Early robot controllers were electromechanical timers for low-technology robots, programmable controllers for medium-technology robots, small as well as large computers for high-technology robots. (*Courtesy of Mack Corporation.*)

Low-Technology Controllers

1. Are not easily reprogrammed and rely on mechanical stops to control their movements.
2. Take a long time to reprogram.
3. Do not have an internal memory for storage of information.
4. Do not have a microprocessor to give command signals for control of axis movement.

Medium-Technology Controllers

1. Are used with a two- or four-axis manipulator.
2. Have a microprocessor and a memory.
3. Have a limited capacity for input/output signals for control of peripherals.
4. Are slow to react to commands and can support movement on only one axis at a time.
5. Can be reprogrammed because they have a memory.
6. Are usually limited in the amount of memory available (usually enough for two programs without reprogramming).

High-Technology Controllers

1. Have a large memory.
2. Have a microprocessor and a comicroprocessor.
3. Have servo control for the manipulator.
4. Can communicate with peripherals with up to 64 input/output signals.
5. Can be reprogrammed quickly.
6. Can manipulate up to 10 axes at a time.
7. Have smooth operation of the manipulator arm (available in all, not just high-technology types).
8. Can store memory on floppy disks, magnetic tapes, and bubble memory cassettes.
9. Can work with computer-aided design/computer-aided manufacturing (CAD/CAM) systems.
10. Can interface with sensing devices.

Controlling the action of the manipulator becomes very important in the operation of any robot. This is where the controller chosen makes the difference in potential uses of the robot. The velocity of the manipulator axis as it approaches the programmed point in the work cell is very important. This is accomplished by speed controls on the motors that operate the manipulator and its end effectors. The motor is stopped at the programmed point by the use of dynamic braking or plugging.

Dynamic braking takes place when a resistor is switched across the armature of the motor as quickly as the electricity is turned off (Figure 3-17). Current is generated by the armature in what is called a *generator action.* This means that the armature acts as a generator once it continues to rotate and a magnetic field is still present for a short period. Putting a resistor across the armature directs this current in the opposite direction from that which caused it, halting the armature because the unlike polarities tend to be attracted.

Plugging takes place when a motor has its electrical power reversed in polarity (Figure 3-18). This reversal of polarity has a tendency to stop the motor immediately. It can be harmful to a motor if done too often. It has to be done quickly, and then the reverse polarity has to be disconnected once the armature has stopped turning. Usually the switching action is such that the polarity is returned to its original direction after the quick application of a reverse polarity.

A stepper motor may be used with medium-technology manipulators. The stepper motor has pulses sent to its windings. These pulses, based on the design of the motor, will cause it to move a certain number of degrees instead of making a full 360-degree turn—it may rotate only 2 degrees with each pulse. By applying

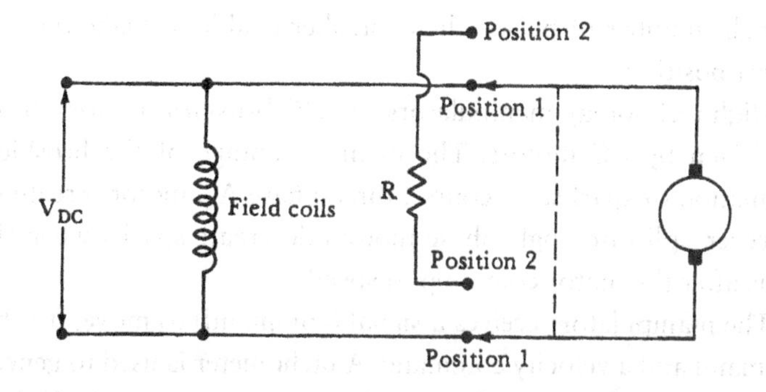

FIGURE 3-17 Dynamic braking circuit (moving the switch from position 1 to position 2 causes a reversal of the DC polarity, and the motor stops quickly). (*Courtesy of PWS-Kent.*)

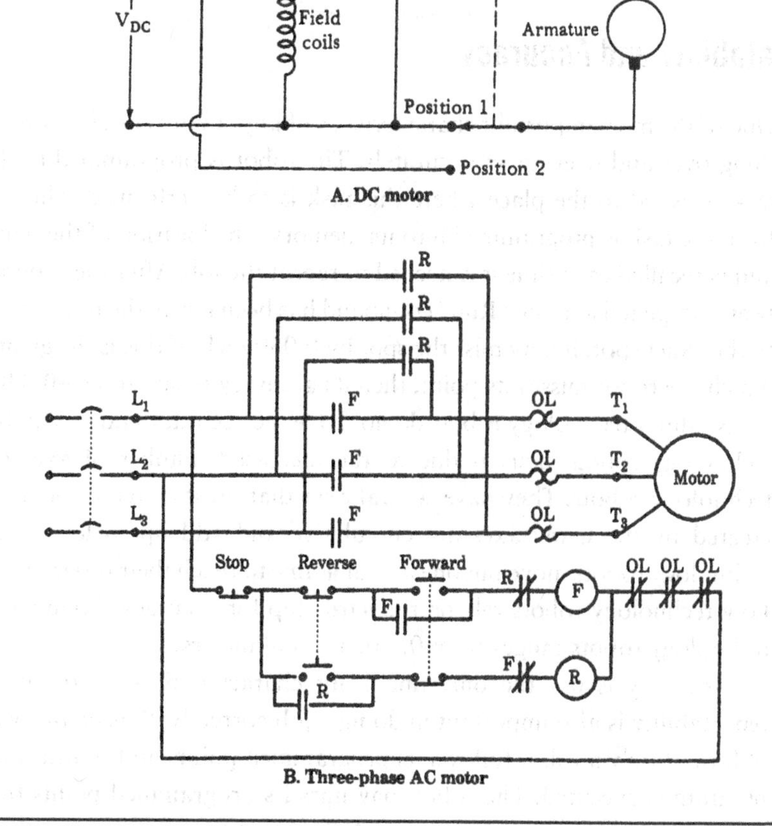

FIGURE 3-18 Plugging circuits (switching action reverses the polarity; three-phase motors can be plugged). (*Courtesy of PWS-Kent.*)

the right number of pulses, the controller is able to make the motor step to the correct position.

High-technology manipulators use DC brushless motors. In some cases, they are driven by AC motors. The main advantage of the brushless motor is the elimination of sparks. Of course, three-phase AC motors create no sparks either. However, split- or single-phase motors do create sparks when their start switch opens after the motor comes up to speed.

The manipulator receives a signal (command) to move. It is both a positional command and a velocity command. A tachometer is used to generate a signal that is fed back to the controller to indicate the speed at which the manipulator is moving. The tachometer is nothing more than a simple DC generator. It resembles a PM electric motor except that it is driven by the turning of the manipulator and thus generates a signal as its armature rotates in the PM field. Tachometers are similar to the PM motors used to drive toys except they are made with a little more precision.

Repeatability and Accuracy

One of the most important characteristics of any robot is its ability to do the same thing over and over again accurately. The robot is programmed to do a specific task. It is led to the place where this task is to be performed. This route to and from the task is programmed into its memory. The location of the points is stored and is recalled each time it is needed to repeat the job. After the programming has been completed and the "Run" command has been given, the robot may not return to the exact spot. It may miss the spot by 0.030 inch. If this is the greatest error by which the robot misses its point, then its accuracy is said to be ±0.030 inch.

Medium-technology robots do not have the accuracy and repeatability of low-technology robots. This is due to the increased number of axes on medium-technology robots. They have several axes that must converge on a point. Errors created by the many axes are cumulative and add up to less accuracy. Low-technology robots move but one axis at a time to reach their programmed position. Low-technology robots rely on the hard stop for accuracy. Accuracy of medium-technology robots ranges from 0.2 to 1.3 millimeters.

Accuracy is not the only important characteristic of a robot. The robot's repeatability is also important in doing a job correctly. *Repeatability* is a measure of how closely a robot follows its programmed points and returns every time the program is executed. The robot may miss its programmed points by 0.030 inch the first time the program is executed. During the next execution of the program, if the robot misses the point it reached during the previous cycle by 0.010 inch, or a total of 0.040 inch from the original programmed point, its accuracy is ±0.030 inch and its repeatability is ±0.010 inch.

Repeatability can change with use and time. The mechanical components have a tendency to wear that increases the inaccuracies of the whole system. Most inaccuracies are easily corrected. In some applications, good repeatability is more desired than accuracy.

Drives

Gears, belts, and chains make up the drive systems of robots. Each has its particular application. The main reason for these drives is to transfer energy. In order to move the manipulator and the end-of-arm tooling, some means of transmitting energy to both must be devised. That is where the chains, belts, and gears become useful. The big task is to transfer the energy from an actuator to the manipulator. The actuator is the motor or drive energy source.

Gears

The most popular method for transferring energy from an actuator to an end effector is gearing. This means that if the actuator is located at the center of the manipulator, some method must be employed to move the energy generated at the actuator to the end effector. Gearing is one method and the one used most often.

Gears are made in many sizes and with different numbers of teeth. They can be designed to give mechanical advantages and to reduce speed. The speed of an electric motor is most often too high for direct application to robot motions. That means that it has to be geared down to the speed needed for operation. Gears can be used to increase or decrease the effective speed of an electric motor.

Gears can be purchased in many different configurations. They can operate at right angles to one another, or they can operate one within another. They have the ability, when arranged properly, to change rotary motion to linear motion by driving a straight bar with gear teeth (called a *rack*). End effectors use different gear mechanisms, and so do the drive portions of robots. There are many different grippers and finger arrangements. They, too, use a variety of gears and levers to convert the power source into the desired motion and power needed to perform the job.

Gear Trains

When several gears are placed together, they form a *train* (Figure 3-19). This is done in order to change the direction of rotary motion or to increase or decrease the speed of rotation of the final gear shaft. There are two types of gear trains: *ordinary* gear trains and *planetary* gear trains. Ordinary gear trains are broken down into two other classifications: *simple* gear trains and *compound* gear trains.

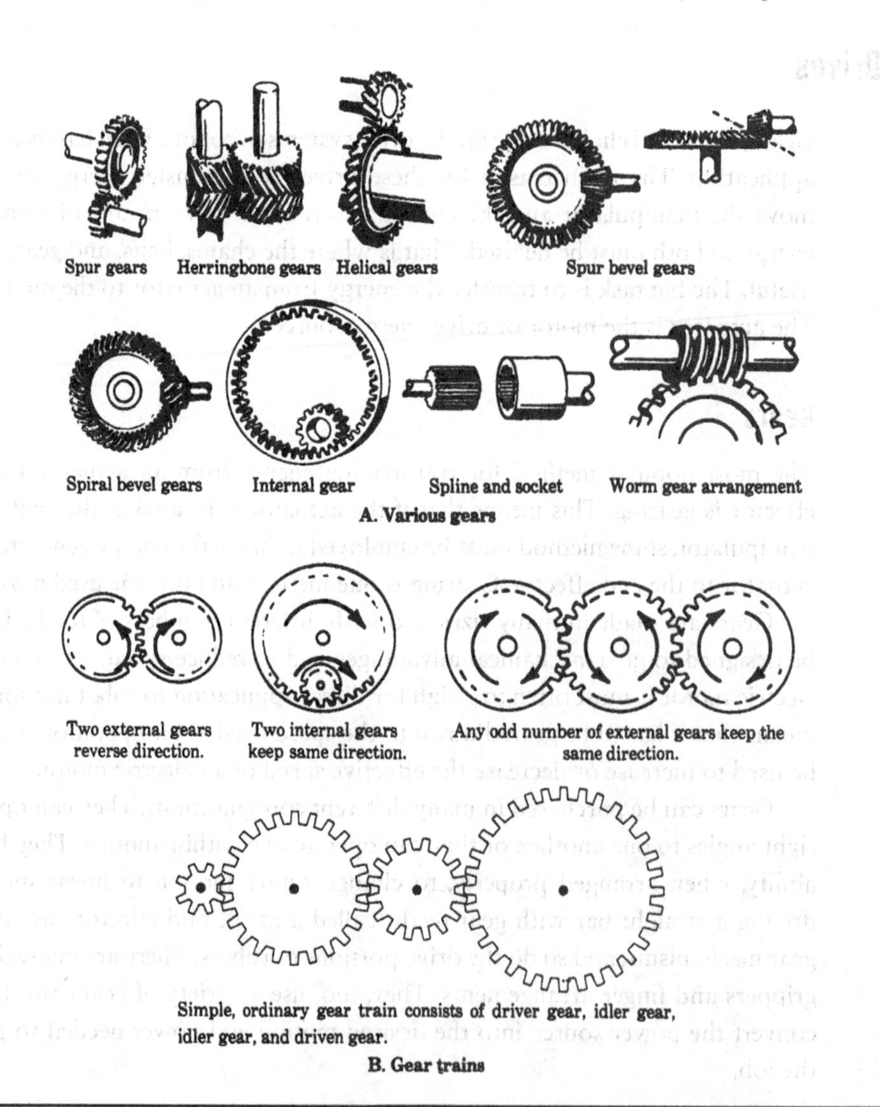

Spur gears
Herringbone gears
Helical gears
Spur bevel gears
Spiral bevel gears
Internal gear
Spline and socket
Worm gear arrangement
A. Various gears
Two external gears reverse direction.
Two internal gears keep same direction.
Any odd number of external gears keep the same direction.
Simple, ordinary gear train consists of driver gear, idler gear, idler gear, and driven gear.
B. Gear trains

FIGURE 3-19 Gear arrangements and gear trains.

Worm Gears

Worm gears are often used for driving the base of a manipulator in robots (Figure 3-20). There are some complex arrangements for making sure that the right speed and direction are obtained. The worm is the screw with either a single thread or multiple threads. The form of the axial cross section is the same as that of a rack. The teeth of the worm wheel have a special form that is required to provide proper conditions for meshing with the worm.

Worm gears are used to convert a circular or rotary motion to a linear circular motion. This is important when it comes to turning the manipulator arm of a robot. An electric motor, a pneumatic motor, or a hydraulic motor may be used to drive the worm gear arrangement to power the manipulator.

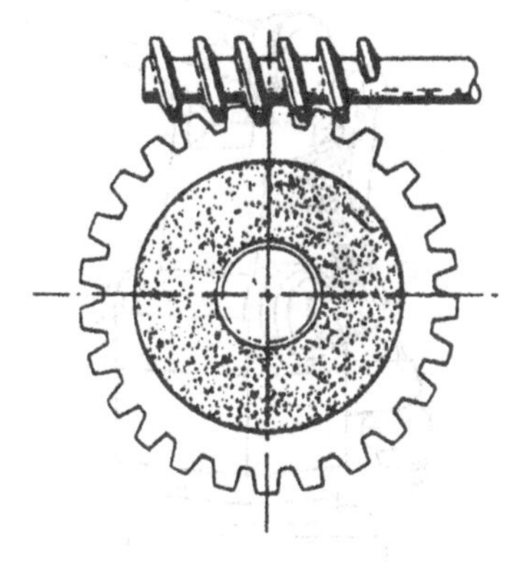

FIGURE 3-20 Worm gear (the wheel part is round, and the worm part is the threaded rod-shaped part).

Ball Screws

When servo motors and stepping motors are used in robots, it is often necessary to convert rotary motion to linear motion. This was done for years by a threaded rod with a nut. If the nut is kept from turning as the threaded rod is rotated, the nut will move along the threaded rod. The major problems with this system were friction and ware. When the threaded rod was turned in the nut, the high friction produced by the tightly fitting screw threads wasted much of the power created by the motor. As the rod turned in the nut, the nut and threaded rod would wear. This wear created inaccuracies that had to be dealt with before this method could be used in robots and more modern automated systems.

These problems have been overcome with improvements in the ball-bearing screw drive (Figure 3-21). The ball screw has been around for many years. The

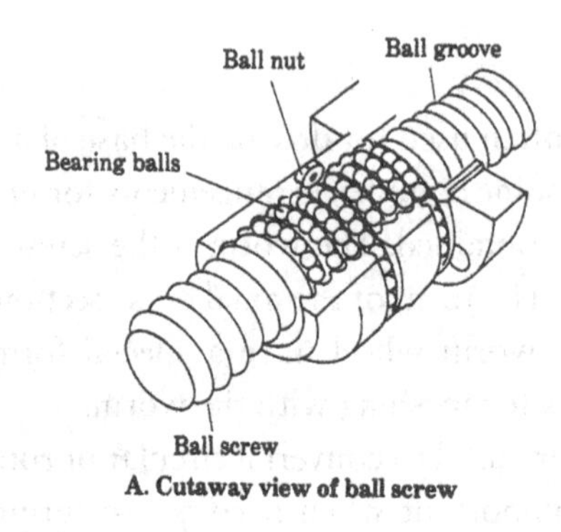

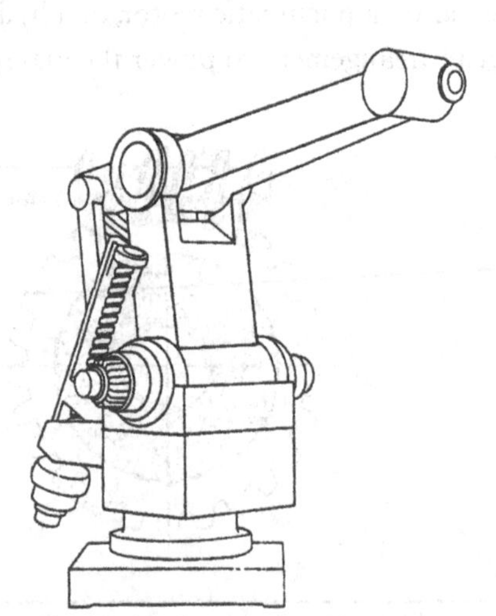

FIGURE 3-21 Ball screws. (*(A) courtesy of PWS-Kent; (B) courtesy of Warner Clutch and Brake Co.)*

groove in the screw is similar to the groove in a standard screw but is cut to allow a ball bearing to roll freely in the groove. Some automobiles now use ball-bearing screw drives in their steering mechanisms.

Bevel Gears

Bevel gears are conical (cone-shaped) gears that are used to connect shafts that have intersecting axes (Figure 3-22). *Hypoid gears* are similar to bevel gears in their general form, but they operate on axes that are offset. Most bevel gears can be classified as either straight- or curved-tooth gears. Spiral-bevel, zero-bevel, and hypoid gears are all classified as curved-tooth gears. Straight-bevel gears are the

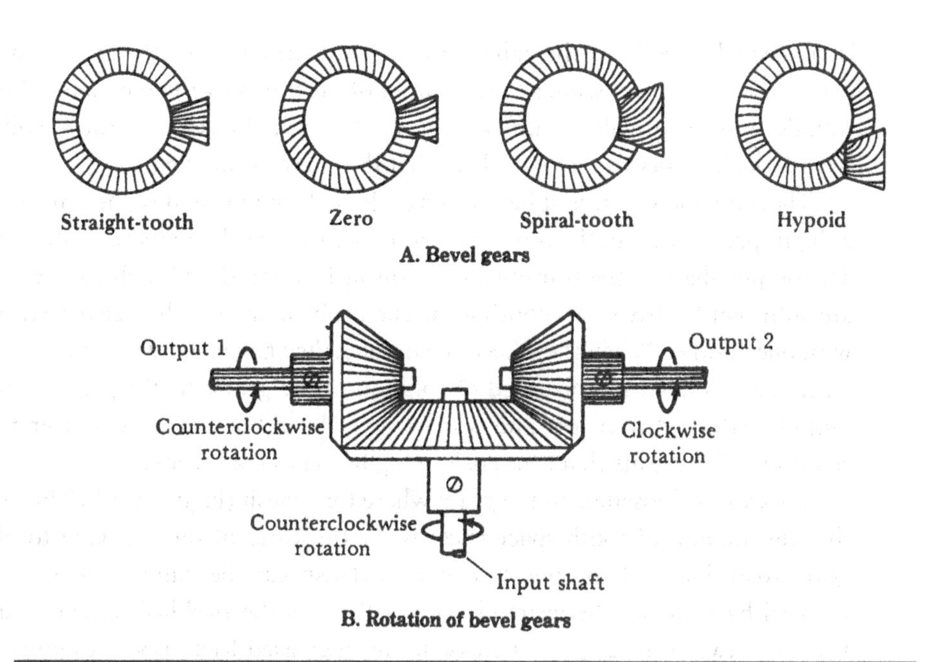

Figure 3-22 Bevel gear arrangements and rotation. (*Courtesy of PWS-Kent.*)

most commonly used type of all the bevel gears. However, most of these are not used on robots. The teeth are straight, but their sides are tapered so that they intersect the axis at a common point (called the *pitch cone apex*) if they were extended inwardly.

Bevel gears are used in robot manipulators when a 90- or 45-degree transfer of motion must take place. The cut of the teeth in a bevel gear determines the gear's ability to withstand heavy loads. Spiral-tooth gears can handle heavier loads than zero-cut gears. The teeth of a hypoid gear and a zero-cur gear are similar. The only difference is that the driver gear is offset from the center contact point of the gear. This gearing arrangement is sometimes referred to as the *miter gear* is because the 45- and 90-degree arrangement appears the same as a miter joint.

Adjusting Gears

Ball-screw drives convert the rotary motion of a servo motor to linear motion. Many robots require both linear and rotary motion. A good example is the joints of a jointed-arm robot. They can be driven with a rotary drive.

Most servo motors cannot be connected directly to a joint of a robot to produce rotary motion. The power developed by most servo motors is not high enough to drive the arm of a robot, especially when the shaft of the servo motor is connected directly to the robot's joint. If the armature of the servo motor is connected to a transmission and the output of the transmission is connected to the robot's joint, there will be sufficient power to move the robot's arm. It will also have enough power to move the load that the robot is carrying,

The servo motor turns at high speed to drive the input shaft of the transmission at high speed. Gears in the transmission reduce the speed and increase the torque. The output shaft of the transmission turns at low speed and high torque, which are sufficient to drive the robot's arm. The problem arises when gear teeth mesh with one another. When they meet one on one, they have a tendency to wear. This wear is accelerated by the speed of rotation of the gears. As the gears wear, the output shaft is driven less and less evenly. The robot's arm no longer moves smoothly. This means that arm positioning becomes inaccurate.

Backlash is looseness in the gears where they mesh (Figure 3-23). This means that the amount of tooth space exceeds the thickness of the engaging tooth. As gears wear, backlash becomes greater. Backlash can be eliminated or at least reduced by bringing the gears closer together. As the backlash is removed, the friction between the gears is increased. This increased friction wastes power and generates heat and noise in the transmission. There must be enough backlash to allow for lubrication to coat the gear teeth. Machinists' handbooks list the correct backlash for gears to run efficiently and quietly.

New drives are being developed and you might encounter some in the latest robots. Whenever there is a demand, there is an answer. New products are constantly being developed.

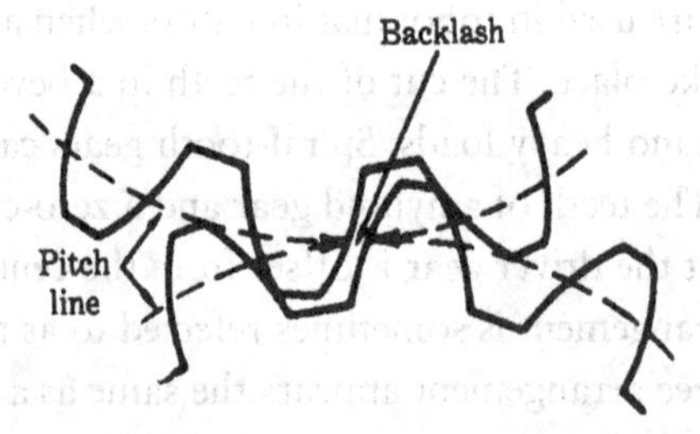

FIGURE 3-23 Backlash—the looseness between gear teeth when they mesh.

Harmonic Drives

Harmonic dives are used in robots to reduce backlash, gear wear, and friction. They have been developed to improve the quality of the transmissions used in robots. A *harmonic drive* consists of three major elements: a circular spline, a wave generator, and a flex-spline (Figure 3-24).

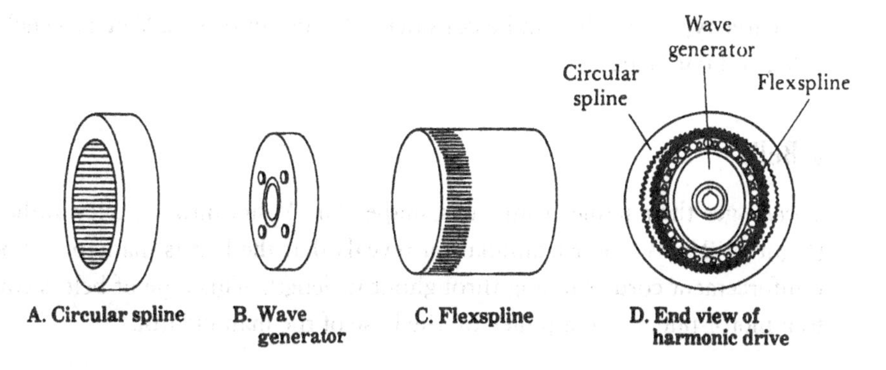

FIGURE 3-24 Harmonic drive. (*Courtesy of PWS-Kent.*)

A wave generator is an ellipse. A flex-spline is a flexible cup with teeth cut on the outside diameter. The wave generator is slid into the flex-spline, distorting it into an ellipse. The circular spline is a nonflexible ring with gear teeth cut on the inside diameter. The wave generator and flex-spline assembly is slid into the circular spline.

Gear reduction is generated when the harmonic drive wave generator is rotated. The ratio of the harmonic drive is controlled by the number of teeth on the circular spline and the difference in the number of teeth between the circular spline and the flex-spline. For example, if the circular spline has 400 teeth and the flex-spline has 398 teeth, the ratio is 400:2, or 200:1. If the circular spline has 100 teeth and the flex-spline has 97, the ratio is 100:3 or 33.34:1. The possibilities are limited only by the number of teeth that can be put on the circular spline and flex-spline.

One of the greatest advantages of harmonic drives over standard transmissions is the total lack of backlash. This alone increases robot accuracy and improves efficiency.

Belts

Sometimes it is not possible to use gears to transfer power from an actuator to a robot. Other methods must be used. Chains and belts are the next best possibilities. Belts are flexible and quiet running. They are also able to absorb some of the shock produced by the stop-and-go operations of a robot. Belts do have some limitations, however, inasmuch as they are flexible and subject to wear. By far the greatest limitation is slippage. Nonetheless, belts can be reinforced and made useful in many operations.

Three types of belts can be considered for use in robots: V-belts, synchronous belts, and flat belts.

V-Belts

V-belts get their name from their shape. The V fits into a pulley rather easily (Figure 3-25). A closer examination reveals that the belt is made of rubber with reinforcement cords running throughout its length. This type of belt is connected to a motor pulley and a pulley on the base of the manipulator.

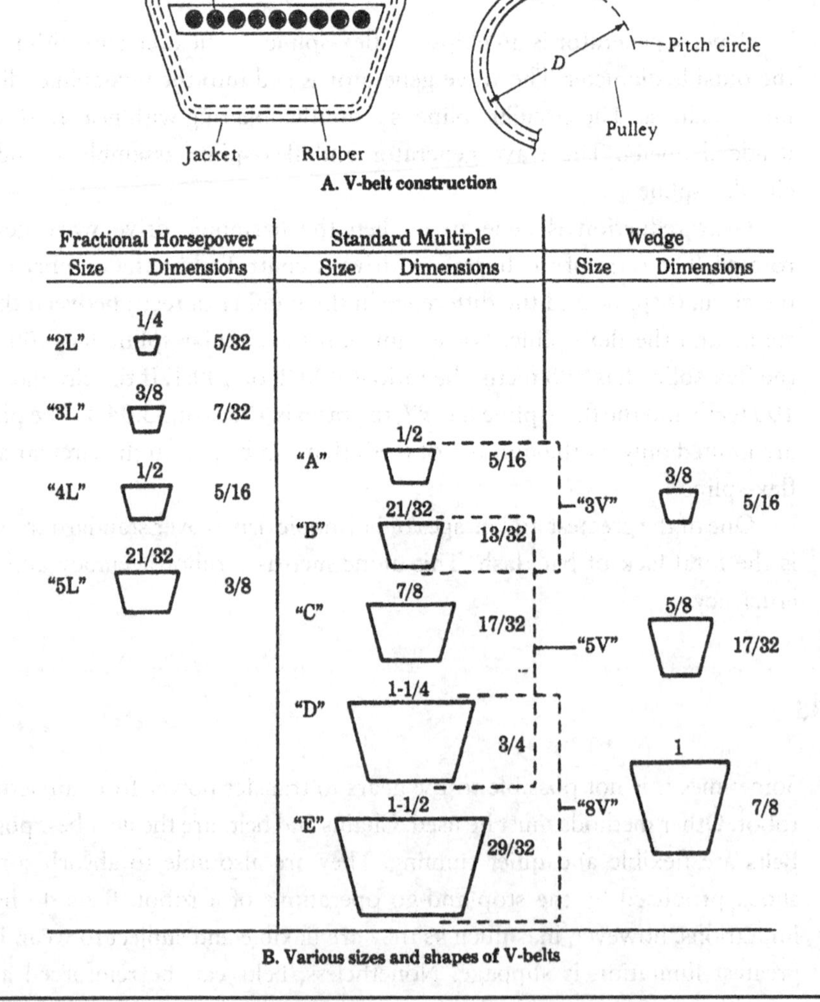

Figure 3-25 Construction details of V-belts. (*Courtesy of PWS-Kent.*)

Synchronous Belts

Synchronous belts are also recognized by their shape. They have evenly spaced teeth where they come into contact with the pulley. The pulley is also specifically designed to go with this type of belt (Figure 3-26). This means that synchronous belts are more expensive than V-belts. The teeth of a synchronous belt mesh with the grooves in the pulley to make sure that there is no slippage. This type of belt is used to provide a positive grip to a robot's manipulator wrist assembly. It is used where there is a constant change in the direction of rotation. Synchronous belts can also be found on some automobile engines, where they are known as *timing belts*.

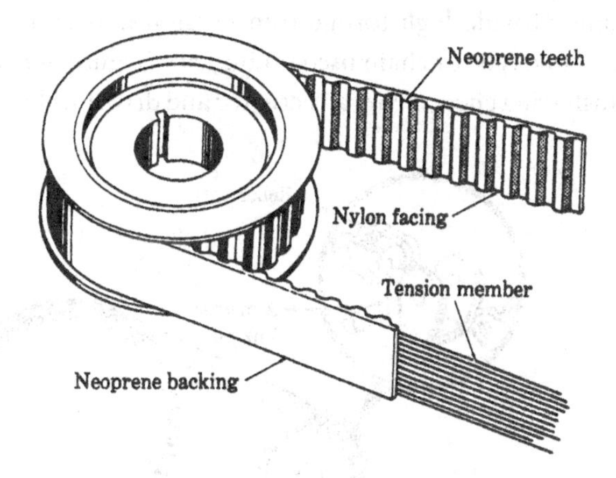

FIGURE 3-26 Construction details of a synchronous belt.

Flat Belts

Flat belts are used in the wrist assembly in small manipulators because they have a tendency to slip when large loads are placed on them. A flat belt is made of rubber and reinforced with cords along its entire length. It is inexpensive and therefore ideal for transferring power from one source to another. It is good for low and moderate speeds to deliver large torque when needed (Figure 3-27).

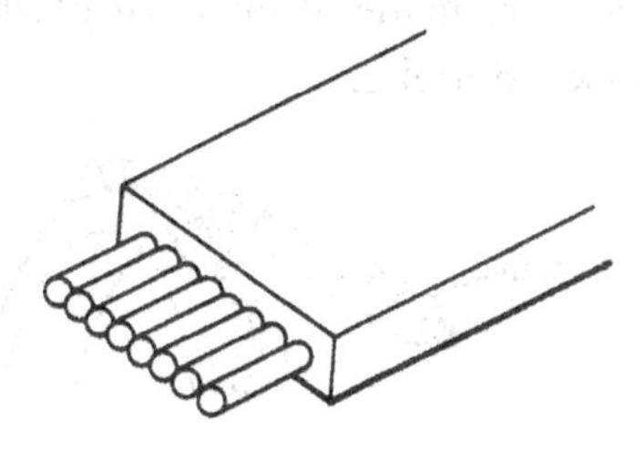

FIGURE 3-27 Flat belt reinforced with cords. (*Courtesy of PWS-Kent.*)

Chains

Chains are used when belts will not do, when you need a source of energy transfer that will not slip and can handle large loads. They come in handy for long-distance (longer than gears and shorter than belts) transfers of energy. Chains do not stretch or slip like belts. Roller chains, not bead chains, are usually used for robots.

Roller Chains

Roller chains provide high torque transmission and good precision. The roller chain is the same type of chain used on bicycles (Figure 3-28). This type is usually used in transferring energy from an actuator and drive mechanism to a manipulator.

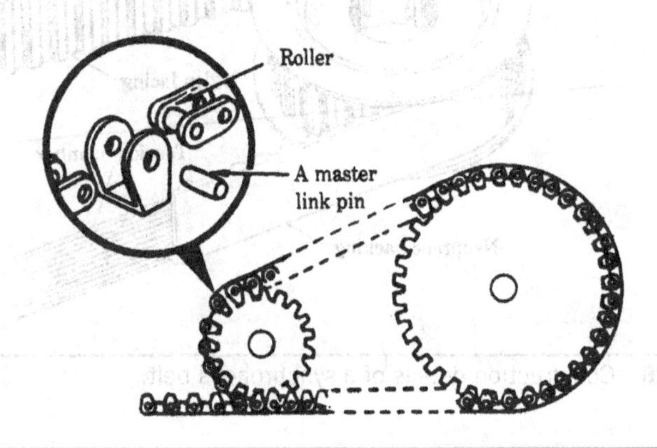

FIGURE 3-28 Roller chain.

Bead Chains

Bead chains are used to close drapes and turn basement lights on and off. They are suited for low-torque applications but are not suitable for robots. Bead chains can be used to drive devices with low torque requirements (Figure 3-29). The chains break easily. The beads are made of metal or plastic. (Note from the figure how they fit into the sprocket dimples.)

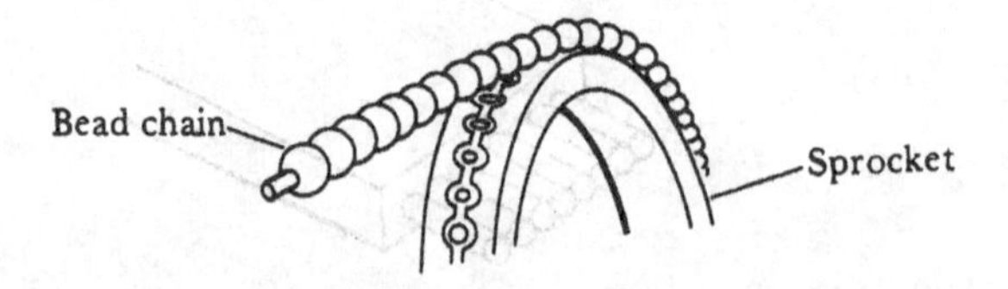

FIGURE 3-29 Bead chain. (*Courtesy of PWS-Kent.*)

Summary

Robots need some type of system to cause them to function. Hydraulic, pneumatic, and electric drive systems are all used to drive robots. Hydraulic systems are used for heavy loads. Pneumatic systems are used for medium- and low-load weights. The electric drive is used for low-load weights. Hydraulic systems use pumps to create the flow needed to do the work at the end of the arm. Several types of hydrostatic pumps are used to make the system operate properly.

Pneumatic systems use air to do work. These systems employ a pneumatic motor on the end of a manipulator to grip or handle the load being processed. A pneumatic system does not need a return system for the air. It exhausts directly into the atmosphere.

Electric drive systems are powered by electric motors. There are many types of electric motors, but the DC types are preferred for precision motion and movement. Permanent-magnet, stepper, brushless DC, and Hall-effect DC motors are used for various functions in robots. AC motors are used for heavy loads and where precision of movement is not necessary. They are classified as induction and squirrel cage motors. Each type has its particular applications. Squirrel-cage motors are further broken down into six classifications according to their starting currents and torque.

End effectors also may be called end-of-arm tooling. The manipulator is used to move the end effector that is mounted on its end. Grippers are used to pick up and hold objects being machined, boxed, or picked up and placed or palletized. There are vacuum-operated grippers and magnetic grippers, as well as a variety of mechanical devices used to grip or hold materials. Many of these grippers are made in the plant where the robot is working.

Positioning is very important in the proper use of a robot. The robot must be able to place an object in the same location over and over again without being too far off the spot. Controllers, and consequently robots, are classified as low, medium, and high technology. The ability of the controller to handle programs for the robot makes the difference in its classification.

Dynamic braking and plugging are both used to promptly stop the movement of a manipulator. They both have advantages and disadvantages and can be used in various locations depending on the application. Repeatability and accuracy are important parts of a robot system. Repeatability is the ability of a robot to place object in the same place again and again. Accuracy is the degree to which the robot can exactly handle the repeat function.

Gears, chains, and belts are the types of drives used in robots. Each has its own applications. Gears are accurate and noisy. Chains have their limitations, whereas belts can be used under circumstances where power is transmitted only short distances. Harmonic drives have been developed to reduce backlash and improve the efficiency of robot operation. They have also eliminated a lot of noise inherent in standard transmissions. The ball screw is a useful adaptation of principles to eliminate backlash and gear looseness.

Key Terms

accuracy The degree to which a robot can place an object in a given spot repeatedly.

backlash The looseness in gears where they mesh.

ball screw A method of using ball bearings to substitute for screw threads (the ball screw changes rotary motion to linear motion).

brushless motor A DC motor that operates without using brushes (an electronic circuit controls its field excitation).

harmonic drive A type of drive that uses a flex-spline, circular spline, and wave generator to accurately position a manipulator with no backlash and little noise.

hydraulics The use of pressure on a fluid to drive an end effector or a manipulator.

plugging A method of stopping an electric motor by reversing the polarity of its power source.

positioning The ability of a robot to place a particular object in a desired location.

pumps Devices, usually electrically driven, used to increase the pressure on a hydraulic fluid.

repeatability The ability of a robot to place an object in the same spot again and again.

roller chain The same type of chain used in bicycles; used to drive manipulators and end effectors.

stepper motor A DC motor whose shaft rotates a specific number of mechanical degrees each time its input is pulsed (the shaft is stepped from one position to the next by the proper combination of fields in the motor).

synchronous belt A belt that has teeth that fit a pulley with grooves so that the belt does not slip.

V-belt A belt used to drive a manipulator; shaped to fit a V-pulley.

worm gear A method of changing linear motion to rotary motion and vice versa.

Review Questions

1. List three types of robot drive systems.
2. Why are filters needed in hydraulic systems?
3. How does a relief valve function?
4. What are the two classifications of hydrostatic pumps?
5. What does pneumatic mean?
6. List the parts of a pneumatic drive system.
7. Describe a PM motor.
8. What is a stepper motor used for?
9. How do DC motors work without brushes?
10. What are the two types of AC motors most often used in robots?
11. Where is a squirrel-cage motor used?
12. List the torque characteristics of the six classifications of squirrel-cage motors.
13. What is slip? How is it used to advantage?
14. How do vacuum grippers work? What are their limitations?
15. How do magnetic grippers work? What are their limitations?
16. What is another name for end-of-arm tooling?
17. What is positioning?
18. What is repeatability?
19. What is meant by robot accuracy?
20. What is the advantage of a roller-chain drive?

CHAPTER 4
Sensors and Sensing

Performance Objectives

After reading this chapter, you will be able to:

- Identify the types of sensors and their classes used in robots.
- Discuss pulsed infrared photoelectric controls.
- Know what a strain gauge is and its use in a robot.
- Identify a Hillis touch sensor.
- List the types of sensing and how each is used in robots.
- Identify and discuss the key terms used in the chapter.
- Answer the review questions at the end of the chapter.

Because robots are made with as many human qualities as possible, they must have the five senses humans possess—sight, hearing, taste, touch, and smell. Some of these senses are used to make robots do special tasks. In most instances, a robot does not need to smell. Of all the senses, sight is the most difficult to perfect in robots. Vision for robots is discussed later.

Sensors and Sensing

One of the tasks most suited for robots is picking and placing. However, in order to pick up things, a robot must be able to sense when it has the object gripped tightly enough to hold. In addition, the robot must not crush the object while trying to pick it up. Therefore, there must be some type of sensing device in the gripper or attached to it that regulates the amount of pressure applied to the object being retrieved. Transducers are used to convert nonelectrical signals into electrical energy. A transducer, then, can serve as a sensor. It converts the pressure applied to it into a signal that is fed to the robot controller. The controller then takes this signal strength and uses it to increase or decrease the pressure being applied by the gripper or end effector (Figure 4-1).

Limit switches are designed to be turned on or off by an object hitting a lever or a roller that operates the switch (Figure 4-2). Robots that do repetitive jobs with little demand for complex operations use limit switches to tell them when to reverse or stop. These switches are either on or off and are very simple in operation. If a robot is to make decisions and assemble products, it needs force sensors,

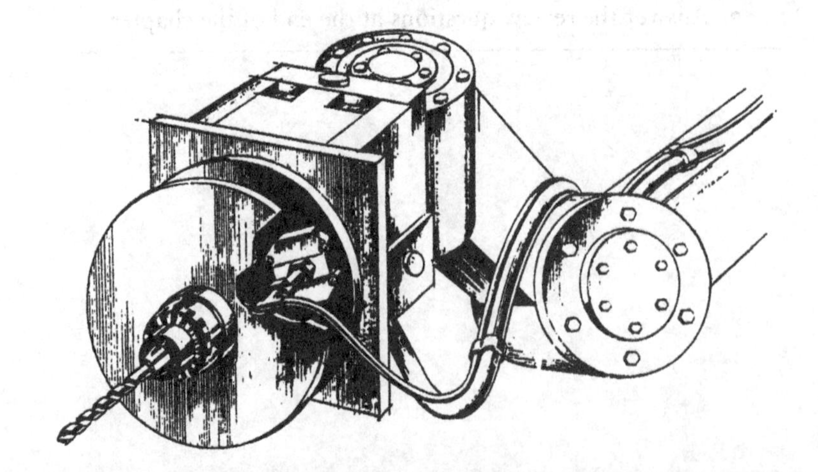

FIGURE 4-1 Piezoresistive transducer on the end effector used to monitor pneumatic pressure for cost-effective control of automated drilling. (*Courtesy of Micro Switch, a Honeywell Division.*)

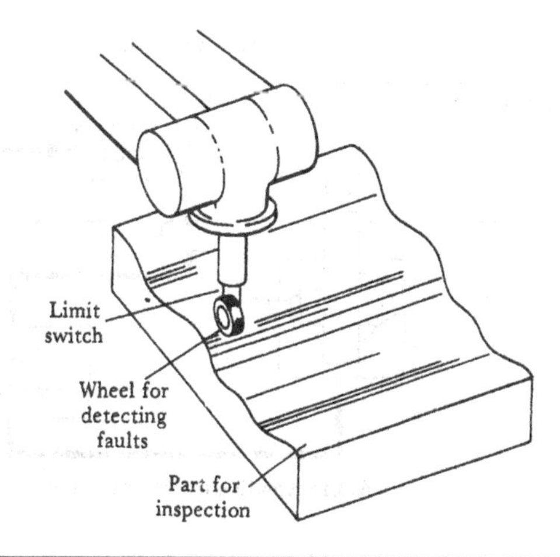

FIGURE 4-2 Limit switch. (*Courtesy of PWS-Kent.*)

vision sensors, and tactile (feel or touch) sensors. As the types of sensors on a robot increase, its ability to do complicated processes increases.

Classes of Sensors

Sensors may be classified as either contact or noncontact. They may be further classified as internal or external and as passive or active.

A limit switch is a contact sensor. This limit switch permits the robot to sense whether an object is present or missing (Figure 4-3). If the object makes contact with the limit switch, then the robot knows that the object is near enough that the robot can begin its operation. If the switch is not closed, this means that the object is missing, and the robot has to react accordingly. This usually generates what is referred to as an *alarm condition*. Force, pressure, temperature, and tactile sensors all respond to contact. They all send their signals to the controller for processing.

Noncontact Sensors

Pressure changes, temperature changes, and electromagnetic changes can all be sensed by noncontact methods. Noncontact sensors usually react to a change in a magnetic field or light pattern. If an electromagnetic field is disturbed, it is sensed and fed to the controller. The same is true of the disturbance of a light beam. Changes in the light beam—its intensity or whether or not it is present—are sent to a controller for processing and then sent back to the robot to react according to the disturbance (Figure 4-4).

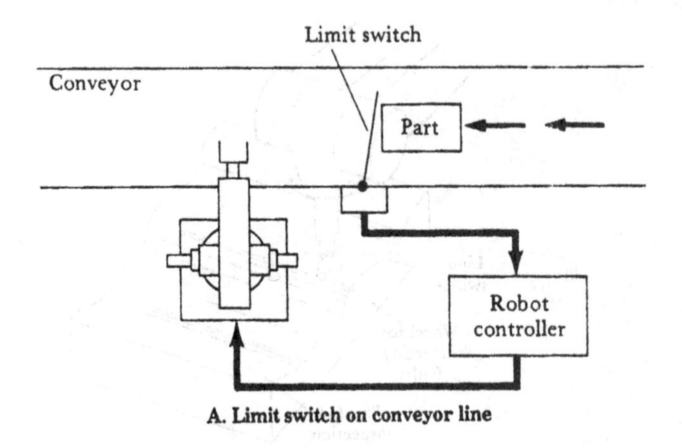

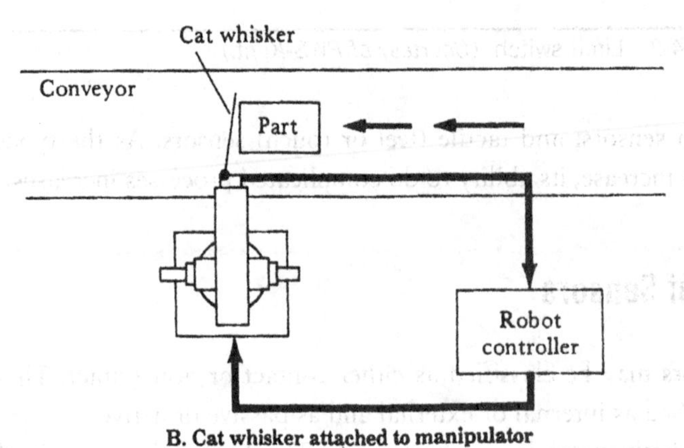

FIGURE 4-3 Contact sensor. (*Courtesy of PWS-Kent.*)

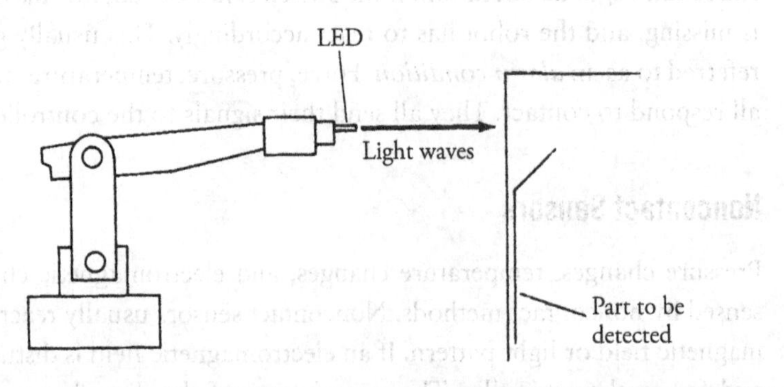

FIGURE 4-4 Noncontact sensor. (*Courtesy of PWS-Kent.*)

A number of light-emitting diode (LED) sensors are used in robotics. The LED produces a low-level light beam that is picked up by another device. If the light beam is broken, the part to be picked up by the robot is present. If the light beam is not broken, then the robot goes into a different mode for which it has been previously programmed (Figure 4-5).

Another noncontact sensor is a television camera mounted on the end of the manipulator. It can see the presence of parts, compare what it sees with what is in the computer memory, and then pick up the right part and move it to a preprogrammed location (Figure 4-6).

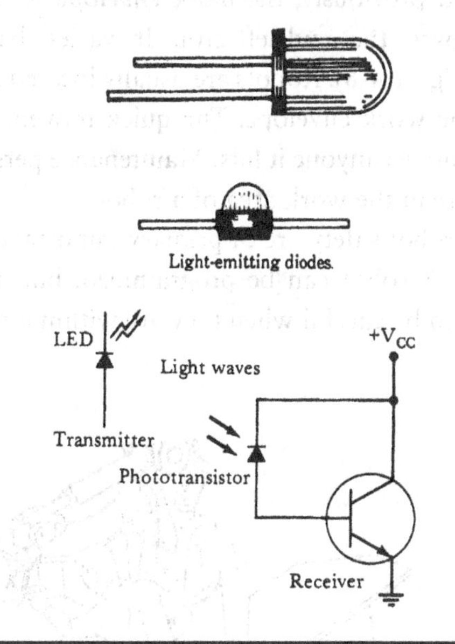

FIGURE 4-5 LED sensor. (*Courtesy of PWS-Kent.*)

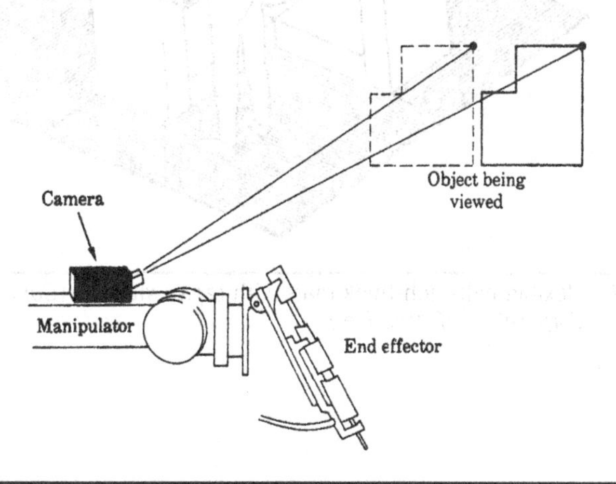

FIGURE 4-6 TV camera on a manipulator. (*Courtesy of PWS-Kent.*)

Self-Protection

A robot must be protected from any damage it may do to itself as well as any injury it may cause to a human worker within its work envelope. Properly mounted sensors make sure that the arm of the robot does not move where it is not programmed to be. If this does occur, the robot comes to an emergency stop and will remain stopped until the operator starts the program again. Figure 4-7 shows a mat that contains switches that prevent a robot from starting while a person is inside the work envelope where the mat is located.

As mentioned previously, the *work envelope* is that area within which the manipulator moves the end effector. It varies based on a robot's design characteristics (Figure 4-8). Robots are usually located in a cage to prevent humans from entering the work envelope. The quick movements of the robot arm may cause serious injury to anyone it hits. Maintenance persons have to be very careful whenever they are in the work area of a robot.

Worker and robot safety are of primary importance in any installation where robots are used. A robot can be programmed, but it takes some time to train human workers to be careful when they are within a robot's domain.

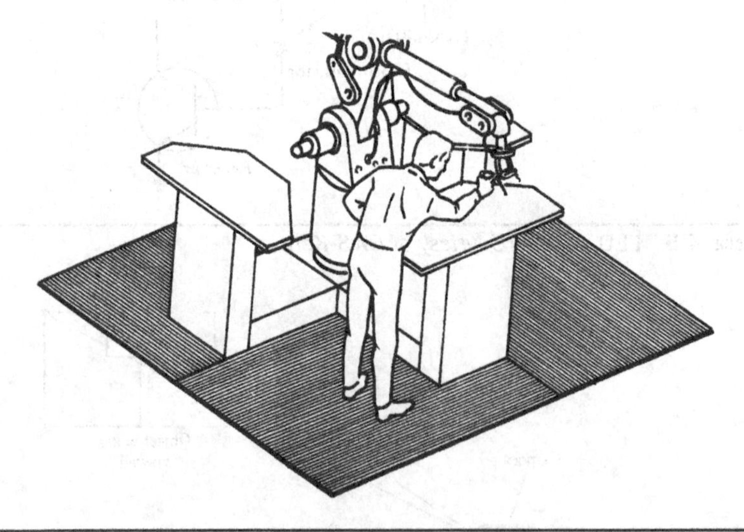

FIGURE 4-7 Flexible half-inch-thick mats with tape switches. (*Courtesy of Tape-Switch Corporation of America.*)

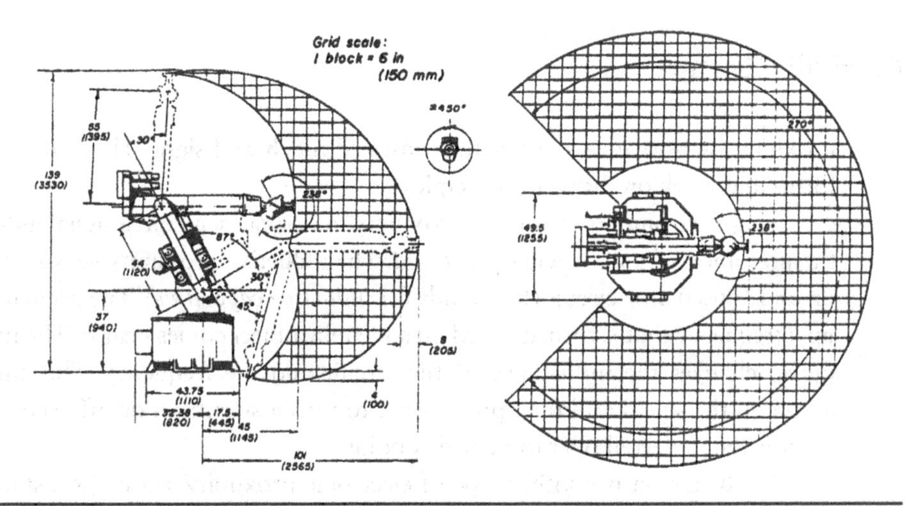

FIGURE 4-8 Work envelope. (*Courtesy of Cincinnati Milacron.*)

Collision Avoidance

The biggest challenge a programmer has in programming a robot is collision avoidance with the object to be serviced. If the gripper hits the part to be picked up and knocks it out of place, then the gripper will not be able to locate it, and the whole operation will have to be stopped until the right program is fed into the controller. Irregular shapes are difficult to program, and it takes time to develop a proper program for them. Proximity sensors can be used to sense the presence or absence of the object (Figure 4-9).

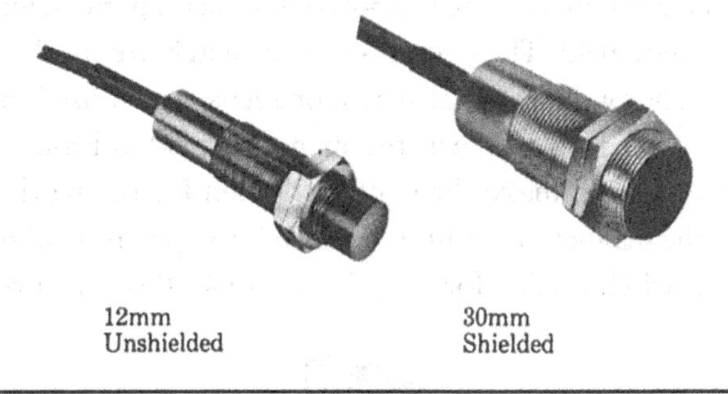

FIGURE 4-9 Proximity sensor for collision avoidance.

Proximity Sensors

Proximity sensors give a robot the senses of touch and sight. They take various forms and work on different principles.

Using electronic commands, a proximity sensor can send out a signal indicating the proximity of a part being processed. The inductive proximity sensor is based on an LC oscillator circuit (LC stands for *inductive capacitive*). The oscillator sets up a frequency that is turned off when a metallic object comes near it. The metallic object changes the inductance of the circuit and the frequency. The amplifier amplifies the signal and then processes it to turn a switch on or off according to what is the preferred operation at this point.

The RC circuit is another type of electronic proximity sensor (*RC* stands for *resistive capacitive*). This oscillating circuit has its frequency changed by the closeness (proximity) of the object being sensed. The signal is processed by an amplifier, and a switch is turned on or off based on the preferred operation at this point. An advantage of a resistive-capacitive sensor is its ability to detect metallic or nonmetallic objects.

Pulsed infrared (IR) photoelectric controls are used in industrial robots to sense the presence of any type of object. These devices can detect the product on the in-feed or out-feed conveyor and the protected robot workstation.

Eddy-current proximity detectors use magnetism to function. They induce a magnetic field in any object nearby. A small coil picks up any change in the magnetic field and sends it to the controller for processing. The reed switch is another magnetic proximity switch that responds to a controlled magnetic field. It makes or breaks contact when exposed to either a permanent magnetic field or an electromagnetic field. The contacts of the switch are inside a hermetically sealed glass tube. The two flat metal strips, or reeds, are housed in a hollow glass tube filled with an inert gas. When the magnetic field is brought near, the reeds are pulled together and make the contact needed for the work to be done. In some instances, the magnet may cause the switch to open instead of close depending on the type of switch needed for the job. Figure 4-10 shows a reed switch.

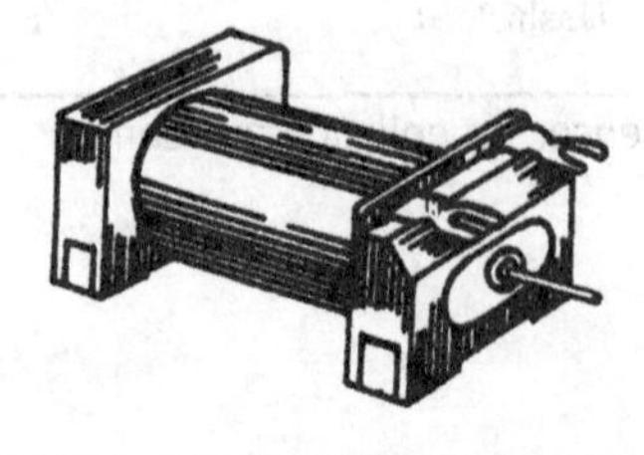

Figure 4-10 Reed switch.

Range Sensors

If precise distance is needed, it is best to use a range sensor. Range sensors are used to locate objects near a workstation to control a manipulator. One range-sensing system is the laser interferometric gauge. It is very expensive and is sensitive to humidity, temperature, and vibration. A television camera is another range-sensing system.

Tactile (Touch) Sensors

Tactile sensors rely on touch to detect the presence of an object. The two types of tactile sensors are *touch sensors* and *stress sensors*. Touch sensors respond only to touch. Stress sensors produce a signal that varies with the magnitude of the contact made between the object and the sensor.

The simplest touch sensor is a microswitch that is turned on or off by the presence of an object (Figure 4-11). A strain gauge is a good example of a stress sensor. Figure 4-1 shows a good use for a sensor in an end effector.

Figure 4-11 Micro-switches are available in hundreds of shapes and sizes.

Strain Gauges

A strain gauge is used to sense mechanical movement. This device can be used with grippers to determine the amount of force being applied to an object with the grasp of the mechanical fingers (Figure 4-12).

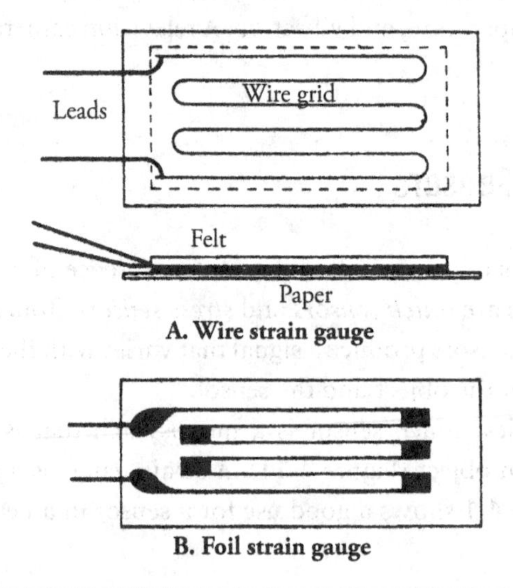

FIGURE 4-12 Strain gauges.

Older strain gauges were made of fine wire (about 0.001 inch in diameter). The wire was attached to an insulating strip of material. As the unit was stressed, the wire would stretch. As the wire got longer, its cross-sectional area was reduced and its length increased. This changed the resistance of the wire, thereby changing the current through the circuit.

Newer strain gauges are made of semiconductor materials and provide greater sensitivity. They change resistance approximately fifty times faster than metallic gauges and have a higher output.

Pulsed Infrared Photoelectric Control

These controls are used to sense the presence of any type of object. They detect a product on the in-feed or out-feed conveyor. They can also be used to detect the presence of an object moving to and from the protected robot workstation (Figure 4-13). These devices will operate in any environment. They disregard ambient light, atmospheric contamination, and thin-film accumulations of oil, dust, water, and other airborne deposits. Low voltage is used from the sensor to the controller.

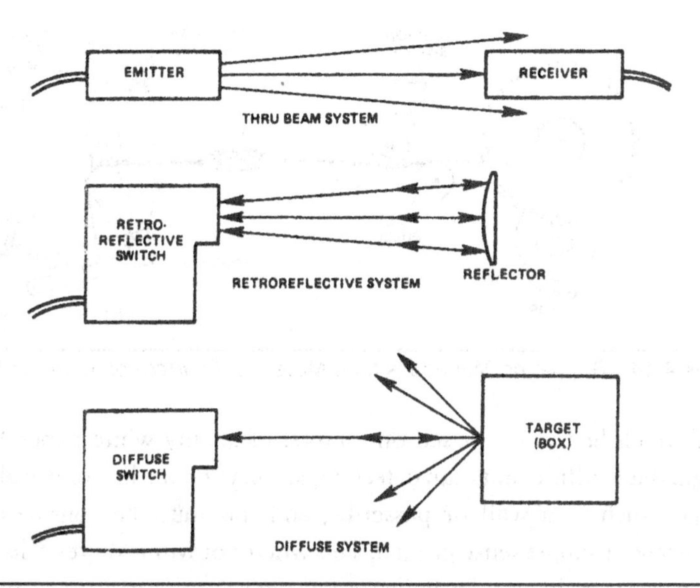

FIGURE 4-13 Infrared photoelectric control. (*Courtesy of Square D Co.*)

About 35 percent of all robots are used for pick-and-place operations, so the robot's gripper is very important. The gripper must be able to pick up an object and hold it until the robot is commanded to release it. There also may be requirements as to when and under what conditions the object should be released, which is where tactile sensing comes into play. The robot must be able to respond to various pressures and act accordingly.

Through-beam units have the source in one enclosure and the receiver in another, pointing at each other. These units offer the longest ranges or highest gains and are generally used in extremely dirty environments because they can tolerate the most attenuation and still function. However, because alignment is critical for peak performance and two devices must be purchased and installed, this system is more expensive than retroreflective or diffuse types.

Retroreflective units have the source and detector in one enclosure and operate by bouncing the light beam off a reflector or reflective tape. These units are less expensive and easier installed than a through-beam unit, but because the beam must travel twice as far, ranges and gains are less. These units are not as good as through-beam units for dirty areas, and care must be taken that the target is not shiny or will look like a reflector to the switch and be ignored.

The construction of diffuse types is similar to that of retroreflective units, but they detect the small amount of light that is reflected diffusely from nonshiny targets such as cardboard boxes. Sensing ranges and gains are lower than those of through-beam or retroreflective types and depend on the reflectivity of the target,

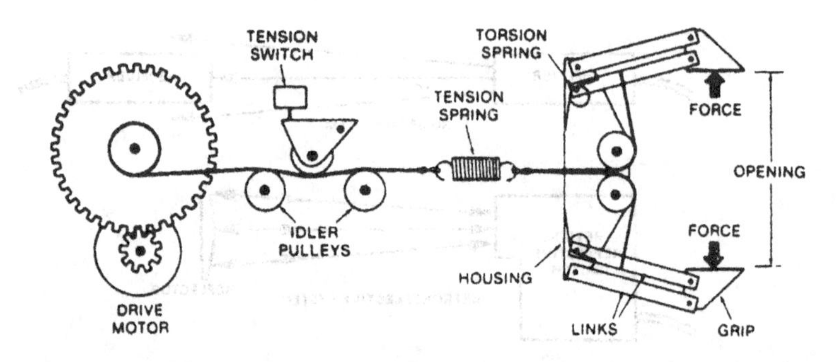

FIGURE 4-14 Gripper on Microbot's Mini-Mover 5. (*Courtesy of Microbot.*)

with published values based on a target-of-quality white paper. While easiest to align, these diffuse units can detect objects beyond the line of travel of the intended target, such as a wall or passer-by, and this must be considered. Figure 4-14 illustrates a simple sense gripper. This Microbot Mini-Mover 5 has a drive motor, a tension switch, an idler pulley, a tension spring, a torsion spring, housing, and links. Force is applied to the end of the gripper. The gripping action is cable driven. A tension spring supplies passive compliance. (*Compliance* is the ability to trade off on position and force.) A small stepping motor powers the gripper, so passive compliance is important. The tension spring provides the passive compliance. The tension switch signals to the computer that the gripper is firmly closed. In other words, as the fingers encounter force, the cable tightens, straightens, and pushes the switch until it closes.

The gripper may require more dexterity if odd-shaped objects are to be gripped. This is provided by pressure-sensitive materials mounted on the gripping surfaces. The surfaces are then divided into a grid. The computer can sense the balance of force over the contacting surfaces. A good example of such a touch sensor is the Hillis touch sensor (Figure 4-15). It is useful in picking up nuts, bolts, flat washers, and lock washers that are identified by touch. The sensor is made of anisotropically conductive silicone (ACS). ACS is electrically conductive only along one axis in the plane of the sheet. The rubber sheet (ACS) and a flexible printed circuit board are separated by a nylon mesh. As increasing pressure is applied to an area of the ACS sheet, more of the sheet contacts the flexible printed circuit board, and resistance in that region goes down. The flexible printed circuit board is etched with parallel conducting lines at 90 degrees to the conductive direction of the ACS sheet. A second circuit board is used as a contact board for driving columns along the conductive lines of the ACS sheet. The result of this arrangement is a sensing grid with a resolution of approximately 1 square millimeter, matching the sensing capability of the human fingertip. The word *anisotropic* means "having unequal responses to external stimuli."

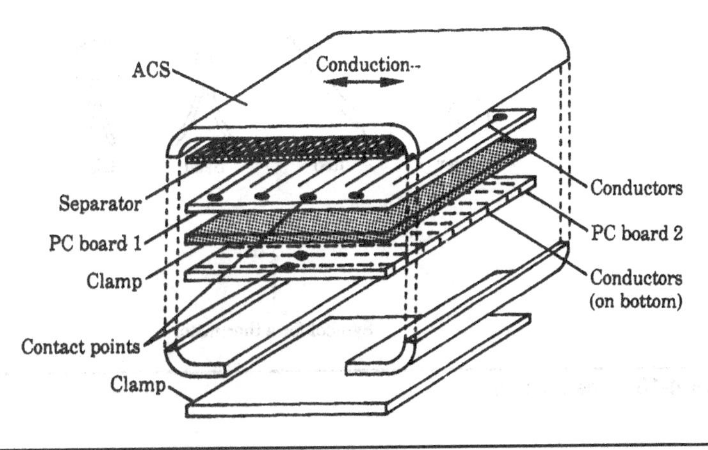

FIGURE 4-15 Hillis touch sensor.

One advantage of touch sensors is that the computer can respond to them better, in some instances, than it can to a TV camera because the information does not have background to contend with. A disadvantage is that these touch sensors can be damaged easily. The piezoelectric touch sensor is another tactile sensor that is becoming more common (see Figure 4-1).

Temperature Sensing

Temperature changes are easily sensed with the devices presently available. They either increase resistance with an increase in temperature or decrease resistance with an increase in temperature. They also may produce a small (usually in millivolts) change in voltage as a result of the application of heat. This very low voltage can be amplified to be used in the computer and fed to the controller.

Thermistors respond to changes in temperature by increasing or decreasing resistance (Figure 4-16). If the temperature is increased, the thermistor decreases its resistance, thereby increasing the current in the circuit. If the thermistor temperature is decreased, the resistance of the thermistor increases, thereby decreasing the current in the circuit. This increase and decrease is usually not linear, so the information fed to the computer must be equated with the response curve of that particular thermistor.

Thermocouples have been used for years as temperature sensors. They are made of two dissimilar metals, such as iron-constantan, copper-constantan, or platinum-rhodium (Figure 4-17). The ends of the metals are fused together. When heated, they give off a few millivolts that can be amplified and fed into a computer for processing. As the temperature increases, the output of the thermocouple increases.

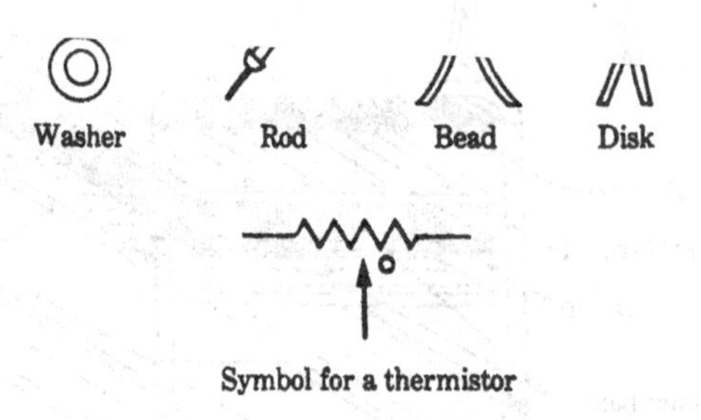

FIGURE 4-16 Thermistors.

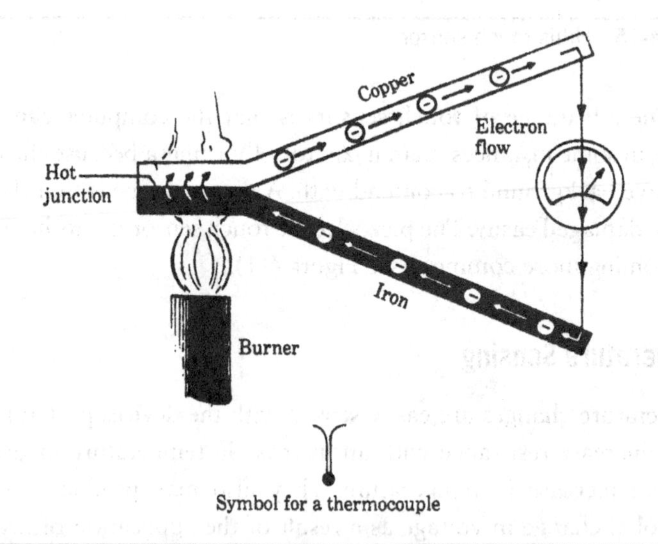

FIGURE 4-17 Thermocouples.

Displacement Sensing

The exact location of the gripper or manipulator is important so that this information can be fed into the computer or controller for action. Many different methods are used to tell the computer or controller where a particular unit is at a specific time. One such method is to use a resistive sensor, which has a fixed resistance—usually a wire-wound resistance—with a slider contact. When you apply force to the slider arm, it changes the circuit resistance. This change in resistance, in turn, changes the amount of current flowing in the circuit. This change can be equated with location, and the computer can act on it to determine the exact placement of the arm or gripper (Figure 4-18A).

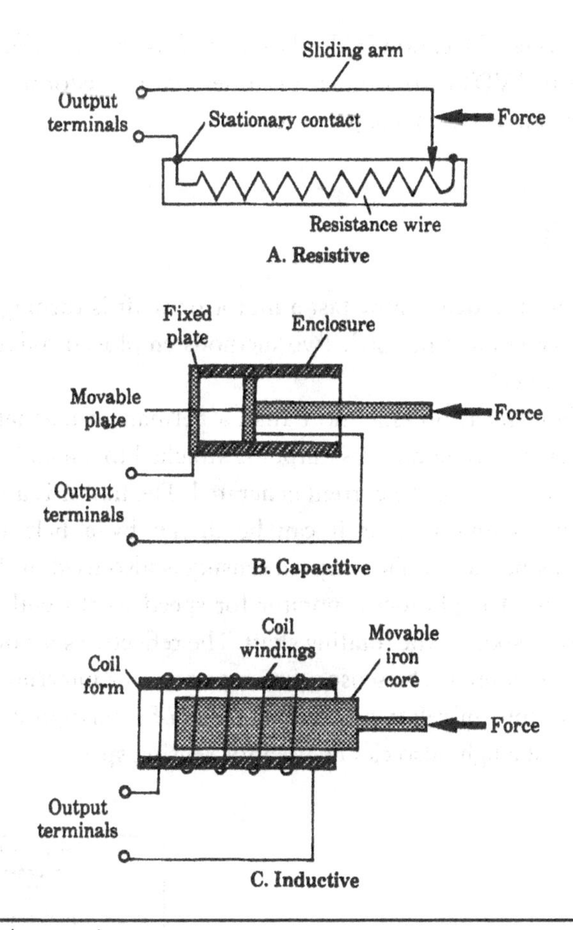

FIGURE 4-18 Displacement sensors.

Another sensing method is *capacitive*. It is capable of detecting location. A capacitor is made up of two plates. If one is held stationary and the other is allowed to move, then it is possible to use this varying capacitance to change the frequency of an oscillator. The change in oscillator frequency can be detected and fed into the computer for processing in regard to the location of the manipulator or gripper (Figure 4-18B).

Another way to detect a change in location of any device is to place an *inductive* sensor on it. The moving force of the manipulator or gripper will cause the movable iron core of the sensor to change its location. As this metal core changes, it changes the inductive reactance X_L of the coil. This inductive reactance, when divided into the voltage applied across the coil, determines the current in the circuit. As the current changes because of the inductive reactance change, the movement that caused the change can be detected, processed by the computer, and fed to the controller for its action (Figure 4-18C).

Other types of location or displacement sensors are linear variable differential transformers (LVDTs), rotary variable differential transformers (RVDTs), encoders, pots, and resolvers, to name just a few.

Speed Sensing

You may need to detect how fast a motor or shaft is turning, so some method of reliable sensing must be used. Two methods employed today are the tachometer and the photocell.

A *tachometer* is nothing more than a permanent-magnet (PM) motor that is being driven as a generator. Its output is attached to a meter circuit, and the speed is read out in terms of the current generated. The motor is attached directly to the device being monitored, or it can be driven by a belt arrangement using a synchronous belt and pulleys. Speed sensing is also used in the control of a robot (Figure 4-19). The photocell monitor for speed works with a light beam that is reflected by a spot on the rotating shaft. The reflective spot on the shaft causes the light to be disrupted. The sensor then reacts to this interruption and produces an electrical output in relation to the number of interruptions per second (Figure 4-20). A strobe light also can be used for sensing speed.

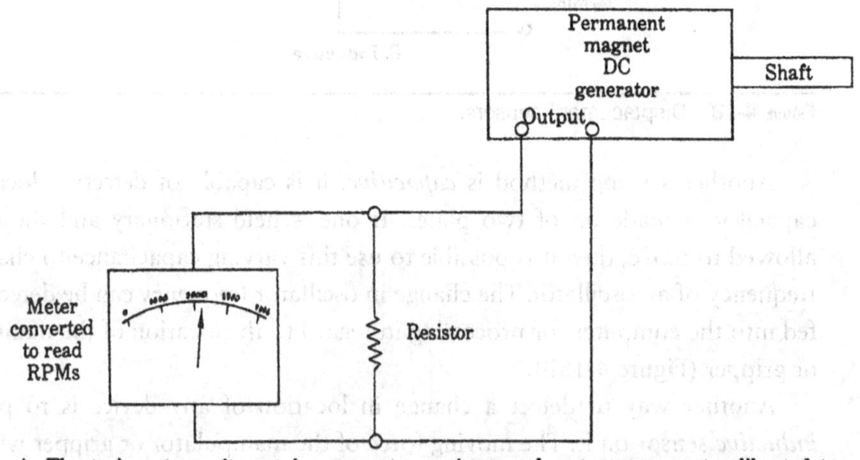

A. The tachometer unit contains generator, resistor, and meter movement calibrated to read RPMs.

FIGURE 4-19 Tachometer.

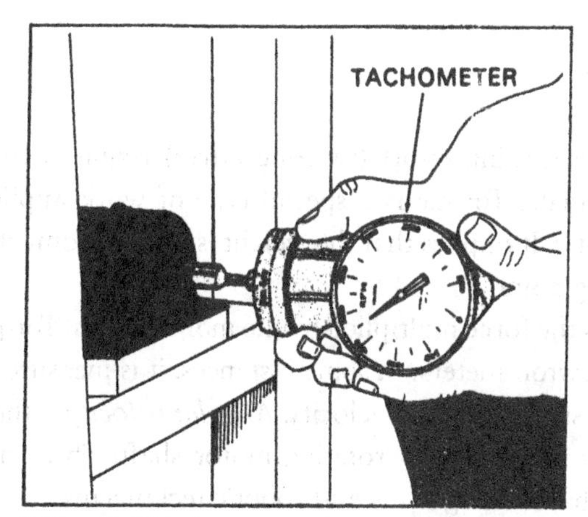

B. Hand-held tachometer can be used to check accuracy of permanently installed unit.

FIGURE 4-19 (*Continued*)

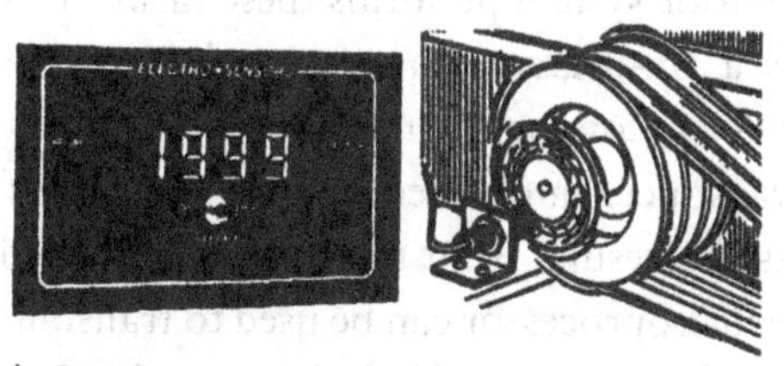

A. Small magnets can be placed on pulley and counted as they cause a pulse to be generated in the stationary coil. Or, a transistorized "Hall effect" can be used to detect the speed of the motor.

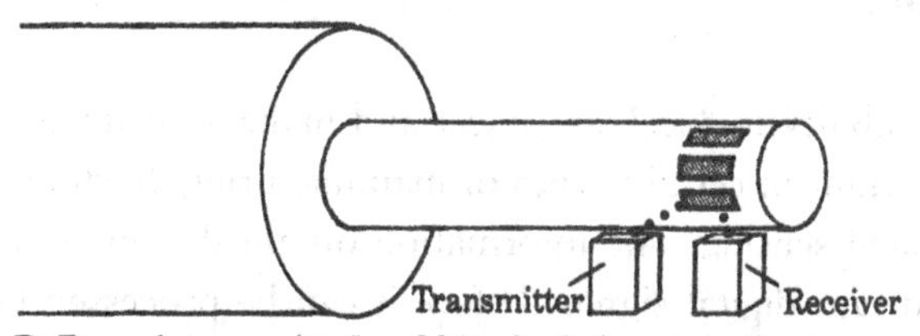

B. Beam is transmitted and hits shaded spots on the rotating shaft; reflected beam is picked up by receiver and counted.

FIGURE 4-20 Magnetic and photoelectric speed sensors.

Torque Sensing

Torque is the turning effort (twisting effect) required to rotate a mass weight through an angle. Torque is a special type of work applied in a rotary manner around a center. It implies that the weight is located some distance from the center of the turning point.

Torque is the force multiplied by the moment arm. Torque is measured in inch-pounds or newton-meters. In some instances, it is measured in inch-ounces. It can also be expressed in angular velocity. *Angular velocity* is the expression used when the torque is taken from a rotating motor shaft. These measurements are dealt with by mechanical engineers and robotic technicians.

Force and torque sensors are used to control robot motion. They may also make go/no-go decisions, adjust task and process variables, detect end-effector collisions, determine required actions to unjam an end effector, and coordinate two or more arms.

One torque sensor system performs these tasks in robots. A basic component of the system is a six-degree-of-freedom solid-state piezoresistive force/torque transducer. The microcomputer automatically resolves the forces and torques applied to the transducer into six equivalent Cartesian force/torque components. It then transmits the results from the robot at better than 100-hertz rates. The serial port of the microprocessor can be used to transmit force/torque information to the robot controller or to a readout device such as a cathode-ray tube.

Vision Sensors

A robot's ability to "see" an object and make adjustments to fit that object are very important in certain types of manufacturing. Such an ability calls for some sophisticated sensing. The information obtained from a video camera has to be converted into digital signals before it can be processed by the computer—and therein lies the problem.

Very intelligent controllers can make real-time adjustments in a robot's program in regard to its positional data. This is done to compensate for variables in the work piece. This ability to compensate for variations in the work piece is sometimes referred to as *adaptive control*. To make these adjustments, the controller needs some sophisticated information about the current state of the process. This cannot be provided by simple go/no-go limit switches; end sensors generally are used with nonintelligent controllers. The type of information provided to the robot controller deals with where the part is located, how it is placed, and the type of part it is.

Machine vision systems (MVSs) are used for recognition and verification of parts. They can be used for inspection, sorting, and making noncontact measurements. MVSs also provide part position and orientation information to the robot controller (Figure 4-21).

A number of imaging devices are used for robot vision systems. They include dissectors, flying-spot scanners, vidicoms, orthicons, plumb icons, charge-coupled devices (CCDs), and others. These devices differ in the way in which they form images as well as in the properties of the images they form. One thing they all have in common is their ability to convert light energy to voltage in much the same way. Research is currently being done on a number of them. The most popular vision system has been the *vidicom tube*. It is the camera tube of a television system. However, the CCD system is becoming more popular and can be used in many industrial locations for robot vision. Fiberoptics are presently being introduced into this field of sensing. A detailed analysis of the various means of identifying parts for processing or inspection will be provided in Chapter 7.

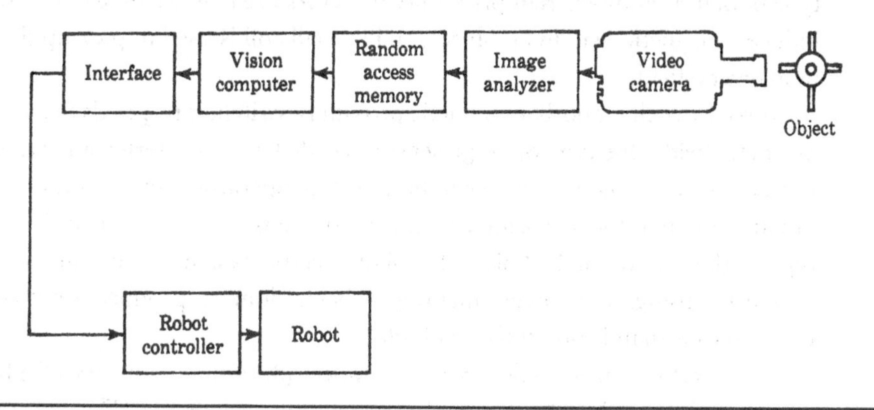

FIGURE 4-21 Machine vision system (MVS).

Summary

In order for a robot to hold and recognize objects, it must be able to sense their presence, size, and shape. Transducers are used to convert nonelectrical energy into electrical energy. A transducer, then, can serve as a sensor. Limit switches are designed to be turned on and off by an object contacting a lever or roller that operates the switch. Some low- and medium-technology robots use this type of sensor. Sensors are classified as contact or noncontact. They may also be called internal or external and passive or active. Most robotic sensors are contact or noncontact. A limit switch is a contact sensor. Touch, force, pressure, temperature, and tactile sensors all respond to contact. Pressure changes, temperature changes, and electromagnetic changes can all be sensed by noncontact methods. They usually react to a change in a magnetic field or light pattern.

A number of LED sensors are used in robotics. The light beam is used in a number of applications. Another light sensor is the television camera mounted on the end of the manipulator. It can see the parts and compare them to what is in the computer memory. The work envelope is that area where the manipulator moves the end effector. This area is usually enclosed by a cage to protect maintenance persons from being near the moving manipulator when it is operating. Proximity sensors are used to give the robot a sense of touch and sight. They take various forms and work on different principles.

Different types of proximity sensors include the inductive sensor, capacitive sensor, and resistive sensor. Some have advantages that others do not. Pulsed IR photoelectric controls are used in industrial robotics for presence sensing of any type of object. Eddy-current proximity detectors use magnetism to function. They induce a magnetic field in an object, and a small coil is used to pick up the change in magnetic field.

A reed switch is another magnetic proximity switch. It responds to a controlled magnetic field. One type of range sensor is called a laser interferometric gauge. It is very expensive and is sensitive to humidity, temperature, and vibration. Another ranging system is the television camera. Tactile sensors rely on touch. The simplest type is the microswitch. Pulsed IR photoelectric systems also can be used for presence sensing. Some experimenting is being done to produce a better-quality tactile sensor than is presently available.

Temperature sensing is done with thermocouples and thermistors. Displacement sensing is done with capacitive, inductive, or resistive devices. The strain gauge is a device used to sense mechanical movement and, in some cases, weight or force.

Speed sensing can be done with a tachometer or photocell. Torque sensing is used to make go/no-go decisions, adjust tasks and process variables, detect end-effector collisions, and determine required actions to unjam an end effector and to coordinate two or more effectors. Machine vision systems are used for recognition and verification of parts, as well as for inspection and orientation. Fiberoptics are presently being introduced into this field of sensing.

Key Terms

collision avoidance The ability of a robot to avoid colliding with the part it is supposed to pick up.

contact sensor A sensor that detects the presence of an object actually making contact.

light-emitting diode (LED) Low-level light beam that is picked up by another device.

limit switches Switches designed to be used with a moving body that should not go past a given point.

machine vision systems Devices that allow a robot to recognize and verify parts.

proximity sensors Devices used to detect how close an object is.

range sensors Devices used to detect the precise distance between an object and a gripper or manipulator.

strain gauge A device made of thin wire that reacts to stretching of that wire (strain gauges are made of semiconductor materials today).

tactile sensors Devices used to detect the presence of an object by using touch.

thermistors Devices that change their resistance in the reverse manner (if the temperature increases, the thermistor lowers its resistance).

thermocouples A union made by two dissimilar metals (when heated, the junction produces a small electric current).

transducer A device used to convert nonelectrical energy to electrical energy.

Review Questions

1. What is a transducer?
2. How does a limit switch work?
3. List the two classes of sensors.
4. List the device used to sense touch, force, pressure, temperature, and vision in a robot.
5. What is an LED?
6. What is a work envelope?
7. What is meant by collision avoidance?
8. What does a proximity sensor do?
9. List three types of proximity sensors.
10. What are the two types of tactile sensors?
11. How many robots are used for pick-and-place operations?
12. Describe how ACS is used in a touch sensor.
13. What are the two types of devices that sense temperature?
14. What is meant by displacement sensing?
15. How would you make a simple strain gauge?
16. What do you use for speed sensing?
17. How is torque measured?
18. What is a machine vision system and why is it needed?

CHAPTER 5

Control Methods

Performance Objectives

After reading this chapter, you will be able to:

- List the characteristics of motors used in robots.
- Know the difference between single- and three-phase motors.
- Know the applications of non-servo-controlled robots.
- Know what is meant by the word *feedback* and how it is used in robots.
- Describe the types of actuators used in robots.
- Describe the difference between an actuator and a controller.
- Understand how robots are programmed.
- Know what a teach pendant is and what it will do.
- List and identify the contents of a work cell.
- Identify and discuss the key terms used in the chapter.
- Answer the review questions at the end of the chapter.

Robots need a source of power to do work. The source of power may be one of three or any combination of three types of power, that is, electricity, hydraulic pressure, and pneumatic pressure. Hydraulic pressure and pneumatic pressure are driven by electric pumps or compressors. Let's take a closer look at these three sources of energy for robots.

Electrical Power

Electric motors were previously mentioned in reference to manipulators and grippers. The direct-current (DC) brushless, DC permanent magnet, and the series, shunt, and compound wound DC motors also were covered. We have not, however, taken a close look at the single- and three-phase AC motors that power manipulators and do a great deal of the heavy work (Figure 5-1).

Single-Phase Motors

Single-phase motors can be classified in a number of ways. Of primary concern here is quick recognition of the general characteristics of the three single-phase motors (split-phase, capacitor-start, and shaded-pole) that are used in robots.

The *split-phase induction motor* is one form of a fractional-horsepower motor. This motor has two sets of windings: one is the *run winding*, and the other is the *start winding*. The run winding is the main workhorse of this motor. The start winding is used only when the motor is started and is therefore called an *auxiliary winding*. Once the motor reaches 75 percent of its run speed, the start winding is taken out of the circuit by a centrifugal switch.

The split-phase motor has the constant-speed, variable-torque characteristics of the shunt DC motor. Most of the motors are designed to operate on 120 or 240 volts. For the lower voltage, stator coils are divided into two equal groups that are connected in parallel. For the higher voltage, the groups are connected in series. The starting torque is 150 to 200 percent of the full-load torque, and the starting current is six to eight times the full-load current. The direction of rotation of the split-phase motor can be reversed by interchanging the start winding leads. Fractional-horsepower split-phase motors are used in various machines and ventilating fans.

The *capacitor-start motor* is a modified form of the split-phase motor. It has a capacitor in series with the start winding. The capacitor produces a greater phase displacement of currents in the start and run windings than is produced in the split-phase motor.

Motor Characteristics

	⊐D	Duty	Typical Reversibility	Speed Character	Typical Start Torque*
POLYPHASE	A-C	Continuous	Rest/Rot.	Relatively Constant	175% & up
SPLIT PHASE SYNCH.	A-C	Continuous	Rest Only	Relatively Constant	125-200%
SPLIT PHASE Nonsynchronous	A-C	Continuous	Rest Only	Relatively Constant	175% & up
PSC Nonsynchronous High Slip	A-C	Continuous	Rest/Rot.†	Varying	175% & up
PSC Nonsynchronous Norm. Slip	A-C	Continuous	Rest/Rot.†	Relatively Constant	75-150%
PSC Reluctance Synch.	A-C	Continuous	Rest/Rot.†	Constant	125-200%
PSC Hysteresis Synch.	A-C	Continuous	Rest/Rot.†	Constant	125-200%
SHADED POLE	A-C	Continuous	Uni-Directional	Constant	75-150%
SERIES	A-C/ D-C	Int./Cont.	Uni-Directional•	Varying‡	175% & up
PERMANENT MAGNET	D-C	Continuous	Rest/Rot.§	Adjustable	175% & up
SHUNT	D-C	Continuous	Rest/Rot.	Adjustable	125-200%
COMPOUND	D-C	Continuous	Rest/Rot.	Adjustable	175% & up
SHELL ARM	D-C	Continuous	Rest/Rot.	Adjustable	175% & up
PRINTED CIRCUIT	D-C	Continuous	Rest/Rot.	Adjustable	175% & up
BRUSHLESS D-C	D-C	Continuous	Rest/Rot.	Adjustable	75-150%
D-C STEPPER	D-C	Continuous	Rest/Rot.	Adjustable	■

* Percentages are relative to full-load rated torque. Categorizations are general and apply to small motors.
■ Dependent upon load inertia and electronic driving circuitry.
• Usually unidirectional—can be manufactured bidirectional.
† Reversible while rotating under favorable conditions: generally when inertia of the driven load is not excessive.
‡ Can be adjusted, but varies with load.
§ Reversible down to 0°C after passing through rest.

Figure 5-1 Characteristics of electric motors. (*Courtesy of Bodine Electric Company.*)

There are three types of capacitor motors: the capacitor-start, the permanent-split capacitor, and the two-value capacitor. Each type has its own characteristics. One of the advantages of the capacitor-start motor is its ability to start under load and develop high starting torque. It can be reversed at rest or while rotating. The speed is relatively constant, and the starting torque is 75 to 150 percent of rated torque. The starting current is normal. This type of motor can be used on air compressors and hydraulic pumps when the voltage rating is available (Figure 5-2).

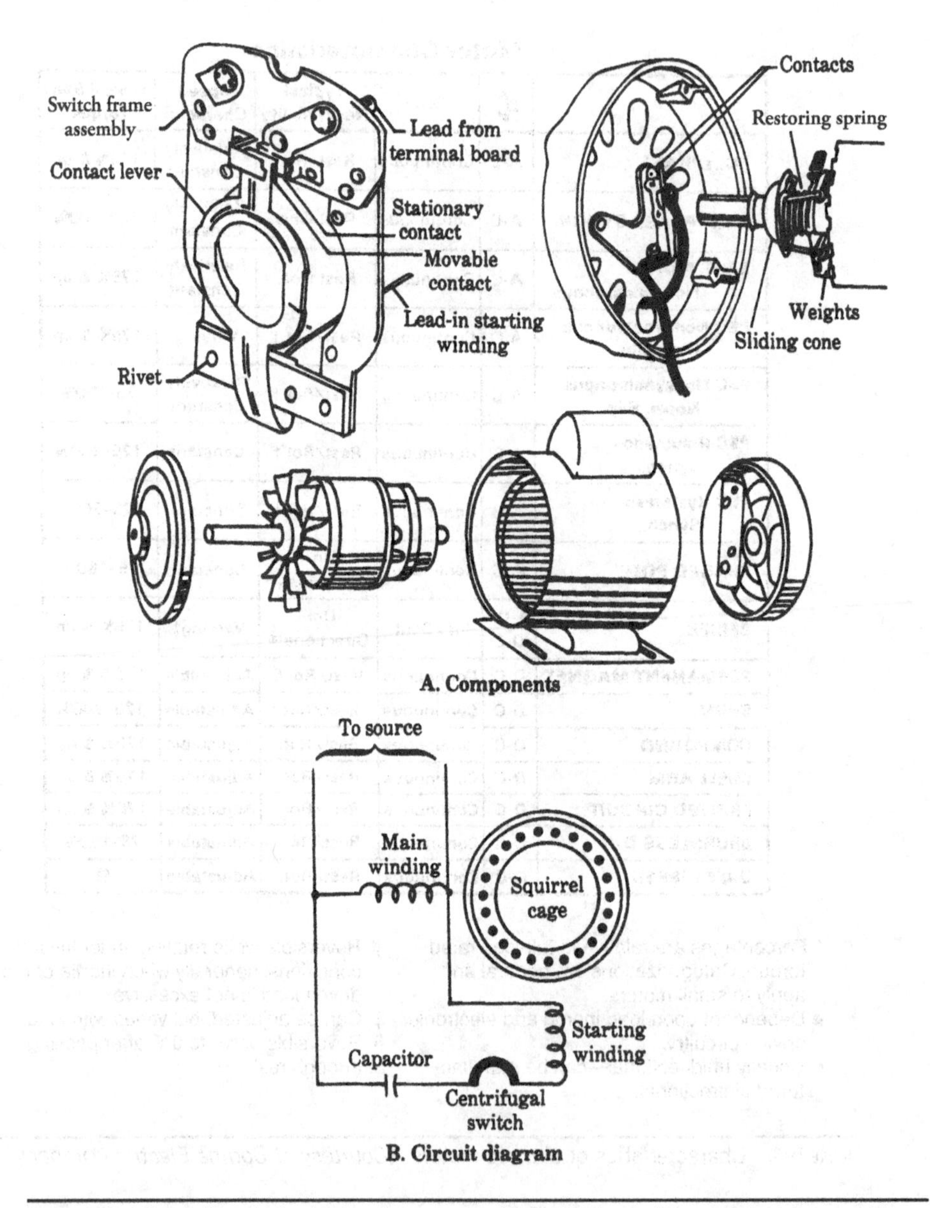

FIGURE 5-2 Capacitor-start, single-phase motor (note the large "bump" that houses the capacitor on top of the motor).

The *shaded-pole motor* comes in three types. It is built in a wide variety of ratings, with any number of performance characteristics. Shaded-pole motors are made in 0.00025 to 0.1 horsepower. Speeds for the shaded-pole motor are stated with no-load conditions. They are not very powerful (Figure 5-3).

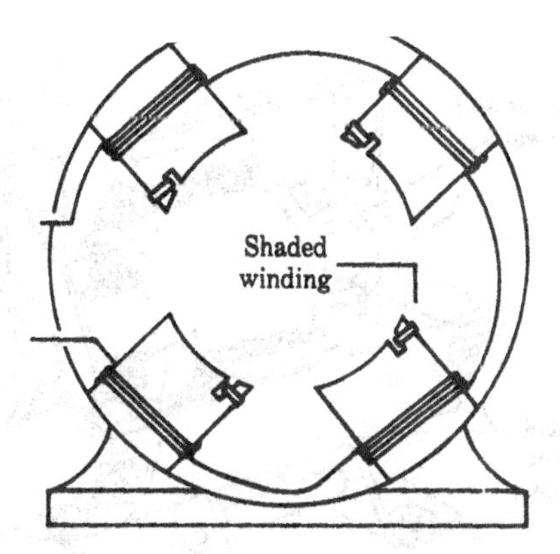

FIGURE 5-3 Shaded-pole motor.

The shaded-pole motor has a low efficiency rating, a low power factor, a low starting torque, high noise and vibration, and is very low in cost. Having small poles shorted by heavy copper rings, it requires no starting mechanism. The shaded-pole motor finds applications in fans and small-horsepower devices such as clocks.

Three-Phase Motors

Most industrial motors are three phase, primarily because maintenance of a three-phase motor is practically nil. Industrial motors do not have the starting devices that single-phase motors have. The three phases of alternating current that supply power for the motor produce the phase shift needed to get the motor started and to keep it running once it is started (Figure 5-4). All commercial power produced in the United States is generated as three-phase current. It can be converted to single phase by dividing the three separate phases and sending them into three different subdivisions or locations. It is cheaper to distribute single-phase AC than three-phase AC. Three-phase power requires at least three, and sometimes four, wires for proper distribution.

Three-phase motors have good overall characteristics. They are ideal for driving machines in industrial uses. They can be reversed while running by reversing any two of the three connections to the power line.

Motor speed, under normal load conditions, is rarely more than 10 percent below synchronous speed. At the extreme of 100 percent slip, the rotor reactance is so high that the torque is low because of a low power factor. The three-phase motor has a high starting and breakdown torque with a smooth pull-up torque. It

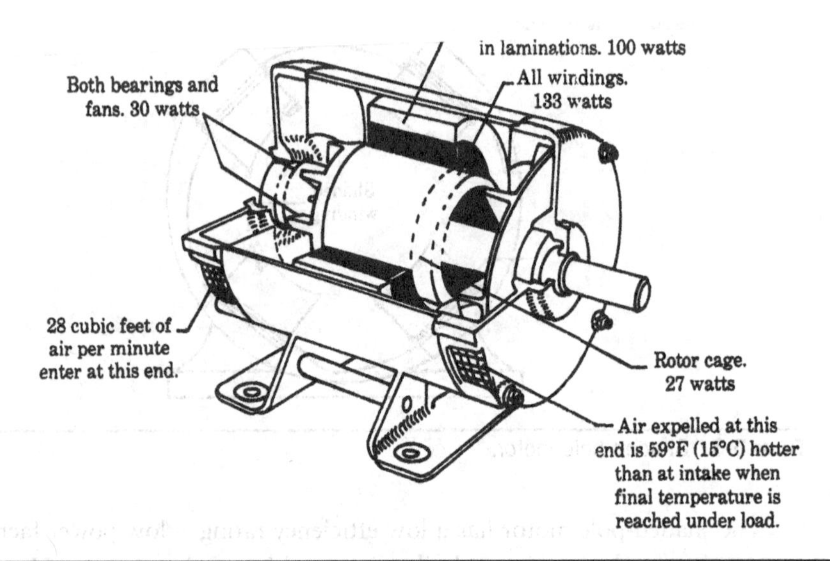
in laminations. 100 watts
All windings.
133 watts
Both bearings and
fans. 30 watts
28 cubic feet of
air per minute
enter at this end.
Rotor cage.
27 watts
Air expelled at this
end is 59°F (15°C) hotter
than at intake when
final temperature is
reached under load.

FIGURE 5-4 Three-phase motor.

is very efficient to operate. It is available in 208–230/460-volt sizes, with horsepower ratings varying from one-quarter to hundreds. It can be obtained for operation at 50 or 60 hertz.

If you classify robots by their power supply control, you get two different types of machines: servo-controlled and non-servo-controlled robots. A closer examination of these two types will produce some understanding of what is inside the large unit called the *controller*.

Servo-Controlled Robots

A *servo-controlled robot* is driven by servo mechanisms. Servo motors are driven by signals rather than by straight power-line voltage and current. This means that the motor's driving signal is a function of the difference between command position and/or rate and measured actual position and/or rate. The servo-controlled robot is capable of stopping at or moving through a practically unlimited number of points in executing a programmed trajectory. In other words, signals are produced that cause the robot to know where it is and where it is going. This is a more sophisticated system than that of a non-servo-controlled robot.

A servo-controlled robot can do more things than a non-servo-controlled type. It can move up and down and back and forth and is able to stop at any point within its work envelope. It can also return to the spot any time the program is

run.
However, it is expensive. Some type of control system must be devised to make sure that the robot can stop and go and reach a programmed point every time.

Feedback is the basic difference between non-servo-controlled and servo-controlled robots. Feedback is needed to tell where a manipulator is in reference to where it is supposed to be. This feedback system is what costs so much and produces so much research into methods of accomplishing it more efficiently with less cost (Figure 5-5).

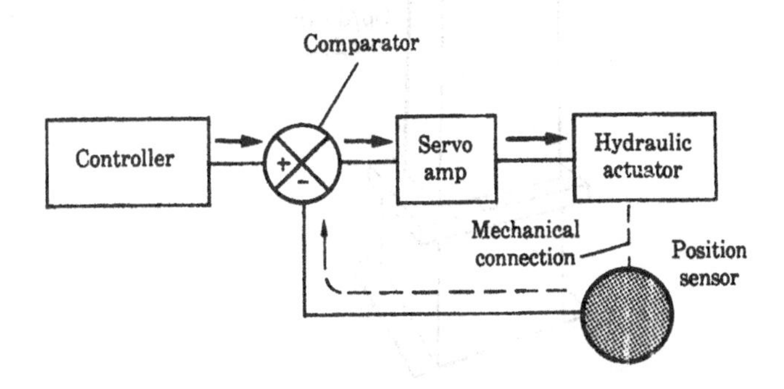

FIGURE 5-5 Feedback system. (*Courtesy of PWS-Kent.*)

To make the arm aware of where it is located within its work envelope, the controller must have some way of determining that the arm is not where it is supposed to be and then making a comparison of where it is with information on where it should be. The controller sends out a command signal. A comparator compares the command signal from the controller with a feedback signal from the position sensor and registers an *error* in the position of the arm. This means that the error has to be corrected. So the amplifier is fed the information, and it drives the hydraulic actuator to cause the arm to move to its programmed location. While the arm is moving, it feeds its positional information back to the comparator for comparison to see how close it is to correcting the error at any given point along the way. Once the error has been corrected and the arm is where it should be, the actuator stops supplying power to the arm.

Non-Servo-Controlled Robots

Non-servo-controlled robots are usually controlled by limit switches or by banging into stops at each side of its swing. They operate on very simple switching or limiting of their arms. Figure 5-6 is an example of a non-servo-controlled robot,

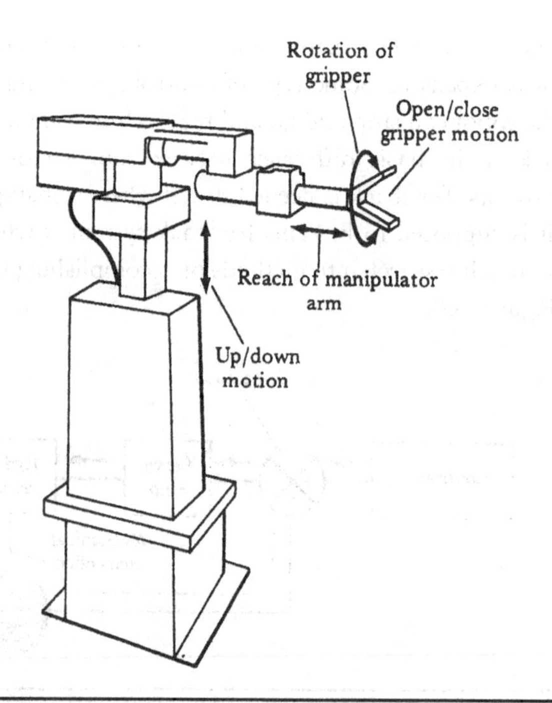

FIGURE 5-6 Non-servo-controlled, low technology robot. (*Courtesy of PWS-Kent.*)

sometimes referred to as a *bang-bang robot*. Non-servo-controlled robots are classified into three types—electric, hydraulic, and pneumatic—based on the means used to drive their manipulators.

Electric Non-Servo-Controlled Robots

This option for powering robots has been eliminated by virtue of the fact that electric motors cannot withstand the stop and go associated with the swing of a robot arm. These motors are easily damaged by the constant and rapid changing of direction. Most non-servo-controlled robots use some other means of power.

Pneumatic Non-Servo-Controlled Robots

Air pressure can be used to power the manipulator arm of a non-servo-controlled robot. Air pressure is fed into the robot drive mechanism, and the arm moves. Once the arm reaches the stop, it is at rest. In order to return to the preceding stop, it must be pushed back by air pressure from a second airline. The second line is activated, and the first line is disconnected. A valve is used to disconnect one line and connect the other. The valve is actuated by using an electric current to energize a coil that draws the sliding bar or valve back and forth inside the unit. This works very well with a pneumatically operated robot because it can exhaust the air into the atmosphere every time it hits a stop (Figure 5-7). (Actuators are discussed later in this chapter.)

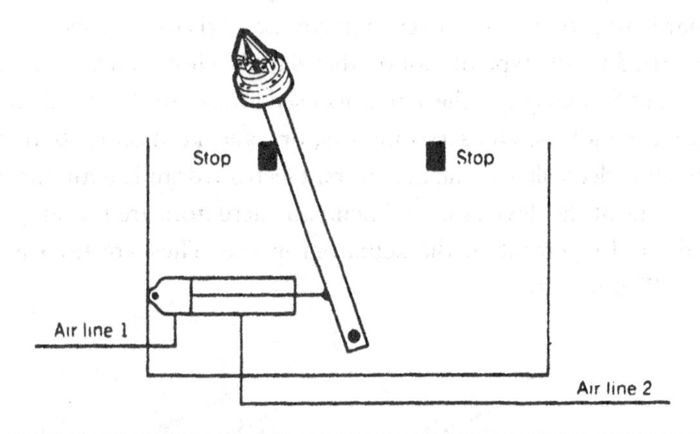

FIGURE 5-7 Pneumatic non-servo-controlled robot.

Hydraulic Non-Servo-Controlled Robots

Hydraulic fluid under pressure can be used to power a non-servo-controlled robot. Because it is very hard on a robot to be constantly hitting a stop at full speed, a switch is added to slow it down. A limit switch is located such that the arm will hit it just before hitting the stop. By opening the switch, the pressure in the hydraulic system is reduced. This allows the arm to coast to a stop against the stop. Shock absorbers also can be used to cushion the shock of sudden stops (Figure 5-8).

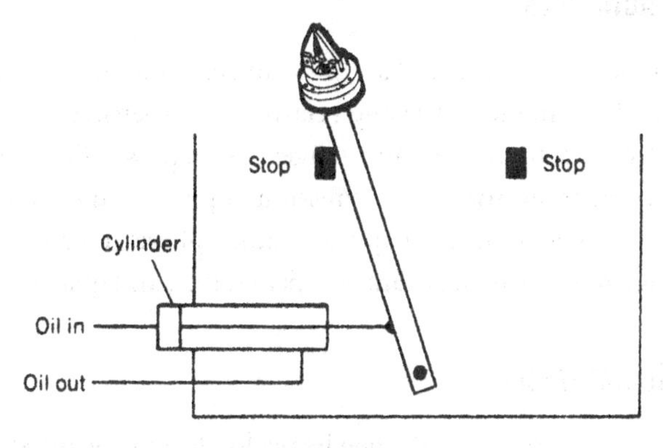

FIGURE 5-8 Hydraulic non-servo-controlled robot.

Actuators

Actuators are motors, cylinders, or other types of mechanisms used to power robots. They convert one type of energy to another. Actuators are employed primarily to provide the power to move each axis of the robot arm. Actuators are categorized by the type of motion they supply, either linear or rotary.

About 50 percent of the actuators used today are electrically driven. They may be stepper motors, DC servo motors, or pancake motors. Both small and large robots use electrohydraulic actuators. Electrohydraulic actuators make up about 35 percent of the devices used. Pneumatic actuators are the simplest and account for about 15 percent of the actuators in use. They are used in pick-and-place robots (Figure 5-9).

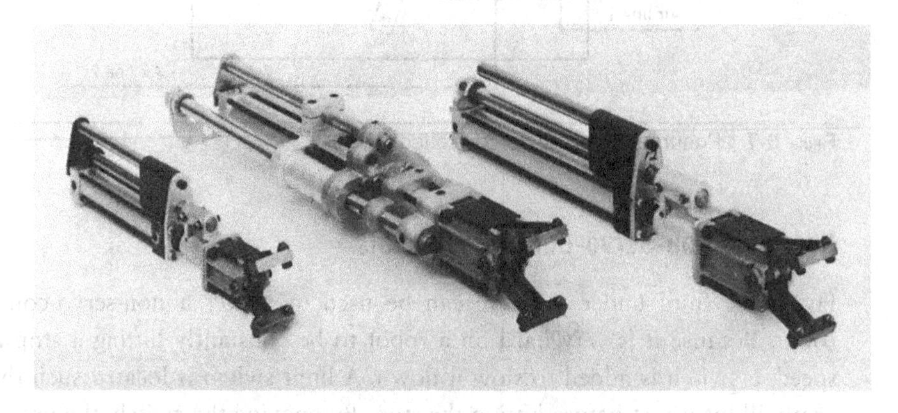

Figure 5-9 Pick-and-place pneumatically operated actuator. (*Courtesy of I.S.I. Manufacturing, Inc.*)

Electric Actuators

An electric actuator is fast and accurate and easily adapted to sophisticated control techniques. It is simple to use and relatively inexpensive.

The electric actuator's disadvantages are its power limits, its backlash (because it requires gear trains for the transmission of power), and its tendency to arc, which may cause problems in an explosive atmosphere. Electric actuators are highly recommended where high accuracy, repeatability, and quiet operation are necessary.

Hydraulic Actuators

Hydraulic actuators are used when heavy loads must be lifted. They have moderate speed. Due to the characteristics of the hydraulic fluid, the joints, once positioned, can be kept motionless, offering accurate control characteristics. The disadvantages of hydraulic actuators are their expense, their tendency to leak and make a lot of noise in operation, their slow speed in cycling, and their stiffness in operation.

Pneumatic Actuators

Pneumatic actuators are cheap, clean, and safe. Their lack of stiffness can be a disadvantage in some uses. They generate high speeds and do not pollute the workspace because they exhaust clean air into the atmosphere (see Figure 5-9). They are often used in laboratory work.

Some disadvantages of the pneumatic actuator are its noise and the problem of air leaks all along the system, especially when the air is under high pressure. This actuator requires good air filtering and drying and has a high level of maintenance and construction. It also has limitations due to the compressibility of air, which does away with its stiffness and, in some cases, its joints creep instead of remaining motionless.

Controllers

Six basic types of controllers are in use today: rotating drum, air logic, relay logic, programmable, microprocessor-based, and minicomputer. Each type of controller has advantages and disadvantages. In fact, some types are no longer widely used because of the rather rapid development of inexpensive electronic controllers.

Rotating-Drum Controller

One of the first controllers to be used was the *rotating-drum controller*. It was made up of a cylinder or mechanical drum that had cams on its surface. These cams would cause plungers to be activated and either release or press against them to open and close valves. These valves, in turn, controlled the air or hydraulic system to produce arm motion. The motion was halted by hitting a stop placed at the desired location. This type of controller had no feedback and no way of knowing where the arm was at any given time. The controller would repeat the same program every time the drum made a complete revolution. It is possible to design a program that takes into consideration the speed with which the drum rotates and the placement of the cams on the surface of the drum. The rotating drum resembles the workings of a music box with its rotating drum causing the various strings to vibrate.

Rotating-drum controllers are used on pick-and-place robots and are very reliable. They do require maintenance from time to time, but they can continue to perform without attention for long periods. Thousands of them are still in operation.

Air-Logic Controllers

Air-logic controllers were used on pick-and-place robots in the sixties and seventies. They are very useful in explosive atmospheres because they do not have electrical switching to create sparks. A series of air valves is placed so that the arm is actuated by the valves at the end of a cylinder's desired travel. The valves are connected together with short pieces of tubing that supply the air pressure needed for operation. In other words, a valve closes, creating pressure that causes the arm to move, and then opens, releasing the pressure, which lets out the air. Another valve closes, and the air pressure causes it to move again. By placing the valves where needed, it is possible to push the arm back and forth over a path desired for its operation.

Relay-Logic Controllers

Relay logic controllers were used with machines before robots were developed. An electromechanical switch, called a *relay*, was used to make sure that switches were closed in the right order to perform a task (Figure 5-10). A ladder diagram was developed for each operation (Figure 5-11). The circuit (ladder diagram) is nothing more than the two lines coming from the power supply. Various relays are connected between these lines. Limit switches and other actuating switches that

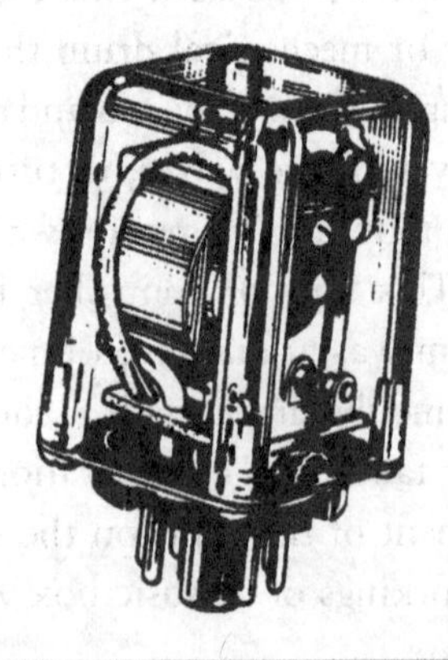

FIGURE 5-10 A relay.

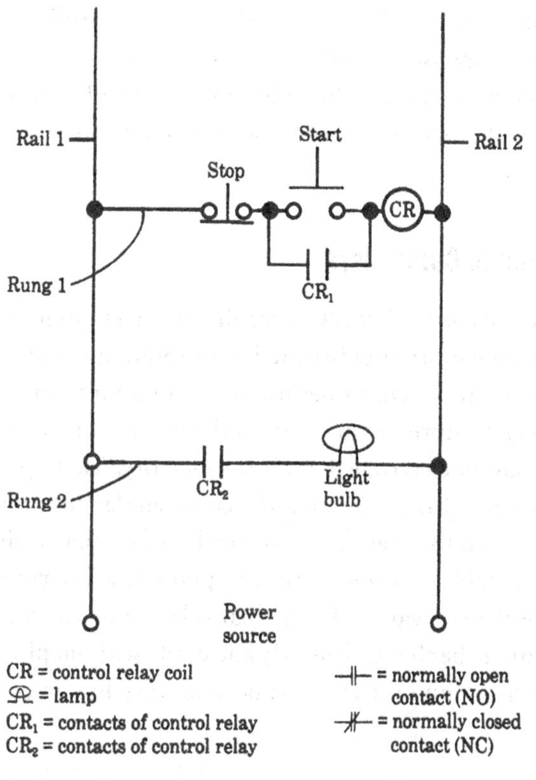
Rail 1
Start
Rail 2
Stop
CR
Rung 1
CR1
Rung 2
CR2
Light bulb
Power source
CR = control relay coil
= lamp
CR1 = contacts of control relay
CR2 = contacts of control relay
= normally open contact (NO)
= normally closed contact (NC)
Push start button and complete the circuit to CR coil. Coil energizes and closes CR1 and CR2. This causes CR to remain energized until stop button is pressed to open the circuit. When the relay is energized, CR2 contacts are also closed, causing the lamp to light and show power on.

FIGURE 5-11 Ladder diagram.

are mechanically closed or opened by the robot's manipulator or cylinders cause the relays to open or close to perform the required operation. The relays cause the solenoids to be energized and to open and close the pneumatic or hydraulic valves. The valves make the robot arm do what it is supposed to do. Ladder diagrams are used today for programmable controllers and other electrical equipment.

Relay switches need a great deal of attention and cause a lot of downtime for their machines. Industry has quickly replaced relays with electronic components, just as the telephone company now uses transistor switching instead of relay switching. There are no moving parts in transistors and no contact points to be cleaned periodically. With relays, the whole circuit had to be rewired to make simple changes. In semiconductor devices, no rewiring is necessary; one needs only to remove a board and substitute another or, in most instances, reprogram without removal.

Rotating-drum, air-logic, and relay-logic controllers are not used today for robots. Less expensive controllers with no moving parts have replaced these maintenance-prone types. The older types had to be reprogrammed along with a lot of physical work. They could not communicate with other robots or with a central location.

Programmable Controllers

Programmable controllers (PCs) are the next step up in controllers (Figure 5-12). They are a little more sophisticated than rotating-drum, air-logic, and relay-logic types. Programmable controller use electronics for timing and sequencing. Instead of using pegs to store the memory as the drum controller did, the programmable controller has an electronic circuit for memory. Many programmable controllers are programmed from a keyboard that is similar to a typewriter keyboard. The keyboard is used to enter the proper order of switch closings.

Programmable controllers are computer-based devices that are nothing more than the electronic way of doing what relays did. The main advantage is that the PC program can be changed quickly and easily without physical work. Maintenance is less expensive, and the cost has become very low in comparison with previous controllers.

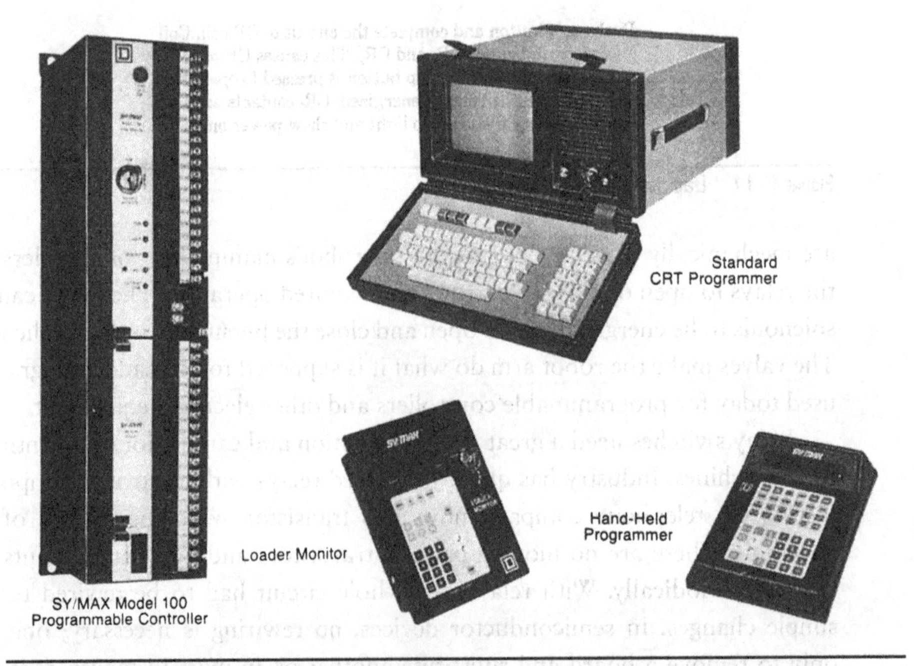

FIGURE 5-12 Programmable controller programming equipment. (*Courtesy of Square D Company.*)

PCs are used in a variety of applications to replace conventional control devices such as relays and solid-state logic. When compared with conventional controllers, PCs allow ease of installation, quick and efficient system modifications, more functional capability, troubleshooting diagnostics, and a high degree of reliability. Typical installations include automated materials-handling control, machine tooling control, assembly machine control, wood and paper processing control, injection-molding machine control, and process control applications such as film, chemical, food, and petroleum.

PCs consist of system hardware and programming equipment. System hardware can be a processor, one or more rack assemblies, power supplies, input/output (I/O) modules, and various other modules that provide additional capabilities. The rack assemblies and associated I/O modules communicate with external I/O control devices such as limit switches and motor starters. System hardware may also be a PC that incorporates all necessary hardware in a single package.

Programming equipment consists of either a hand-held programmer or a deluxe cathode-ray tube (CRT) programmer that is used during program entry or for monitoring operation of the system. A separate loader/monitor for monitoring and changing data or printing messages also can be obtained.

PC processors provide relay logic, latch/unlatch relays, data transfers, timers, counters, master control relays, synchronous shift registers, data comparisons, bit read and control, transitional output, and register fencing; respond to communications from other processors; and have I/O forcing capability.

As you can see from this terminology, a number of new terms apply to this rapidly advancing field. It is difficult to do justice to the topic of programmable controllers within the confines of this chapter. An entire book is needed to develop the technical expertise associated with PC devices and their capabilities. However, this brief introduction to PCs indicates how they are connected to the advancement of robots and the field of robotics.

Microprocessor-Based Controllers

The microprocessor is the building block of the microcomputer. Microprocessors are designed to do a specific job. In this book, we are concerned with how microprocessors are designed to work with robots and give them certain capabilities to perform tasks (Figure 5-13).

Microcomputers are used with all types of robots. The microprocessor is specially designed for a specific robot. It has a memory and the circuits needed to cause the robot to function and be easily programmed or edited. In some cases, the microcomputer will have a CRT to display the contents of its memory and the program that has been placed in it.

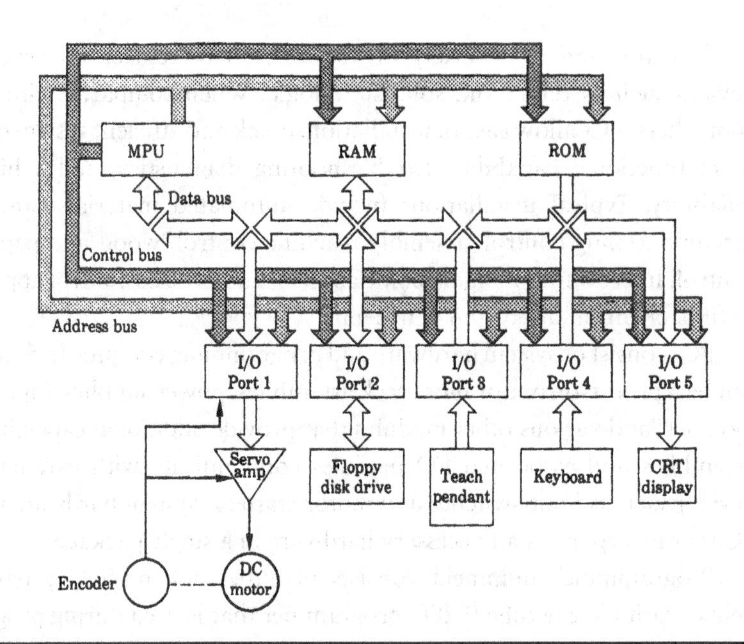

FIGURE 5-13 Block diagram of a microprocessor.

The output of the program must be able to power the solenoids that control the manipulator arm. This calls for an amplifier to take the very weak signals from the computer and boost them sufficiently to energize solenoids that control hydraulic or pneumatic valves.

Minicomputer Controllers

Minicomputers are used with some robots to give them a greater level of sophistication and to enable them to do tasks that could not be done otherwise. Robot manufacturers have mated computers with their robots to expand their applications. The output of the computer is fed into circuits to be amplified and then fed to the control devices for the manipulator, wrist, and gripper (Figure 5-14).

Programming a Robot

Robots have to be "taught" to do jobs. They are easily programmed, or taught, once their electronics packages are understood. Electronics makes it possible for robots to do so much.

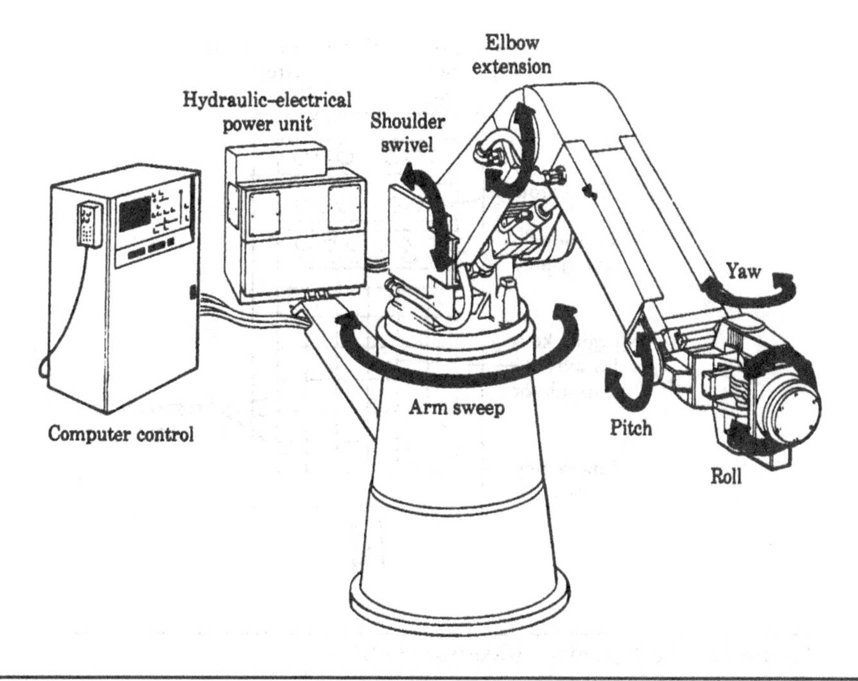

FIGURE 5-14 Robot system with a minicomputer. (*Courtesy of Cincinnati Milacron.*)

Putting pegs in a drum and changing air hoses as a means of programming a robot have already been discussed. This technology is obsolete. Now, besides keyboarding as a way to teach a robot a task, the teach-pendant method, lead-through programming, and computer terminal programming are used commonly.

Teach Pendant

A *teach pendant* is the most popular device for programming a robot because it requires very little mathematical figuring and program editing. The desired position for the manipulator is reached by "walking it" to the spot where it is needed. This walking is done by pressing buttons on the teach pendant that cause the arm to move where you want it. A teach-pendant button is then pressed to store the information in memory (Figure 5-15). The manipulator is led to the next desired position by pressing the right buttons. Once it is in the correct position, that too is entered into the memory of the robot as the next point in the program. None of the moves made by the programmer between the first and second points is remembered by the robot; the robot remembers only the final points. The program is played back, and the robot's manipulator moves smoothly from the first point to the next point stored in its memory.

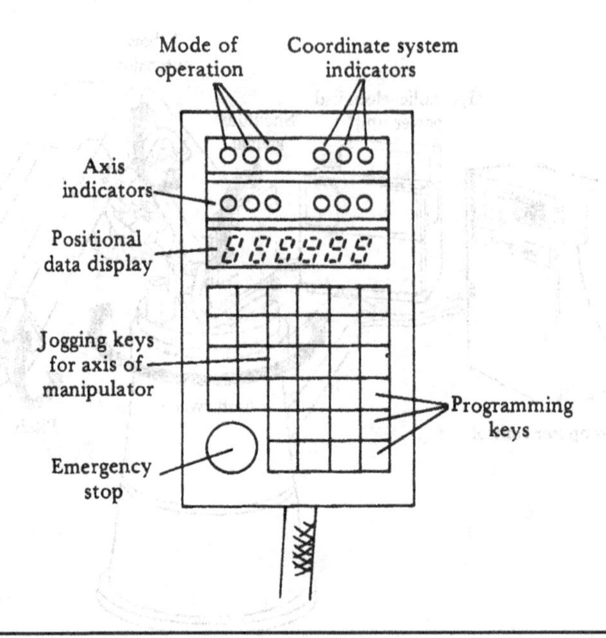

Figure 5-15 Teach pendant. (*Courtesy of PWS-Kent.*)

The program is tested before the robot is put into production runs. The speed from one point to the other also can be programmed with the teach pendant. This allows the operator to make sure that the arm moves smoothly and comes to a stop without abrupt action. It is also possible to program the acceleration and deceleration of the arm by using the pendant. Smoother operation is then possible. The programmer steps through the program after it has been stored, recalling all the information and putting the manipulator through its motions. If errors are detected, they can be edited out with the teach pendant before the robot is put into operation on the production line.

Once the program has been checked for correctness, the robot can be put to work, and it will repeat the motions for as long as needed. You will see why this is the easiest way to program a robot once you have looked at other methods of programming.

Lead-Through Programming

Lead-through programming is usually done with continuous-path robots. They can be programmed to follow a circle, an arc, or a straight line. This type of robot is programmed by grabbing the arm and moving it through the actions needed to do the job. The robot will memorize the path and then play it back when called for. The speed at which the arm moves can be adjusted later after the exact strokes or movements are committed to memory. Adjustments also can be made for small errors that are detected before the robot is put into production. This type of robot is useful as a painter or welder. For example, an experienced painter leads the arm through the same motions he or she would make if he or she were doing the job. The robot memorizes the path and then repeats it over and over again. This type of robot can remember thousands of points between start and stop. Figure 5-16A shows a spray-painting robot that is easily programmed by using the lead-through teach method. Figure 5-16B is an editing module used to correct errors in speed and movement.

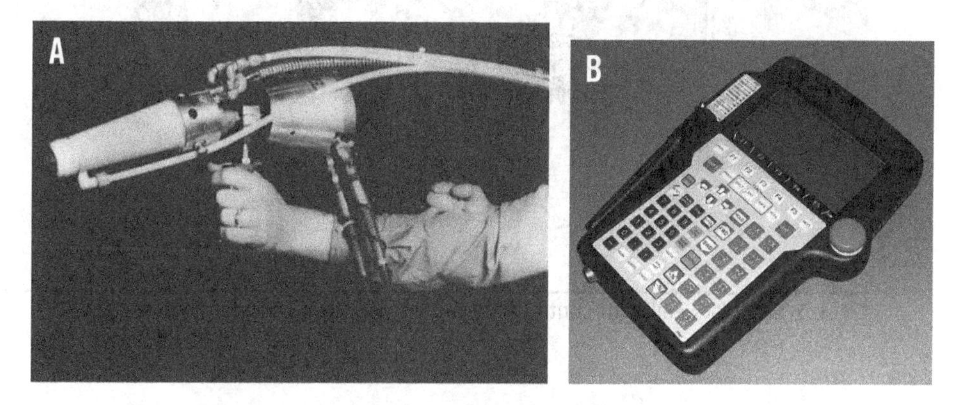

FIGURE 5-16 (A) Lead-through programming and (B) a hand-held editing device similar to a teach pendant. (*Courtesy of (A) Binks Manufacturing Company and (B) FANUC.*)

Computer Terminal Programming

Using a computer to program a robot can be difficult. Robots use a number of languages to move their manipulators from one place to another. These languages can be used to call up the program that was memorized by a walk-through and modify it for smoother and more precise operation. A computer language also can be used to write a program from scratch. The computer program comes in handy when the robot is to work in concert with a welder, conveyor belt, or other types of equipment that make up a work cell (Figure 5-17).

Computer controllers were developed in the 1960s when integrated circuits (ICs) were introduced. An IC is a complete circuit with its resistors and transistors on a single silicon chip (Figure 5-18). These chips may have hundreds of transistors and diodes in a single package, which make possible microcomputers in very small packages. The chips also allow for the design of a computer for each robot. The price of electronic components has decreased so much that it is now cheaper to

Figure 5-17 Work cell with controller and computer. (*Courtesy of FANUC.*)

buy an electronic controller than it is to buy rotating-drum or air-logic controllers. The extended capabilities of computer terminal programming are also much more economical when it comes to keeping robots operational.

Inexpensive computer terminal programmers make it possible to have vision, touch, and hierarchical control for almost all robots. Desktop computers such as the Apple can be used to program a robot. The Apple communicates with the robot through the robot's controller. The controller actually operates the robot with the information furnished by the computer. The controller has a microprocessor that has to be *interfaced* (properly connected) with the computer.

Summary

Robots need a source of power to do work. The power may be from a single source or from any combination of three sources, that is, electricity, hydraulic pressure, or pneumatic pressure. Single- and three-phase motors are used to provide the energy to move heavy loads. Single-phage motors may be either split phase, capacitor start, or shaded pole. The shaded-pole motor is usually employed to power fans and ventilation devices. The split-phase motor does not start well

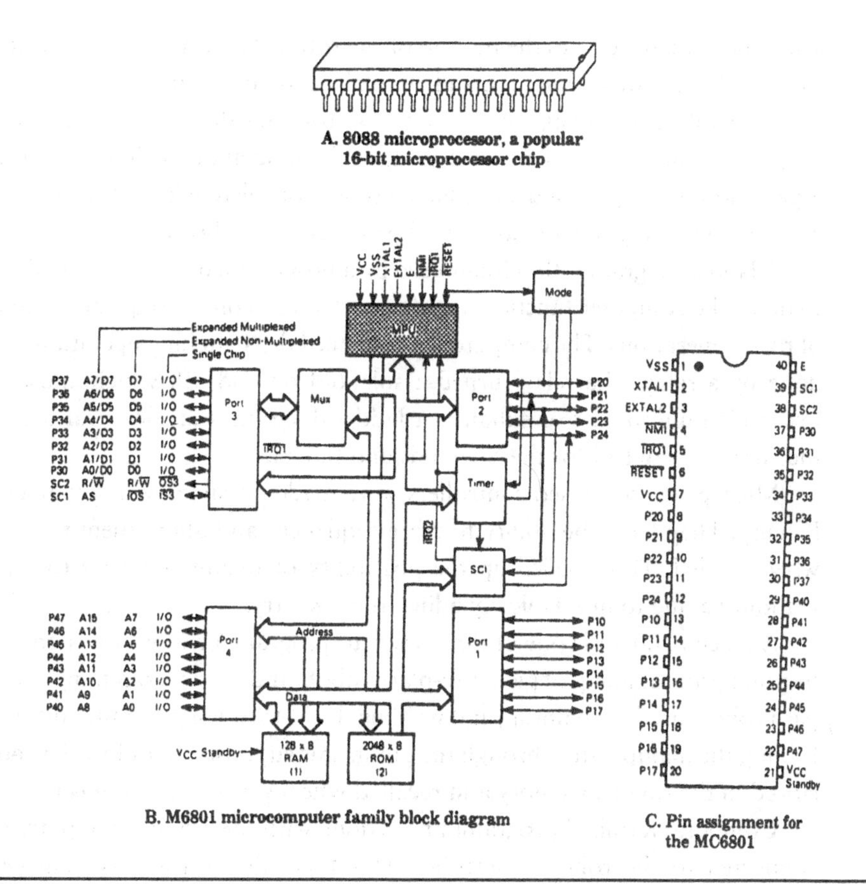

A. 8088 microprocessor, a popular 16-bit microprocessor chip

B. M6801 microcomputer family block diagram

C. Pin assignment for the MC6801

FIGURE 5-18 Integrated circuit chip. (*Courtesy of Motorola, Inc.*)

under load, but the capacitor-start motor does. The capacitor-start motor can be used to power compressors and similar devices where lower voltages (120/240) are available. The three-phase motor is one of the most reliable of electric machines. It is the workhorse of industry.

A servo-controlled robot can do more things than a non-servo-controlled robot. It can move up and down and back and forth and is able to stop at any point within its work envelope. Non-servo-controlled robots usually are controlled by limit switches or banging into stops at the end of each swing. There are electric, pneumatic, and hydraulic non-servo-controlled robots.

Feedback is the main advantage that servo-controlled robots have over non-servo-controlled robots. Feedback tells the controller where the manipulator is located at all times.

Actuators are motors, cylinders, or other mechanisms used to power robots. They are employed primarily to provide the power to move each axis of the robot arm. The actuator causes the motion of the robot. There are pneumatic actuators, electrically operated actuators, and electrohydraulic actuators.

Controllers are available in six types: rotating drum, air logic, relay logic, programmable, microprocessor based, and minicomputer. Rotating-drum, air-logic, and relay-logic controllers have become obsolete with the advent of the IC chip and its ability to store and recall programs for robots.

A ladder diagram is the circuit used by a programmable robot. It is also needed to make the computer function as a device that can control sequencing and timing of robot operations. The computer has replaced the switching operations normally done by a relay. The relay turned a solenoid on and off, which allowed air or hydraulic fluid to pass or exhaust. A ladder diagram is an electrical schematic of the circuit of control for the solenoids and timers.

Microprocessor-based controllers were made possible by development of the IC chip. They have the ability to store sequences and allow them to be recalled when needed. This opened up the possibility of making changes to a program without having to mechanically adjust the circuitry.

A teach pendant is one of three ways to program a robot. It is used for point-to-point programming of pick-and-place robots and for programming continuous-path robots used for painting and welding. Lead-through programming is done by leading the manipulator through the points it is supposed to follow. The points are stored in the robot's memory and recalled whenever the program is repeated.

Computer terminal programming is done with the aid of a computer properly connected to the robot's controller. This then allows for easy changes in the program if needed to change the robot's job.

Key Terms

actuators Motors, cylinders, or other mechanisms that are used to power robots

chips Integrated circuits that contain many transistors, chips, and resistors to make up various electronic circuits on a small (about 8 millimeters square) piece of silicon; components that store a computer's program and act as its memory.

controllers Units needed to control a robot by preparing the proper signals and timing to cause the robot to operate with some degree of accuracy and repeatability.

feedback The ability of a device to feed a signal back to its controller to aid in keeping track of the position of the manipulator or gripper.

interface Proper connections between a robot and its computer or microprocessor.

ladder diagram Electrical circuit drawing of the sequence of switches needed to cause a robot to perform programmed activities.

lead-through programming Taking the continuous path of a robot's manipulator and leading it through a sequence of motions to accomplish a task.

microcomputers Small-scale computers.

microprocessor The brain of a microcomputer; an IC that performs arithmetic, logic, and control operations within a microcomputer.

minicomputers Midsize computers, slightly larger than microcomputers and smaller than mainframes.

relays Electromechanical devices that are energized and that close or open switches.

servo motors Motors driven by signals rather than by straight power-line voltage and current; motors whose driving signal is a function of the difference between command position and/or rate and measured actual position and/or rate.

solenoids Coils of wire with plungers that can turn on or off a fluid or air line.

teach pendant Device used to teach a robot's memory a new program.

Review Questions

1. What type of power does a power supply have to run a robot?
2. Name three types of single-phase electric motors.
3. Which type of single-phase motor has good starting torque?
4. Name three types of non-servo-controlled robots.
5. What is the advantage of the servo control over non-servo control in a robot?
6. What is the basic difference between a non-servo-controlled and a servo-controlled robot?
7. What does an actuator do?
8. How are the greatest number of actuators driven today?
9. Name six types of controllers in use today.
10. Why have the rotating-drum and air-logic controllers become obsolete?
11. What is a ladder diagram?
12. Why are relays used in a ladder diagram?
13. How is a programmable controller programmed?
14. Describe a microprocessor controller.
15. What is a teach pendant?
16. How do you program a pick-and-place point-to-point robot?
17. How do you use a teach pendant to program a welder robot?
18. How is lead-through programming done?
19. What is an IC chip?
20. What does the term *interface* mean?

CHAPTER 6

The Robot and the Computer

Performance Objectives

After reading this chapter, you will be able to:

- Understand how the robot-computer interface is used to control robots.
- Identify what is meant by the terms *VAL*, *HELP*, *MCL*, *RPL*, and *DAR*.
- Describe software for use in robot control.
- Explain what ASCII code is.
- Explain what parts are and how they are used with ASCII code.
- Describe the interfacing of robot and computer.
- Explain how a robot can see.
- Explain how object recognition is possible with robots.
- Describe the clustering process.
- Identify and discuss the key terms used in the chapter.
- Answer the review questions at the end of the chapter.

Constant supervision is necessary to ensure that robots do their assigned tasks the way they are supposed to do them. Many robot manufacturers are now building in this supervision capability. The controller is built into the control system, and it works much the same as the programmable controller. Inasmuch as a robot is always interacting with conveyors, transfer lines, and other materials-handling equipment, it is important that the robot be able to function under these conditions. To make the robot a safe machine, it needs switches interconnected to the other equipment to stop it in case a dangerous situation arises. All this inputs to the controller or microprocessor.

Robot-Computer Interface

A number of methods or systems are used to accomplish the mating of computer and robot. Figure 6-1 is a schematic diagram of the microcomputer system that is often used for this purpose. As shown, the three major parts of the microcomputer are memory, central processing unit (CPU), and input/output (I/O) ports.

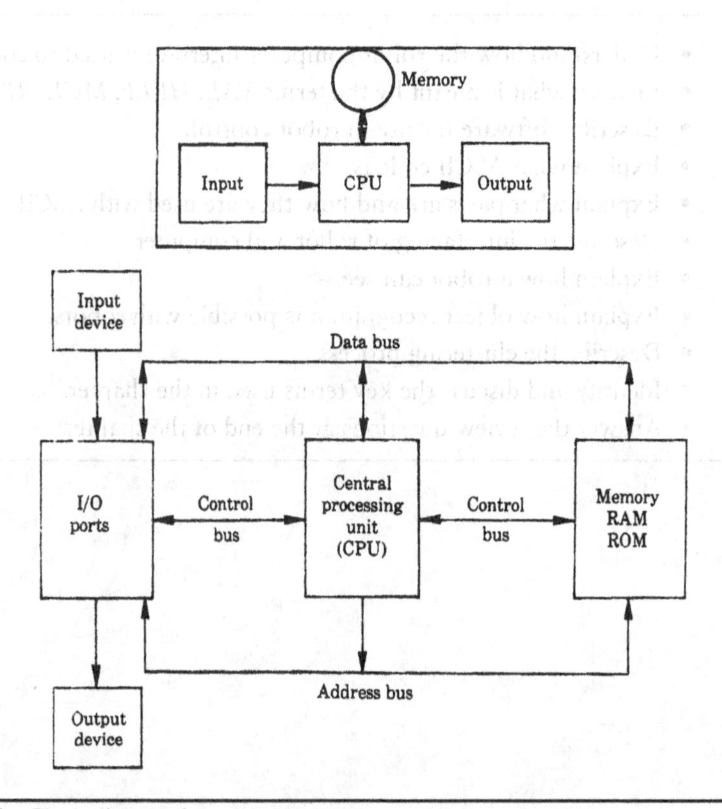

FIGURE 6-1 Block diagram for a robot computer system.

Memory

The computer's memory is used to store data to be processed by the CPU or data resulting from processing.

Central Processing Unit

The CPU, sometimes called the *microprocessing unit* (MPU), contains the control circuitry, an arithmetic logic unit (ALU), registers, and an address program counter. This may be a large computer in larger installations.

The instruction code comes to the CPU from memory and gets decoded and executed. The CPU works in millionths of a second, enabling it to recall an instruction and process it without a slowdown in operation. Instructions are given in many computer languages. Each manufacturer specifies which language or languages the machine will use.

Input/Output

The connections that interface with the outside world of the microprocessor are called *input/output ports*. The input port allows data from a keyboard or other input device to be taken into the CPU. The output port is used to send data to an output device such as a motor. Bus lines carry the signals to and from the major parts of the microprocessor.

Languages

Intelligent robot controllers have the ability to understand high-level languages. Almost every robot manufacturer has its own controller language, making the language unique for a particular controller from a particular manufacturer. Robots are usually preprogrammed for the benefit of the user. Some of the more frequently used languages are VAL, HELP, AML, MOL, RPL, and RAIL. These are relatively simple languages. BASIC, COBOL, and other languages also can be used.

VAL was developed for Unimation. HELP was developed for General Electric's assembly robots. AML, which is a manipulator language, was developed for IBM's assembly robot. MCL stands for "Manufacturing Control Language" and was developed by McDonnell Douglas for an Air Force project. Cincinnati Milacron's T^3 robot uses it. RPL stands for "Robot Programming Language" and was developed by SRI. It is similar to languages such as Pascal and FORTRAN and is useful for communicating with intelligent vision sensors. RAIL was developed by Automatix for robots and vision systems.

Following is an example of a program written in VAL that will give you an idea of how a robot language works.

VAL

The robotic language VAL was developed by Unimation for the specific purpose of controlling the robots that Unimation manufactures. VAL is also used by Adept and Kawasaki. Because Unimation robots are widely used, Rhino chose to emulate the Unimation languageVAL for use with the XR Series robotic system. Following is a Unimation program written in VAL used to palletize products:

```
1.          SETI  PX = 1
2.          SETI  PY = 1
3.    10    GOSUB  100.0.0.0
4.          IF PX = 3  THEN 40
5.          SHIFT PALLET BY 100. O, C
6.          GO TO 10
7.    20    IF PY = 3  THEN 40
8.          SETI  PX = 1
9.          SETI  PY = PY + 1
10.         SHIFT PALLET BY --900.0, 100.0, GO10
11.         GO TO 10
12.   100   APPRO CON, 50
13.         WAIT CONRDY
14.         MOVES CON
15.         GRASP 25
16.         DEPART 50
17.         MOVE PALLET: APP
18.         MOVES PALLET
19.         OPENI
20.         DEPART 50
21.         SIGNAL GOCON
22.         SETIPX = PX + 1
23.         RETURN
24.   40    STOP
```

This VAL program causes the robot to go through the following procedures to pick up an object from a conveyor belt and deposit it on a pallet:

- Lines 1 and 2 keep track of how many parts have been loaded onto the pallet in both the *X* and *Y* directions.
- Line 3 calls a subroutine (**GOSUB**) that will unload one part from the conveyor and place it on the pallet.
- Line 5 redefines a coordinate frame named **PALLET**.
- **SHIFT** on line 5 implements a translation of $X = 100.0$ millimeters, $Y = 0$, $Z = 0$.
- Inside the subroutine, the robot approaches the frame named **CON**.
- The **APPRO** function causes motion to a point displaced 50 millimeters out the *Z* axis of the named frame and oriented such that the *Z* axis of the hand is aligned with the *Z* axis of the named frame. The **APPRO** command also specifies joint interpolated control.
- Line 13 tells the robot to wait for a signal (**WAIT CONRDY**) from an outside source such as a limit switch. The limit switch indicates that the conveyor is ready.
- Line 14 moves the manipulator to grasp position and closes the gripper. **MOVES** indicates Cartesian straight-line motion.
- Line 15 halts the program if the hand closes to less than the minimum anticipated distance of 25 millimeters.
- Line 16 specifies Cartesian motion along the *Z* axis of the hand frame to the departure point. The **50** indicates 50 millimeters out.
- Line 17 shows the alternate form used in VAL. It could have been **APPRO PALLET, 50.MOVE** specifies joint interpolated motion. (Note the difference between **MOVES** and **MOVE**.)
- Line 21 indicates how VAL can output signals to other devices—in this case, the command **SIGNAL GOCON** starts the conveyor in motion.

Unless a person is well versed in computer languages and the programming of computers, it is almost meaningless to go through a program in any greater detail than has been done here. It is significant to note, however, that a more thorough knowledge of programming is needed in order to become a robotics technician.

Software

Computers are used to program robots, and the programming is done off-line. This means that the robot is not directly involved when programming is taking place. The software can be recorded on floppy disk, magnetic tape, or other means and then checked on a robot and debugged or edited. Once the program is proven on a robot system, it can be duplicated and sold to others who want to use robots for similar tasks.

By using off-line programming, the programmer has greater flexibility to carry out complex operations, and the time spent in programming is reduced. In addition, the robot can remain in service while the programming is taking place, thereby increasing its productivity.

Interfacing

Interfacing links allow a robot to communicate with its controller and other parts of the workstation. Parts that do cause the robot to do its task directly are called *peripheral components*. Figure 6-2 shows a simple interfacing link.

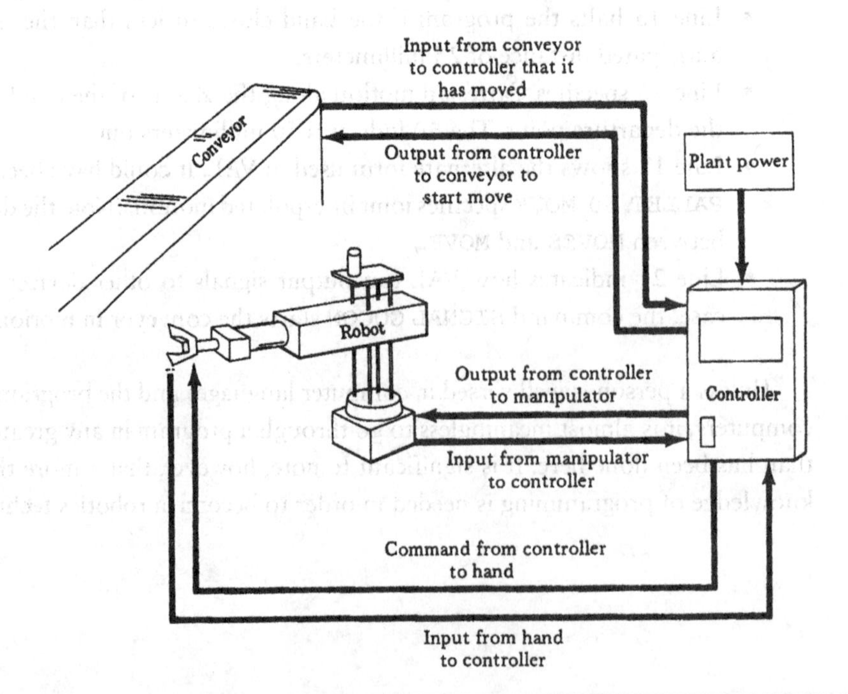

FIGURE 6-2 An interfacing link. (*Courtesy of PWS-Kent.*)

A microprocessor-controlled robot can communicate with the equipment around it through connections called *ports*. For example, it is necessary for a robot to input information to the controller, and it must be able to receive information from the controller. Hence the controller is connected to the robot through a port.

ASCII Code

In order to communicate with the controller, you must be able to input instructions to its microprocessor. A special code has been designed that allows a regular typewriter keyboard to be used to type in instructions. However, the keys of the keyboard are actually switches that send electronic pulses to a decoder, which generates a special binary code. The code most often used is the American Standard Code for Information Interchange (ASCII).

The ASCII code is made up of seven binary bits, with 128 possible combinations (obtained when you take 2 to the 7th power, 2^7). The 128, or 2^7, possible combinations of ones and zeros represent the letters of the alphabet, both upper and lower case, as well as the numbers 0 through 9 and several special codes that include punctuation and machine control information.

Parallel Ports

Parallel ports are the outputs of the microprocessor or computer that have flat cables connected to them with eight conductors. Seven of these wires or conductors carry the information mentioned earlier. The eighth conductor carries the strobe line, the one that prevents switch bounce. When a switch is closed, it bounces and can allow more than the on and off information to be given. This noise or incorrect signal information must be prevented from being transmitted from the keyboard to the controller and then to the robot. It could cause the robot to make an incorrect move.

Serial Ports

The serial format also may be used to transmit data in the ASCII code. Serial ports allow the information to be transmitted along two wires. This makes transmitting over long distances feasible. The parallel-line format is very good for short distances or connection between machines within the same work cell, but if the information has to be sent for a greater distance, it is better to send it by serial formatting.

in
ıch.
The two voltage standards are known as RES-232C and TTL. The two current standards are the 60-milliampere current loop and the 20-milliampere current loop.

The RS232-C standard says that the voltage of the signal will be between −3 and −25 volts to represent the logic 1 or the "on" condition. A voltage between +3 and +25 volts is used to represent the logic 0 or "off" position This standard was developed by the Electronics Industry Association (EIA). The advantage of this standard is that the noise will have to be very high to make any fake signals, and the voltage losses along the line will not affect the signal level as much as lower voltages do. Its disadvantage is that it has to be converted to transistor-transistor logic (TTL) at the port of the computer.

The TTL standard is compatible with transistor-transistor logic and interfaces directly. There are problems with any transmission of data over a distance. If there is a line voltage of at least 0.5 volt, then there is the possibility of receiving incorrect data. Because the peak is only 5 volts, there is always the possibility of picking up a noise signal when a wire is spread over a distance in an electrical noise-generating environment.

The 60-milliampere standard says that a current of 60 milliamperes is logic 1, whereas zero current represents logic 0. The main advantage is that the noise usually encountered over long-distance transmission lines does not affect the quality of the data being transmitted. However, the main disadvantage of this standard is that it has to be converted to voltage variations if used as inputs to a computer port.

The 20-milliampere standard is basically the same as the 60-milliampere standard except that it is 20 milliamperes. The 20-milliampere level represents logic 1, and zero current represents logic 0. The same advantage is experienced with this standard as with the 600-milliampere standard. It is also necessary to convert the current variations to voltage variations if used as inputs to a computer port.

Interfacing Robot and Computer

The controller has input ports for interfacing with various computer controls. The RS-232C and RS-422 formats are used for this interfacing. R-422 is an improvement on RS-232C. A serial format is used for *inputting*. The information fed to the computer has a start bit, a parity bit, and a stop bit, as well as the data information.

Ful
sen

... computer interface allows the programmer of the robotic system to program off-line. This means that the service codes, geometric moves, and velocity rates can be programmed off-line. They are then downloaded to the controller.

The RS-232C format can send information for about 50 feet maximum. An amplifier is needed to ensure that a lot of noise and incorrect information is not picked up. The amplifier boosts the signal sufficiently to keep the signal voltage to the level required to overcome noise. Noise is everywhere in an industrial plant. Motors turning on and off generate spikes that can be interpreted as data pulses.

Sensors

Sensor information is converted to digital code so that the computer can handle it. Controllers have an interfacing port that provides for connecting sensors. The RS-232C port is usually used with full duplexing. Information from the sensors is sent to the controller on its data bus and stored in internal memory.

Program Control

Program control of the periods when data are transmitted and received by the interfacing ports is important to the operation of peripheral components. The interfacing is such that program control of the sequence of events becomes very important if the robotic cell is to function properly.

There are two types of program interfacing: service requests and robot requests. *Service requests* deal with the interfacing operation of peripheral components and provide program control of the periods where the data are transmitted and received at the input ports (Figure 6-3). In order to control the sequence of events in a work cell, the person doing the programming must take into consideration the fact that the signals will be outputted from the controller and what types of signals will be needed by the work cell.

Robot requests also require signals for operation (Figure 6-4). The input and output signals come from either the MPU or the controller. In this case, the controller has been programmed for the sequence of operations that the robot has to perform to open and close the gripper and do other tasks.

The controller must also have some method of finding out if the gripper actually picked up the object intended. As you can see, the input and output signals keep the robot working. The signals also provide communication between and

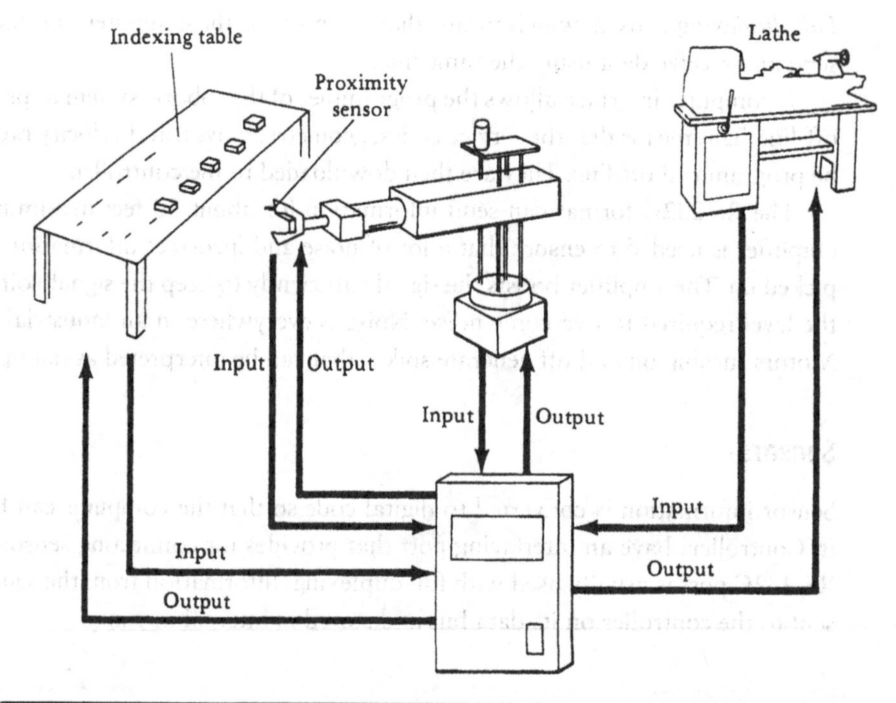

FIGURE 6-3 Signal processing for the operation of a work cell. (*Courtesy of PWS-Kent.*)

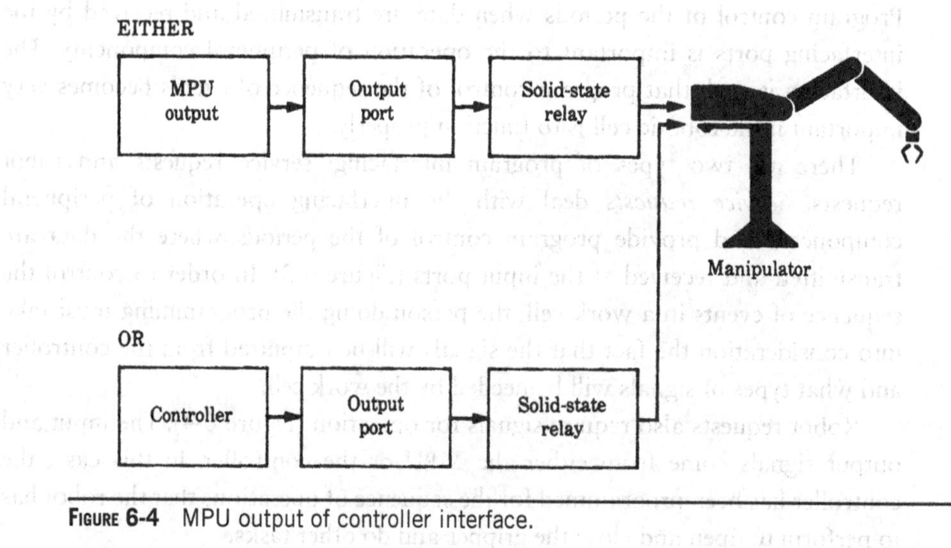

FIGURE 6-4 MPU output of controller interface.

ar ... any automated
pr ... robot has to work in sequence with the conveyor belt and do its job at the right moment. Timing is very important in any group effort. This is where the computer is at its best. It has the ability to organize many operations in a timely fashion and then to check to see whether they are done in sequence. The computer is also capable of keeping track of vast amounts of information for both the robot and the peripheral components.

Vision for the Robot

Providing a robot with vision is one of the major challenges today. A number of different approaches have been tried, and some degree of success has been achieved with several of them. In fact, the auto industry is using vision inspection for a few tasks. Very intelligent controllers have the ability to make real-time adjustments in a program's positional data. This is done to compensate for variables in the work piece. This ability is sometimes referred to as *adaptive control.* To make adjustments, the controller needs more sophisticated information about the current state of the process than can be provided by go/no-go limit switches and sensors used by nonintelligent controllers.

Visual systems provide the robot controller with information about the location, orientation, and type of part to be handled. A machine vision system (MVS) may be used for recognition verification of parts, for inspection and sorting of parts, for noncontact measurements, and for providing part position and orientation information to the robot controller (Figure 6-5).

The video camera can be used to take photographs, and the human brain can react to variations in shades of gray in those photographs (Figure 6-6). However, in order for a computer to be able to react to the information, it must be converted to digital format. This means that some method of shade interpretation must be used. A gray scale is developed, and its parts are given numbers that correspond to digitization. The computer can then handle the information but must be programmed in some acceptable fashion to be able to compare what it is looking for and what is in front of it. Here a problem arises. There is no standardized method for making machine vision. A number of different approaches have been tried and are available on the market for use in limited processing operations.

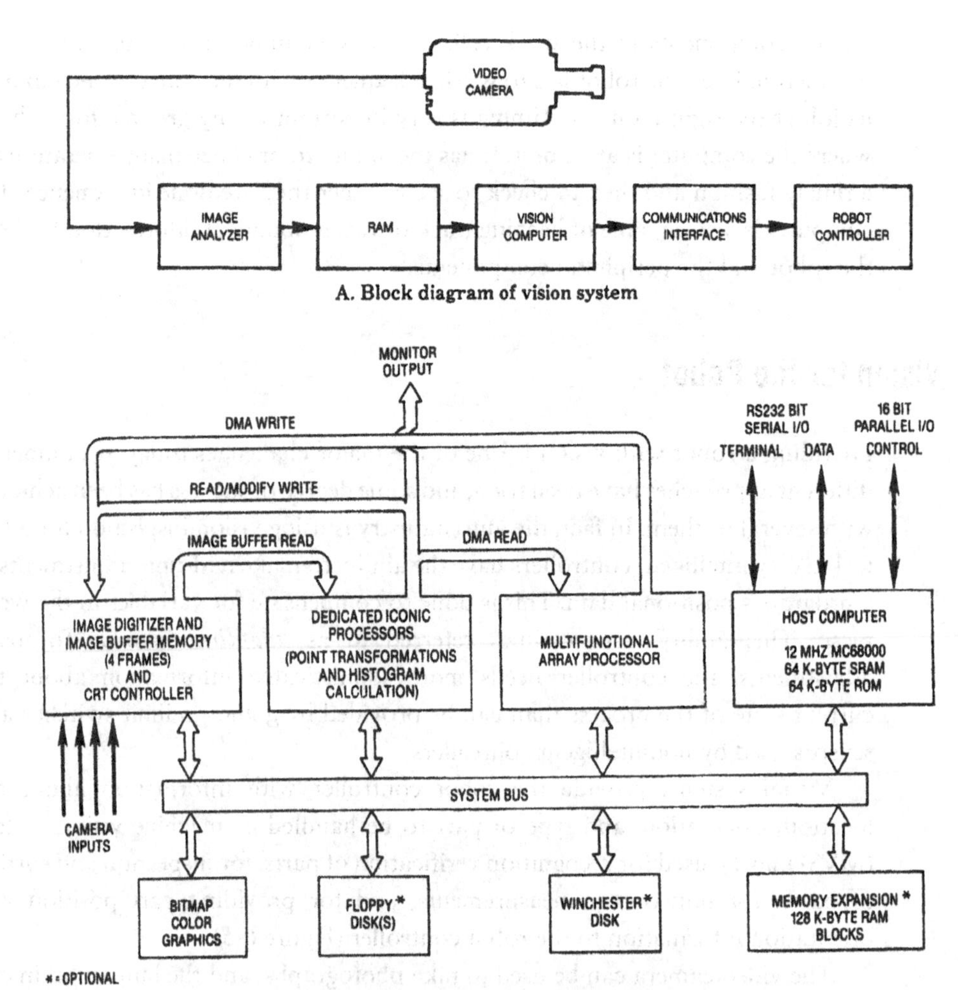

Figure 6-5 Machine vision system for robots. (*(B) courtesy of International Robomotion/Intelligence Group.*)

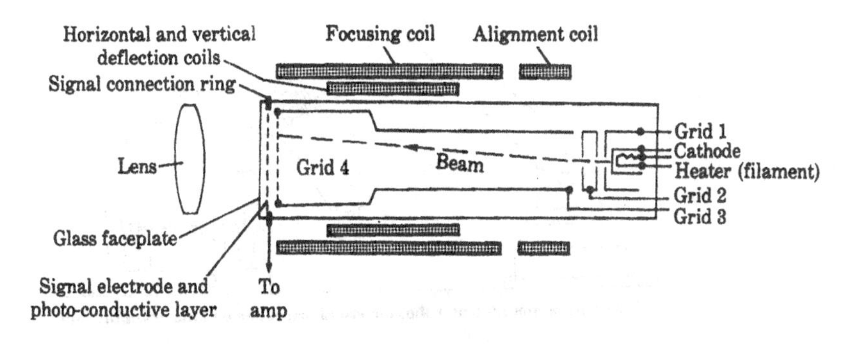

Figure 6-6 Video tube used in a robot vision system.

Object Recognition

Object recognition is the main reason for putting vision on a robot arm, that is, to make the robot capable of locating and distinguishing objects or parts. Images of the objects to be recognized are identified by the video camera. The information is then converted to digital signals and stored in the computer's memory. This is the training phase of the operation. Then the robot, through the vision sensor, distinguishes one part from many different parts grouped together. There are two methods for distinguishing an object: the edge-detection process and the clustering process.

Edge detection uses the difference between light and dark areas of an object to make distinctions (Figure 6-7A). A camera, located on the end of the manipulator arm or somewhere along the conveyor belt, locates an object—in this case, an automobile window. The image of the object is broken down into digital pulses. The pulses are sent to the vision computer. The vision computer searches through its memory and compares the different gray areas of the image with the grayscale already stored in its memory. When the vision computer finds a match between areas that contain the correct grayscale variations, it has located the edge of the part.

Figure 6-7B shows how the information from the video camera is converted to digital information in an analog-to-digital converter before it is fed to the computer. Note that the pixel clock generates a series of pulses to be fed into the analog-to-digital converter, with the video providing the gray level in digital form.

The *clustering process* is similar to the edge-detection process. It processes the camera signal and feeds it to the computer. The computer compares the various contrast levels to what it has stored in memory, and if they match, the object has been recognized. Both the edge-detection and the clustering processes have a problem with signal interference from any industrial environment. A lot of voltage

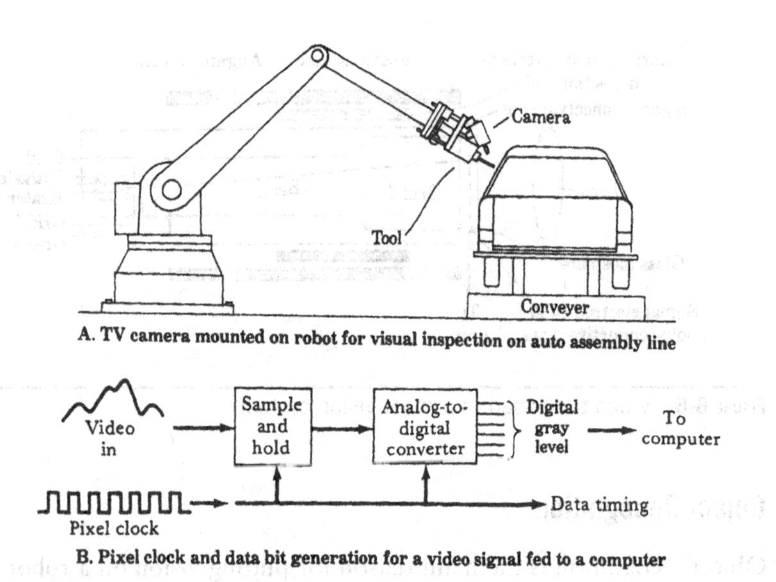

FIGURE 6-7 Example of the edge-detection process. (*(B) courtesy of PWS-Kent.*)

spikes (noise) are generated by the motors and other equipment being cycled on and off. These voltage spikes produce pulses that can be misinterpreted by the computer. The cluster and edge-detection processes both require a high light level for the camera to work properly and to be able to pick up the contrast in object shapes.

The template method of part recognition shows some merit. It takes an image and rotates it until it matches the template stored in the computer's memory. The computer must have a large memory capacity to be able to handle the rotation and comparison process (Figure 6-8).

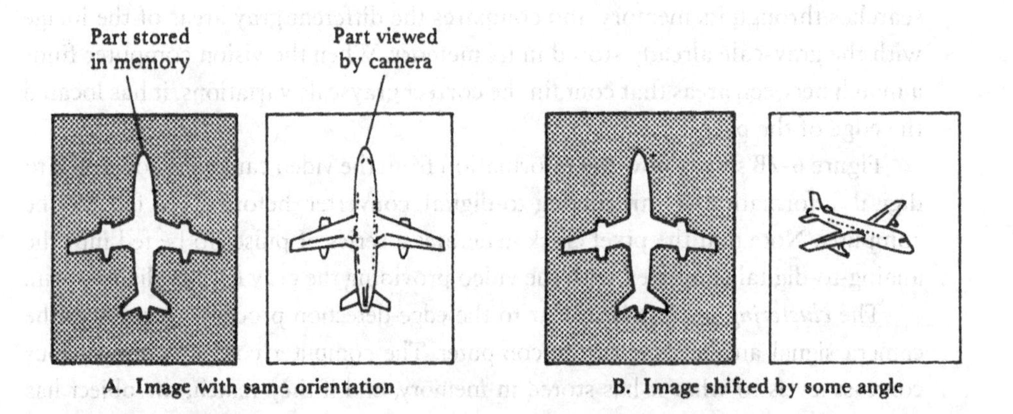

FIGURE 6-8 Template approach for object recognition. (*Courtesy of PWS-Kent.*)

Programs can be written to aid the computer in doing its job. The programs can also be written to cause the robot to reject a part or put it in a different bin or container than that of the desired object. Much is being done on robot vision systems. However, there is much to be done before robots have the ability to recognize objects and identify them in a short time.

Summary

A number of methods or systems are used to accomplish the mating of computer and robot. The microprocessor is one device used to control robots.

Very intelligent robots have the ability to understand high-level languages. Almost every robot manufacturer has developed its own controller language. Some of the more frequently used languages include VAL, HELP, AML, MCL, RPL, and RAIL. BASIC and COBOL are also used for some controllers.

A robot can be programmed by a computer. The program software (disk, magnetic tape, or other means) can be written and then adapted to the robot so that the robot does not have to be taken off the line during programming. Interfacing is the means used to enable a robot to communicate with its controller and other parts of the workstation. A microprocessor-controlled robot is able to communicate with other equipment around it by connections through ports.

ASCII code is the means by which the keyboard can be used to communicate with robot computers or microprocessors. Parallel ports are used when the computer and the machine it is controlling are separated by less than 50 feet. Serial ports are used when long-distance communication is necessary between units.

Information may be transmitted as changes in voltage or changes in current. The RS-232C standard and the TTL standard rely on voltage variations. The 60- and 20-milliampere standards rely on current variations.

The controller has input ports for interfacing with various computer controls. RS-282C and RS-422 formats are used for this interface. A computer interface allows the programmer of the robotic system to program off-line.

Sensor information is converted to digital so that the computer can handle it. Controllers have interfacing ports that provide for connecting sensors. There are two types of program interfacing: service requests and robot requests. Each deals with interfacing with peripheral components and provides control during the period when the data are transmitted and received at the input ports.

Vision systems can provide a robot controller with information about the location, orientation, and type of part to be handled. Machine vision systems are used for recognition and verification of parts, for inspection and sorting of parts, for noncontact measurements, and for providing part position and orientation information to the robot controller. Edge detection and clustering are the two processes used for identification by the computer of the parts viewed by the vision camera. Much has to be done before vision is available for all robots at a reasonable price and with the ability needed for daily production runs.

Key Terms

ASCII Code used to transfer information from a keyboard to the processor MPU.

CPU Central processing unit.

interfacing Matching up of a device with a computer or microprocessor so that they operate as one unit.

language The coded information fed to a computer to make it respond to commands and do certain operations.

MPU Microprocessor unit.

MVS Machine vision system; method of giving a robot the ability to see.

object recognition Ability of a robot to recognize certain shapes and to choose the right one for processing.

off-line Refers to the programming of a robot by use of a computer and then placing the program in the robot's controller for action (the robot is not involved in the actual programming).

parallel ports Method for connecting computers and peripheral devices so that they can share data (uses eight or more wires).

RS-232C standard A standard that uses –3 to –25 volts for logic 1 and +8 to +25 volts for logic 0.

serial ports Method for connecting computers and peripheral devices so that they can share data over distances of 50 feet or more (uses two wires).

60-milliampere standard A standard that uses 60 milliamperes for logic 1 and zero for logic 0.

software information Programmed on floppy disks, hard disks, magnetic tape, or drums.

TTL standard Transistor-transistor logic standard that uses a 5-volt signal for logic 1 and 0 for logic 0.

20-milliampere standard A standard that uses 20 milliamperes for logic 1 and 0 for logic 0.

Review Questions

1. What is the purpose of a controller?
2. What is an MPU?
3. Where is the MPU located?
4. List six languages used by robots.
5. What is software?
6. Identify interfacing.
7. Where is the ASCII code used?
8. What is the difference between the TTL and RS-232C standards?
9. What is the difference between the 60- and 20-milliampere standards?
10. What is the strobe line?
11. What is the difference between a parallel port and a serial port?
12. How can you tell by looking whether a port is parallel or serial connected?
13. Explain the difference between service requests and robot requests.
14. What is the main reason for putting vision on a robot?
15. How is the template method used for robot vision?

CHAPTER 7
Uses for Robots

Performance Objectives

After reading this chapter, you will be able to:

- Understand how loading and unloading are done.
- Identify the types of loading used with robots.
- Know the different types of loading, materials handling, and die casting using a computer.
- Know how a pick-and-place robot works.
- List the various jobs robots can do.
- List how robots can weld, paint, and inspect a finished job.
- Know what CIM is and what it can do.
- Understand how robots of the future will produce needed functions.
- Understand the social impact and new uses for computers and robots in the future.
- Identify and discuss the key terms used in the chapter.
- Answer the review questions at the end of the chapter.

Robots are used for a variety of industrial purposes. Of course, we have to understand that *robot* means a device that can operate on its own without direct human supervision. Pick-and-place robots, the simplest type used in industry, can accomplish a number of jobs that humans do not especially enjoy doing. Humans tire easily when it comes to lifting and doing the same thing over and over. This can cause high absenteeism and affect the quality of the product being produced. Pick-and-place robots can go on all day loading and/or unloading without tiring.

Pick-and-place robots work well in handling materials. In most cases, they do not add value to the product by machining it or performing any other operation. They do the lifting, moving, stacking, and packing of parts or finished products.

More sophisticated machines are needed to do value-added work to a product or part. *Value-added work* is a task such as painting, welding, buffing, or assembling. More intelligent robots are employed to perform these tasks.

Loading and Unloading

Work handling must be fast, accurate, smooth, and dependable if productivity improvements are to be achieved in modern industry. Even the simple picking up of a part and placing it on the line, as shown in Figure 7-1, must be efficient. Manufacturing efficiency increases in direct relationship to controlled workflow rates, good work piece orientation, and precise part positioning. These essential factors allow plant processing equipment to be operated at full capacity for prolonged periods, with downtime limited to scheduled tool changes and service-maintenance periods.

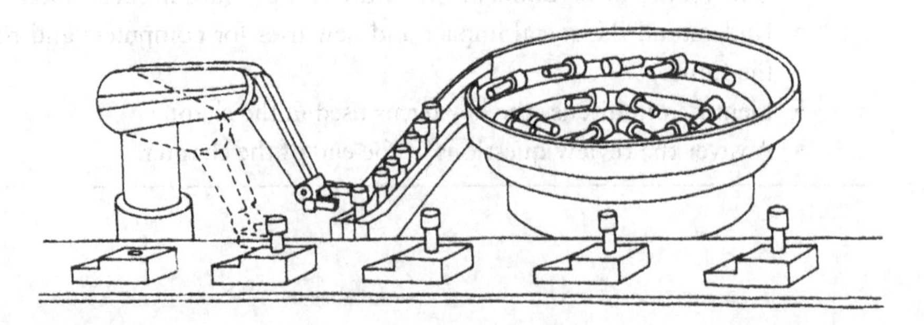

FIGURE 7-1 Pick-and-place robot used to perform simple, tedious materials-handling tasks. (*Courtesy of Feedback, Inc.*)

Lane Loader

A *lane loader* is another application for pick-and-place robots. Lane loaders can be used to load and unload multiple-part fixtures and balance flow rates between a fast machine and several slow machines or several slow machines and fast machines (Figure 7-2).

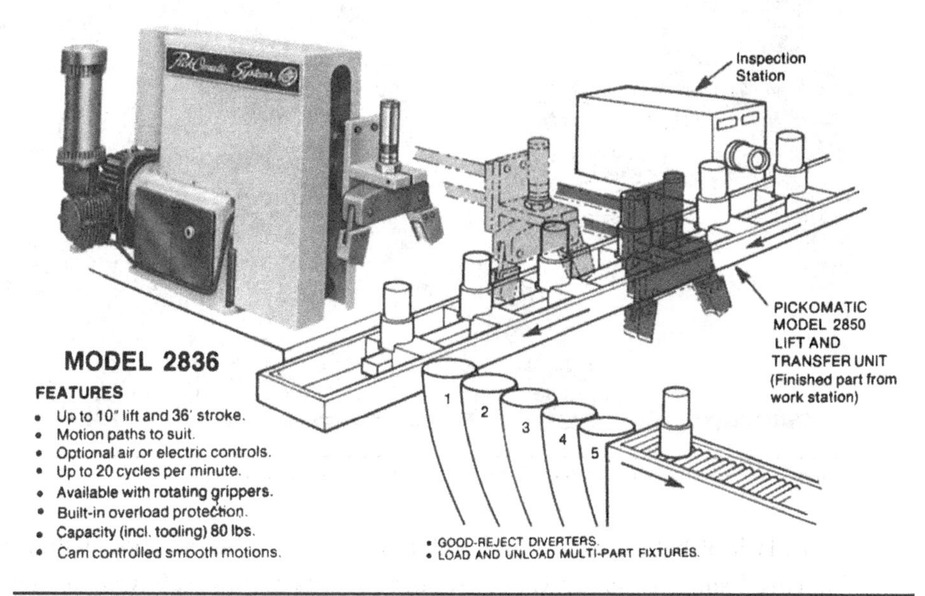

FIGURE 7-2 Lane loader. (*Courtesy of Pick-O-Matic Systems.*)

Flow-Line Transfer

Flow-line transfer enables robots to be programmed to pick up two or more pieces at a time and transfer them from a machining line onto a second transfer line located parallel to the first line. Human effort in picking up and moving heavy loads is thereby eliminated (Figure 7-3).

The main items to be concerned with here are the ability of the manipulator to move over the objects to be lifted and to swing over to put them down where they are needed. Another consideration is the weight of the package or part. The weight of the manipulator has to be factored into the total weight when you look at specifications for purchasing a robot to fit the job at hand.

Loading and unloading are the second largest use for robots. These machines do the jobs humans do not care to do because they may be in locations that are less than favorable for human habitation. The use of a robot to load and unload reduces the personnel injury rate and increases productivity. Excessive heat and

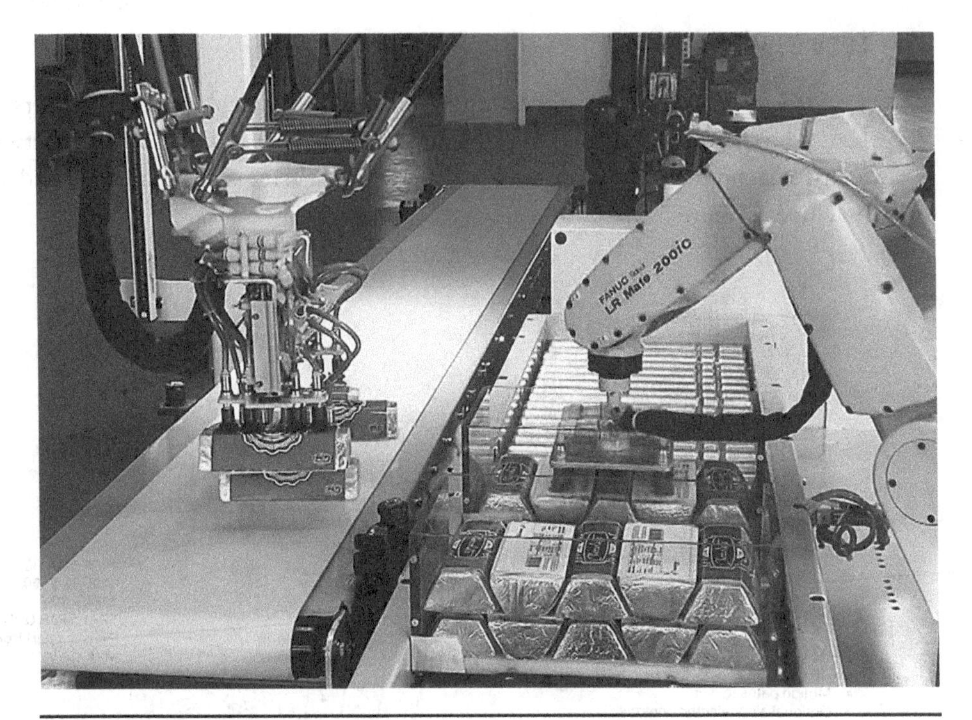

Figure 7-3 Robots used to transfer wrapped food from one line to the other.

noise, lack of light, and dirt and pollution do not bother a robot as they would humans. Using robots can lower production costs because safety equipment does not have to be purchased and maintained.

Machine Loading

One of the most important jobs the robot can do is *machine loading*. Most high-speed machines are numerically controlled (NC) via tape. The ability to make sure that the machine is loaded properly is therefore of concern to any production planner. NC machines work much faster than human operators and can improve the quality of the product if fed properly with the parts they are machining. The cam-operated machine in Figure 7-4 shows how parts handling can be a smooth operation with no human involvement.

Materials Handling

Materials handling is the moving of materials about a manufacturing plant and can refer to raw materials as well as parts of a product that must be moved from

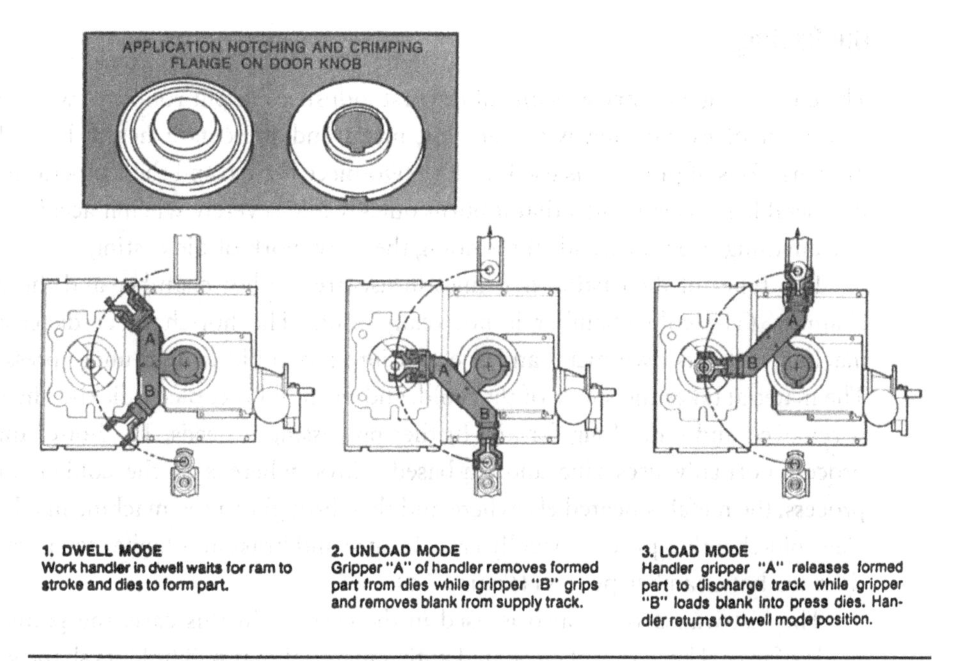

APPLICATION NOTCHING AND CRIMPING FLANGE ON DOOR KNOB
1. DWELL MODE
Work handler in dwell waits for ram to stroke and dies to form part.
2. UNLOAD MODE
Gripper "A" of handler removes formed part from dies while gripper "B" grips and removes blank from supply track.
3. LOAD MODE
Handler gripper "A" releases formed part to discharge track while gripper "B" loads blank into press dies. Handler returns to dwell mode position.

FIGURE 7-4 Automatic press loader. (*Courtesy of Pick-O-Matic Systems.*)

one machine to another or from one part of the plant to another for further processing. Moving the material adds no value to the product, yet it has to be done. This means that industry must keep its materials-handling costs down.

Manufacturers depend heavily on conveyors to move parts and materials. Conveyors are synchronized with robots so that parts or materials are where they are supposed to be when they are needed. The conveyors are connected to controllers so that their operations are synchronized with the machines doing the value-added work (Figure 7-5).

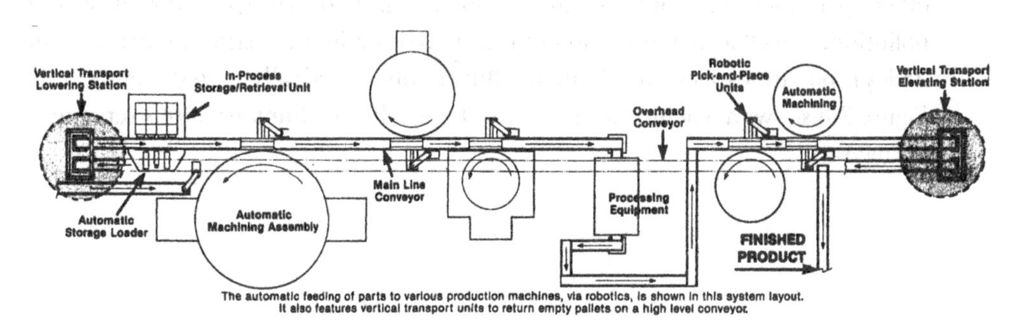

The automatic feeding of parts to various production machines, via robotics, is shown in this system layout.
It also features vertical transport units to return empty pallets on a high level conveyor.

FIGURE 7-5 Typical materials-handling system using vertical transfer units (VTUs). (*Courtesy of Dorner Manufacturing Corp.*)

The die-casting industry was one of the first industries to put robots to work. *Die casting* involves working with very hot metal and presents a health hazard to humans. Tons of pressure is used to close two-piece molds, and the temperature of the metal being cast is such that it burns quickly and severely when it accidentally hits anything. Robots are ideal for doing the dirty work of die casting.

Two types of die-casting machines in use are the hot chamber and the cold chamber. (The cold chamber is not really cold.) The hot-chamber die-casting machine melts its own metal and injects it into a two-piece mold under pressure. The material takes the shape of the mold, and the mold is cooled quickly. The part is removed and sent along for any further processing it needs. The hot-chamber process normally uses zinc and tin-based alloys, whereas in the cold-chamber process, the metal is heated elsewhere and then brought to the machine in a ladle. The cold-chamber process usually casts bronze and brass at a higher temperature than the hot-chamber process (Figure 7-6).

The lost-wax process also is used in die casting. In this case, the pattern is made of wax. The wax is then coated with a material that will hold its shape when dried. The wax and its surrounding mold are heated until the wax melts, leaving a cavity in the mold the shape of the object that was made of wax. The cavity is filled with molten metal and spun around at high speed to cause the molten metal to cling to the inside of the mold. Once the metal has cooled sufficiently, the mold is broken and the metal part removed. This process is used for jewelry making and for very delicate work where fine details must be present in the finished casting (Figure 7-7).

Robots can do a number of operations that may be hazardous if done by humans. A robot can reach in and lubricate the mold when it opens, allowing the pieces to fall out. It can take the piece and quench it or perform any of a dozen other operations that would endanger a human operator. The gases, dirt, dust, and pollution, as well as the heat and hot metal involved in die casting, all produce an undesirable atmosphere for humans. But a robot is ideally suited for the job. Figure 7-8 shows a machine feeding a machine, relieving humans of this repetitive, boring task.

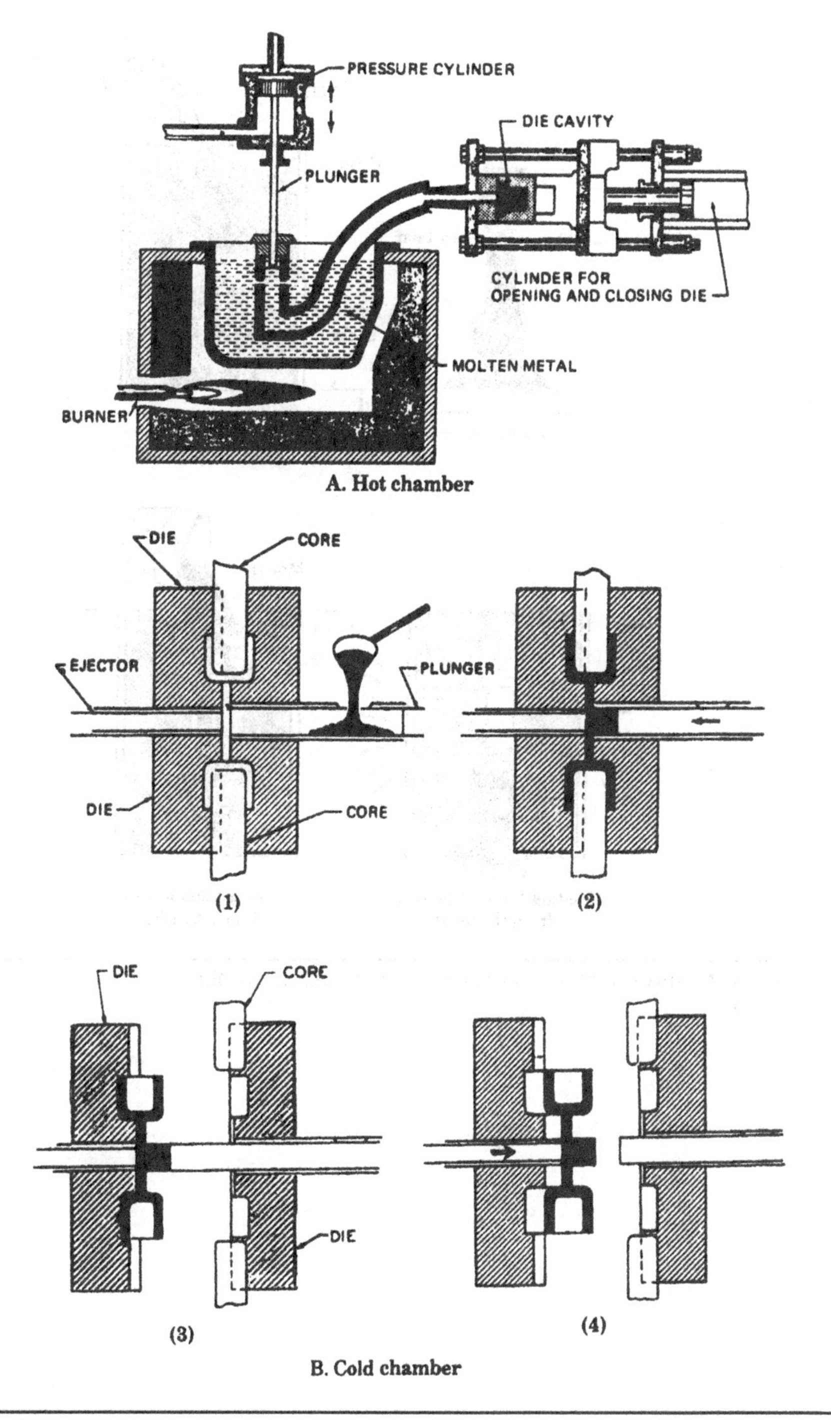

FIGURE 7-6 Die-casting processors. (*Courtesy of New Jersey Zinc, Inc.*)

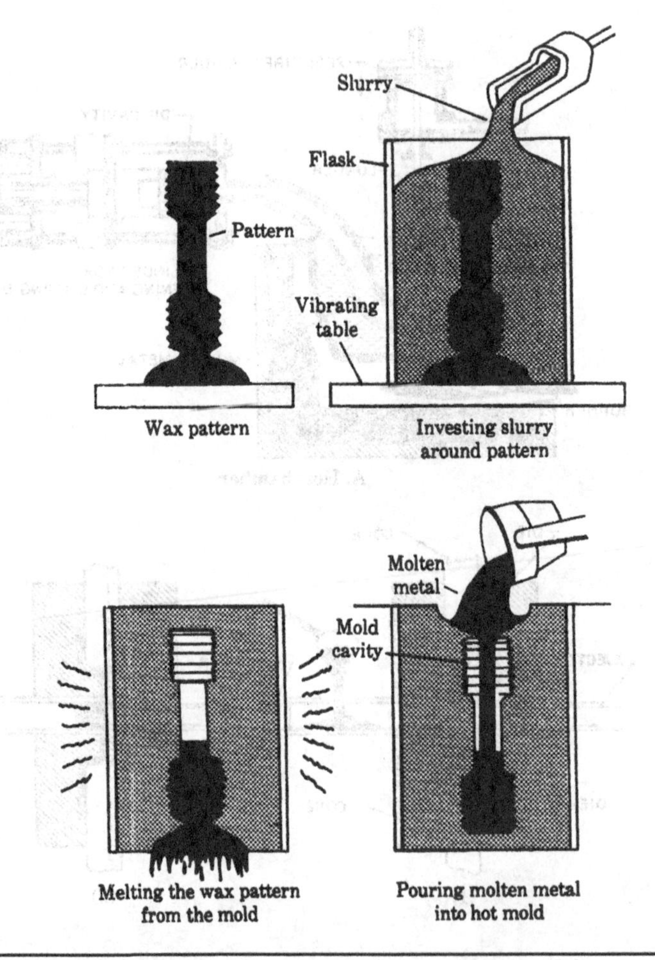

FIGURE 7-7 Investment mold (making a mold over a wax object).

Figure 7-8 Machine fed by a machine. (*Courtesy of I.S.I. Manufacturing, Inc.*)

Palletizing

Stacking parts is easily automated, and robots were quickly used to do this type of work. *Palletizing*, or placing a series of boxes in a given pattern on a pallet, makes it easier for forklifts and other devices to pick up the pallet and move it to a shipping area or to another part of the manufacturing line. Picking up and placing boxes and parts into a specified arrangement can be very boring and tiring for humans. A robot is easily programmed to do such work all day long, never tiring. A good example of a robot palletizing is shown in Figure 7-9, where the robot is placing bottles into a carton.

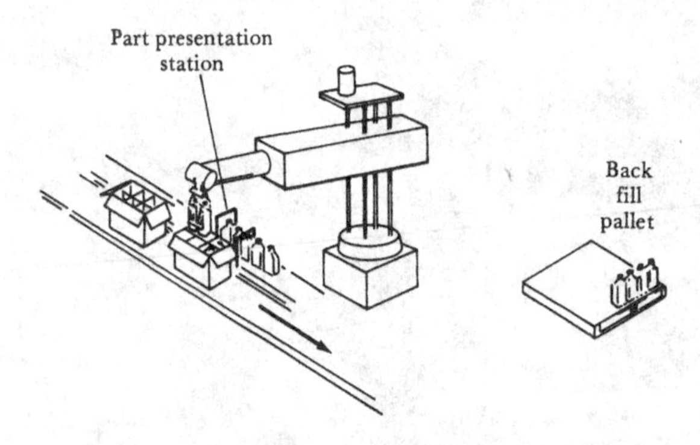

FIGURE 7-9 Robot used to place bottles in a carton for shipping. (*Courtesy of PWS-Kent.*)

Just as parts can be taken from an assembly line and placed into a bin or box, they also can be taken out of a box or bin and placed on the line. When they are removed from a box or bin and placed on the line for processing, the process is called *depalletizing* (Figure 7-10).

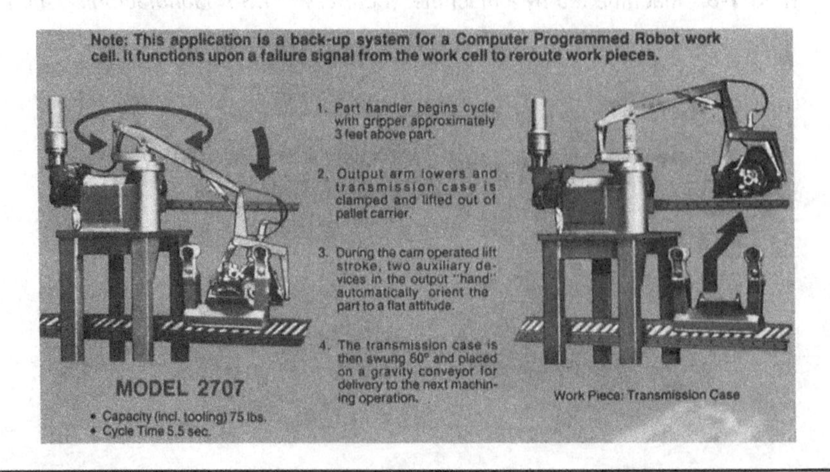

FIGURE 7-10 Pallet carrier unloader. (*Courtesy of Pick-O-Matic Systems.*)

Line Tracking

In some instances, it is necessary for a robot to move with the production line. For instance, it may have to keep up with a part in order to inspect it. Stopping the line for such an inspection would increase production costs greatly. Thus the robot must be placed in a parallel path with the moving part so that it can do its job as the part moves to another location and operation. This is called *line tracking.* Figure 7-11 shows how a robot travels along with a car, inspecting the weld joints with machine vision.

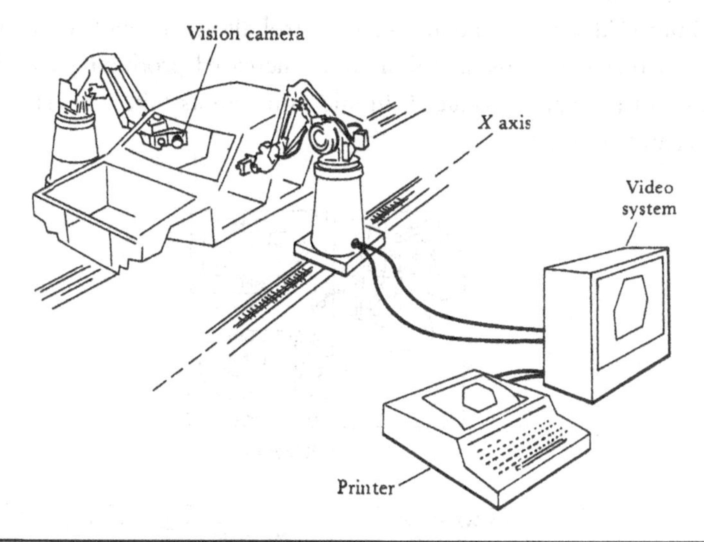

FIGURE 7-11 Line-tracking robot with camera used to inspect car body weld joints. (*From Malcolm Robotics.*)

Process Flow

In order for the production of any product to progress smoothly, you must have a regulated flow of parts and materials, and everything must be at the right place at the right time. This is referred to as *process flow*. A flowchart has to be made up so that the locations of all parts is known at any given time. Then you have to arrange the timing so that the parts arrive at a given location when needed. The demand for new parts also must be met by having raw materials available when needed. Keeping tools operational also has to be factored into the operation. All this is done by a flowchart for the entire plant operation. Then the flow of parts to a given area of supervision is drawn up for either robotic or human supervision. Good planning makes for a good productivity record.

Fabricating

A robot can do a number of the required operations involved in fabricating parts or a product. Some of the more prevalent manufacturing processes being performed by robots are routing, milling, drilling, grinding, polishing, deburring, sanding, and riveting. Figure 7-12 shows a complex double-gripping process in which a hole is drilled and then deburred perpendicular to the main axis of the part. The robot also can change tooling, can change fixtures, and has an increased accuracy and repeatability.

Improved sensory and adaptive control allows a robot to be used in different manufacturing processes. Robots have increased productivity and improved the quality of the parts produced. In addition, robots take workers out of the dust, noise, and polluted air.

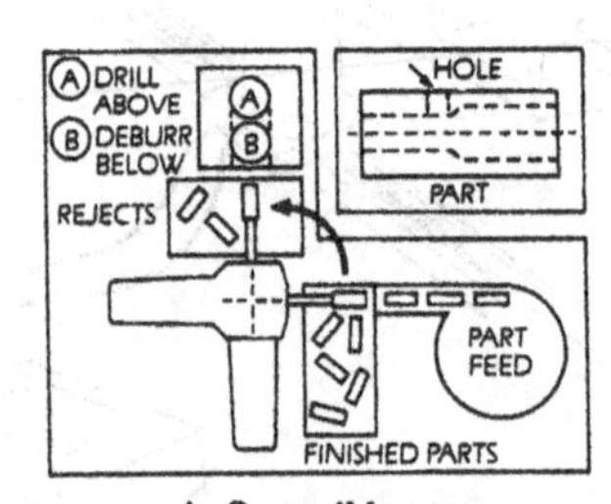

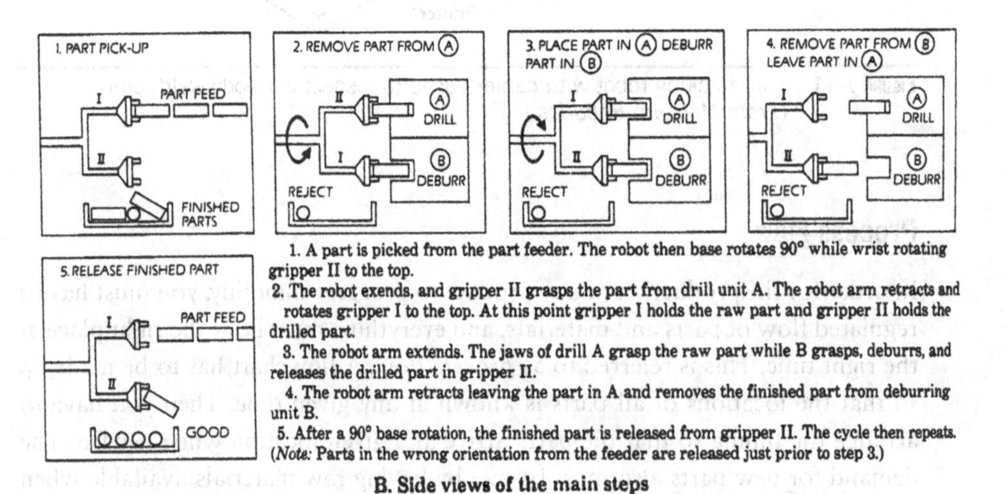

1. PART PICK-UP
PART FEED
FINISHED PARTS
2. REMOVE PART FROM A
DRILL
DEBURR
REJECT
3. PLACE PART IN A DEBURR PART IN B
4. REMOVE PART FROM B LEAVE PART IN A
5. RELEASE FINISHED PART
GOOD
1. A part is picked from the part feeder. The robot then base rotates 90° while wrist rotating gripper II to the top.
2. The robot exends, and gripper II grasps the part from drill unit A. The robot arm retracts and rotates gripper I to the top. At this point gripper I holds the raw part and gripper II holds the drilled part.
3. The robot arm extends. The jaws of drill A grasp the raw part while B grasps, deburrs, and releases the drilled part in gripper II.
4. The robot arm retracts leaving the part in A and removes the finished part from deburring unit B.
5. After a 90° base rotation, the finished part is released from gripper II. The cycle then repeats. (Note: Parts in the wrong orientation from the feeder are released just prior to step 3.)
B. Side views of the main steps

FIGURE 7-12 A double-gripping application used to speed up a manufacturing process. (*Courtesy of Schrader-Bellows.*)

Assembling

Robots have been used in assembling calculators, watches, printed circuit boards, electric motors, and alternators for automobiles. The future will see robots assembling anything that can be broken down into simple step-by-step operations, and this includes almost everything. More and more attention is being paid to the possibilities of using robots for the tedious jobs of assembling small and/or large parts. Figure 7-13 provides an illustration of how a robot can be used to fabricate printed circuit boards.

Figure 7-13 Robot used to fabricate printed circuit boards. (*Courtesy of Fared Robot Systems.*)

One of the problems with converting a product to automated assembly is its design. Some products are not designed to be assembled by machines (Figure 7-14). Designers must take a close look at each product and be aware of the ability of the machines that do the assembly work. By studying the design requirements of a product and the ability of robots or other machines, it is possible to design a better-quality device and have it assembled by machine. The ability to assemble an electric motor without a human hand touching it is worthy of note. The ability of machines to make alternators for automobiles is another application that maintains quality and productivity and keeps cost down. Robot assembly relieves workers from performing boring, repetitive tasks day after day. The quality of the product can be improved by having the robot work at a constant pace. The quality of the product improves because the tolerances are closer when a machine is used to handle the parts.

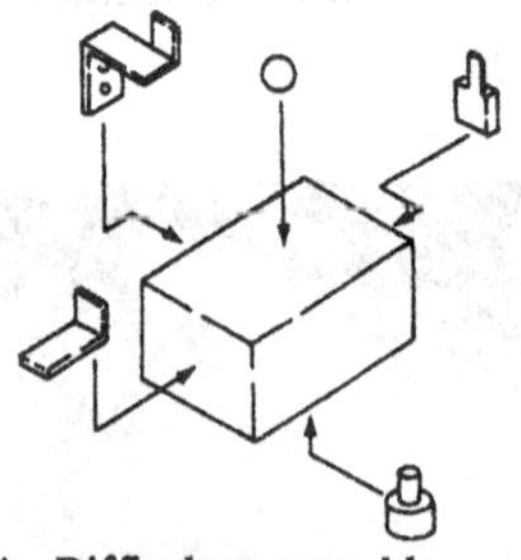

A. Difficult to assemble because of part orientation

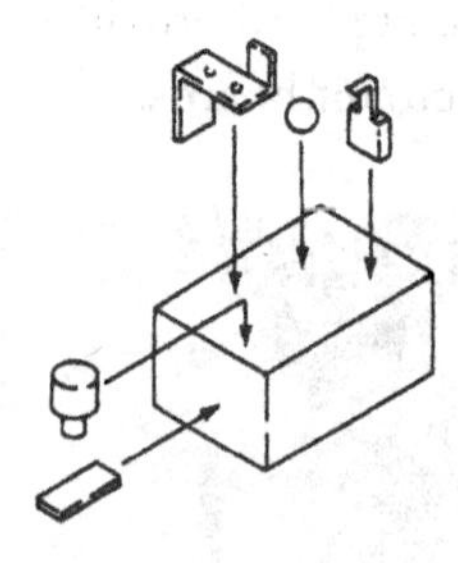

B. Preferred assembly because of new orientation

C. Difficult to grip because no lip provided

D. Preferred assembly because lip provided

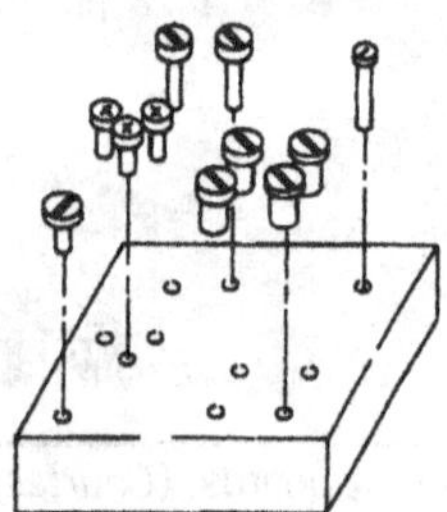

E. Difficult to automate because of too many screw variations

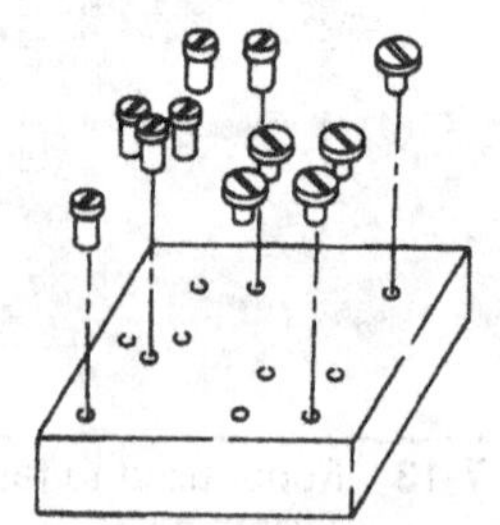

F. Preferred assembly because of fewer screw variations

FIGURE 7-14 Product design—an important factor in automated assembly work. (*Courtesy of PWS-Kent.*)

Painting

A robot's wrist is the usual location for a spray-painting gun. A human operator takes the robot arm through the motions of spray painting. A continuous-path robot with lots of memory is usually used. The robot memorizes the paths and then repeats them. It has to be debugged after a few sweeps because both the correct and incorrect moves of the human operator were memorized and stored. The program has to be checked often before putting the robot to work (Figure 7-15).

FIGURE 7-15 Spray painting a production line item. (*Courtesy of FANUC.*)

Spray painting by robot has several advantages. One of them is removal of the human operator from the fumes, making it possible to use different types of paints. Many of today's coatings are toxic and need to be applied by the spray arm of a robot because humans would be injured if they were to work inside the enclosure with the fumes. Another advantage is that robots can spray paint in very thin lines with accuracy. A robot can paint pinstripes on automobiles all day without faltering. Areas that could not be reached by a human painter can be reached by high-technology articulate robots because they have six (some up to nine) axes.

Welding

One of the largest uses for robots is welding. A number of industries use robots for spot, stud, stick, metal–inert gas (MIG), and tungsten–inert gas (TIG) welding.

(TIG is a form of stick welding.) Spot and stud welding are resistance welding processes. Stick, MIG, and TIG welding are arc welding processes.

Spot welding is the most common type of welding because it is the easiest and is in the greatest demand. When the assembly operation begins on automobiles, spot welding is used to hold the pieces of sheet metal together. Hundreds of spot welds are made to hold the pieces of metal together and to provide a tight body without rattles and squeaks. Spot welding can be performed as the automobile body moves along the assembly line, or the welder-robot can be stationary.

The most common type of arc welding is MIG welding. The welding station consists of a wire feeder, an inert-gas supply, a gas flowmeter, and a welding pan. The welding is a continuous process once it starts. An inert gas is brought to the spot where the welding is taking place to shield the metal from oxygen in the air while it is very hot and in a molten state. The current or amperage is controlled by how fast the wire is fed to the joint (Figure 7-16). Remote welding by robot has been perfected to such an extent that the wire feed speed can be controlled within ±1 inch per minute and the welding voltage within ±0.1 volt.

One of the biggest problems with machine welding is the butting together of the two materials to be joined. The groove must be accurate and fit smoothly. The robot does not, in most instances, have the ability to adjust the welding path or wire feed or to change the current based on demands, as a human operator would.

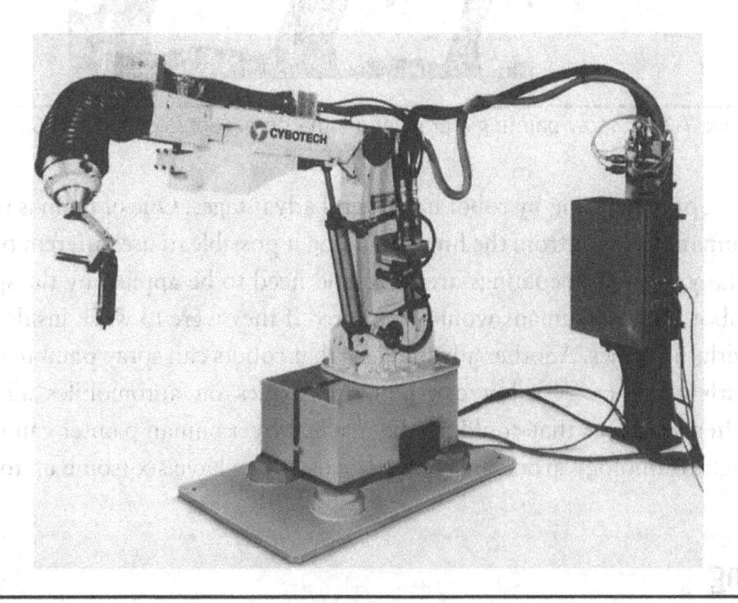

FIGURE 7-16 Arc-welding robot. (*Courtesy of Cybotech Industrial Robots.*)

Inspecting and Testing

Other tasks a robot can perform without tiring are testing and inspecting. Humans have a tendency to tire after doing a repetitive job for some time. Their efficiency and objectivity are both compromised. Thus a robot with the ability to test products without tiring would, in most instances, produce a better-quality product. Testing and inspecting can be done with a number of robots. Robots can pick up a part and place it into an inspection gauge or a more complicated device for go/no-go testing. Robots also may have vision that will allow them to test up to 1,000 points on a part in 1 minute.

Inspection is less complicated than testing. It involves taking a look at a part, gauging it, and then discarding it if it does not fit the established limits. Robots have a number of advantages when it comes to inspections. They can perform the inspection as the part moves along, doing the same thing over and over without tiring, and still be able to test 100 percent of the product line and obtain consistent results (Figure 7-17).

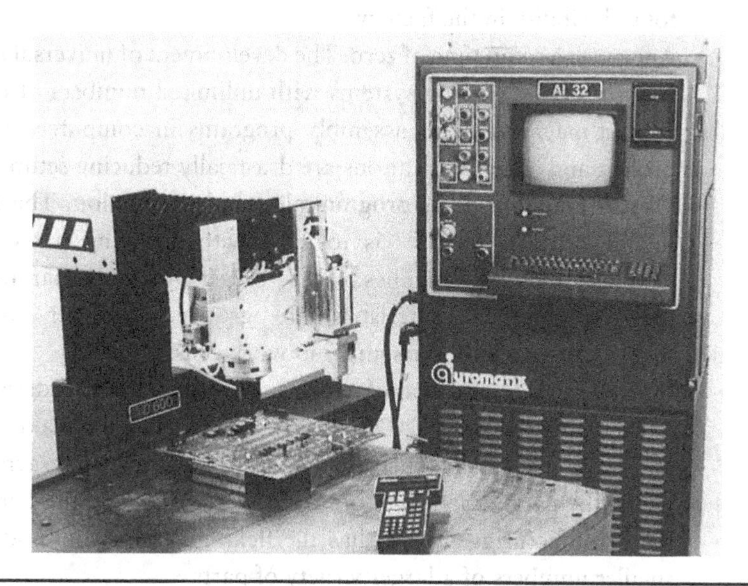

Figure 7-17 Inspection station. (*Courtesy of PWS-Kent.*)

The Future of Flexible Automation

The computer-integrated manufacturing (CIM) factory of the future will look quite different from the factories of today. It will be based on the integration of traditional or process-based technology with the emerging software- or systems-based technology of today and tomorrow. Figure 7-18 shows an example of the factory of the future.

Objectives of CIM

Seven objectives in setting up a CIM process or method of organization for making a product are as follows:

- **Obtain an economic order quantity approaching one.** While it probably will be some time before an economic order quantity of one is feasible for manufacturers, it should be their ultimate goal. It is the key to controlling the cost of inventories of finished and in-process goods and will allow total flexibility in the factory.
- **Approach a setup time of zero.** The development of universal multipurpose fixtures, tool-changing systems with unlimited numbers of tools, storage of part-machining and assembly programs in computer memory, setup probes, and other innovations are drastically reducing setup time.
- **Obtain family-of-parts programming and production.** The rationale for manufacturing processes is to start with grouping parts with similar characteristics into families. Parts related by design parameters and/or manufacturing characteristics are grouped together for optimized manufacturing and/or assembly in work cells.
- **Integrate design and manufacturing.** A computer-aided design/computer-aided manufacturing (CAD/CAM) system is needed to make sure the CAD and CAM package is such that it is an integrated, interactive system in which the parts designs are optimized to be manufactured efficiently, and the manufacturing process has the flexibility to accommodate relatively smaller numbers of a larger variety of parts.
- **Establish inventory integrity and just-in-time parts delivery.** Materials procurement and handling are typically the largest manufacturing cost items (50 percent versus 10 percent for labor and 25 percent for indirect labor). Just-in-time (JIT) production minimizes handling, and work-in-progress (WIP) inventory can dramatically reduce capital requirements, improve cash flow, and reduce the breakeven point.

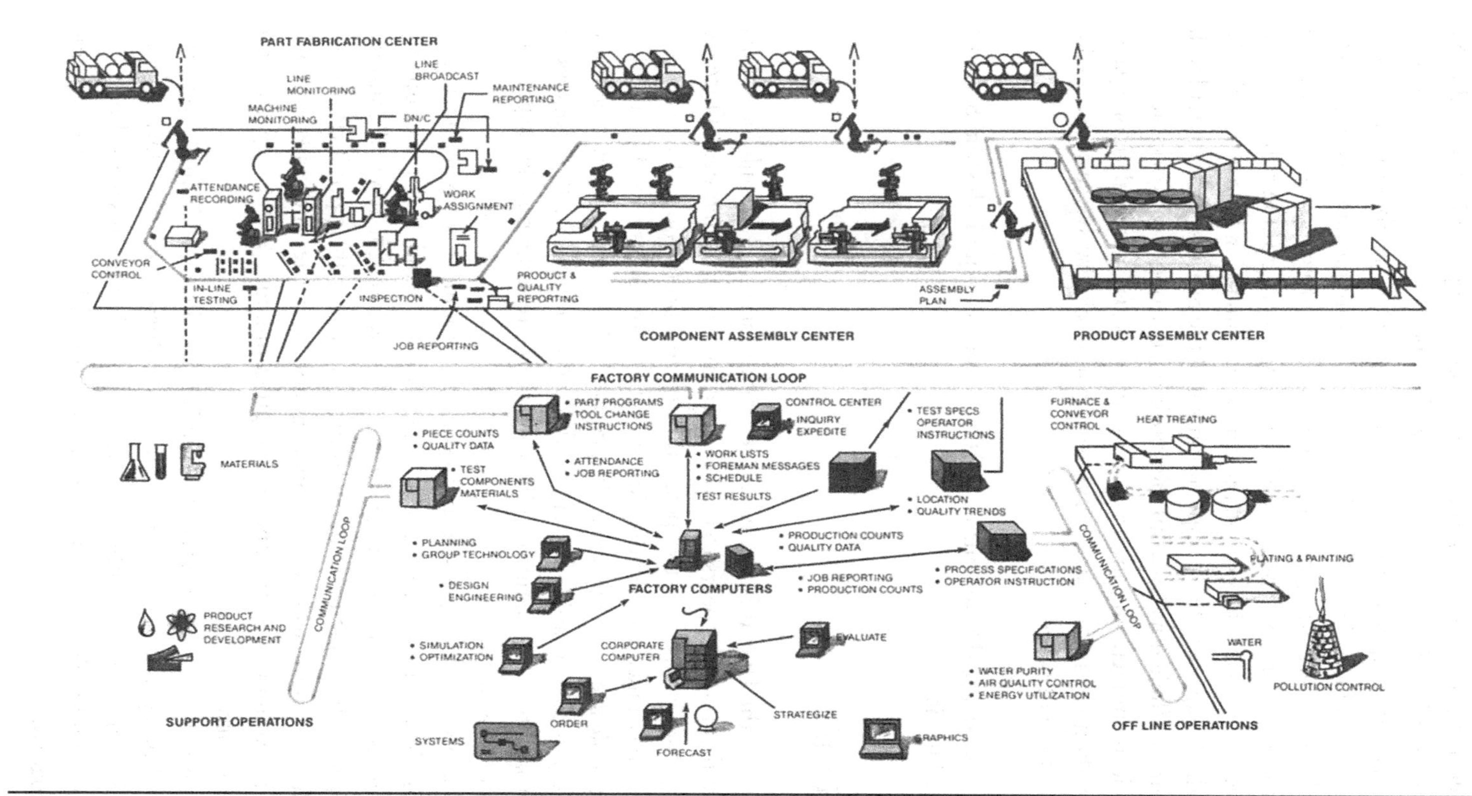

FIGURE 7-18 Factory of the future. (*Courtesy of GCA Corporation/International Systems Group.*)

- **Establish absolute control of the total process.** In a properly designed CIM system, work cells are integrated with other shop floor and factory management computer systems, as well as the central database management system. Interfaces between all subsystems communicate with proven, portable software. All gaps between islands of automation are bridged so that the end user can replace, reschedule, and reprogram in real time. In a properly designed CIM system, the data communication loop is completely closed. Management has complete control of the process at the center level.
- **Maximize efficient use of available workspace.** Robotized workstations in flexible manufacturing and assembly systems use minimal floor space. Pallet shuttles and vertical, gravity-fed parts delivery systems also make efficient use of spacc.

These objectives are listed here courtesy of GCA Corporation/ Industrial Systems of Naperville, Illinois.

The Future of Robots

Manufacturers in the United States have been more successful in building robots than in selling them. This appears to be the sum of things at the moment. Once we get geared up for production, the possibility of overproduction always exists. There is some resistance to the adoption of robots for factory work and other tasks. Some people are very suspicious. If it comes to being replaced by machines, therefore, every type of roadblock may be thrown up to impede the progress of the new approach to manufacturing. However, there are some positive aspects to be gained from robotizing.

Vision systems are too expensive and slow at the moment but will improve as research and development progress to meet the demand for better-quality recognition by robots. Tactile sensing is in need of development and will continue to be developed as manufacturers move to make sure that their products are the best. In the future, robots will become smaller, more accurate, and faster and will have better repeatability. Better and easier programming can be expected for robots of the future.

Research projects are under way at a number of colleges and universities: the University of Michigan, Stanford University in California, and the University of Rhode Island, to name a few. Others are also involved in improving robots. The vision of robots and their contribution to the social health of the nation are some studies in progress.

Social Impact of Robots

Union and industry spokespersons report that the march of industrial robot assembly lines will cost hundreds of thousands of jobs in the next few years but may save some companies from going under.

However, the use of robots has created jobs not only in their manufacture but also in their operation, as well as new jobs keeping them operating at top efficiency. In most instances, these jobs pay more than the manual labor jobs that required mind-bending boring devotion for much less pay.

New Uses and New Forms

The U.S. Army is developing a six-legged robot for walking over various types of terrain and going places that tanks cannot venture. Some of the research for this project will aid in making a robot with legs instead of wheels. Balance problems and movement of the legs will be solved with more applied research. Better vision and tactile sensors will also be developed for this project.

Brain surgery is already being aided by robots in the operating room. When lasers and robots, along with CT scans, are teamed up, it will mean a new approach to solving some surgical problems. The future holds much promise for robots and for those who work with them or on them.

Summary

Work handling must be fast, accurate, smooth, and dependable if productivity improvements are to be achieved. Pick-and-place robots efficiently accomplish many loading and unloading jobs that are slow, tedious, and boring for humans. A lane loader is another application for a pick-and-place robot. Lane loaders can be used to load multiple part fixtures and balance flow rates between fast and slow machines.

Flow-line transfer is also important in keeping the line moving and production going. Twin-armed or even three-armed robots can handle this situation well. Loading and unloading constitute one of the largest uses for robots. Machine loading is one of the important jobs for a robot. It prevents humans from getting their fingers caught in a press or forge. Materials handling, that is, the moving of materials around a manufacturing plant, is one of the expense items in the manufacture of any product. If the cost of handling materials can be reduced, so can the price of the product. Die casting was one of the first industries to use robots. Robots can handle the hot materials without having to react to the heat, dirt, pollution, and lack of lighting. Robots can do things in the casting process that humans are unable to do.

The stacking of parts is easily automated, and robots do the job well. Parts can be taken from an assembly line and placed in a bin or box (palletizing), or they can be taken out of a box or off a pallet by a robot (depalletizing). Line tracking is done by a robot moving along with the line and performing its job as it moves along. Process flow is the moving of parts and materials in an orderly and timely manner.

Fabricating also can be done by robots. Everything from drilling, riveting, sanding, deburring, and grinding to polishing can be done by robots. They can also assemble products partially or totally without human supervision. Painting and welding are two of the value-added processes that robots can do. They are easily adapted to welding and spray painting. Inspecting and testing are also easily automated, with robots designed to do the job without tiring. Robots have a tendency to improve the quality of finished products because they are capable of 100 percent inspections.

The compute-integrated manufacturing (CIM) plant of the future will look quite different from the factory of today. It will be based on the integration of traditional or process-based technology with the emerging software or systems-based technology of today and tomorrow. The seven objectives of CIM provide for cost control, innovative and integrated procedures, and complete and efficient design and control systems.

The future holds promise for six-legged robots and for robots that aid in brain surgery. The movement toward robotization has only begun. Your imagination is the only limit.

Key Terms

assembling Putting together.

CIM Computer-integrated manufacturing; one of a number of proposed organizational methods for manufacturing products in large quantities.

depalletizing The use of a robot to unload parts or boxes off a pallet.

die casting The use of dies to form hot metal into desired shapes.

fabricating Making something.

flow-line transfer The use of a robot to pick up two or more pieces at a time and transfer them from a machining line onto a second transfer line located parallel to the first one.

lane loaders Pick-and-place robots used to adjust the feed between fast and slow or slow and fast lines.

line tracking The process of having a robot move along with the production line to do its work as it moves with the line.

MIG Type of metal-in-gas welding in which an inert gas surrounds the welding spot or area while it is in the molten state.

palletizing The use of a robot to stack parts or boxes on a pallet.

process flow The orderly flow of parts and materials to keep production going.

spot welding Welding a small spot between two electrodes of a spot welder; melting metal with a surge of high current through the metal.

TIG Type of stick welding; tungsten in gas.

Review Questions

1. What are the essential factors for work handling?
2. What is a lane loader?
3. What is flow-line transfer?
4. What does NC stand for?
5. What is a conveyor?
6. What is meant by the term *value-added*?
7. How is die casting done?
8. What are two types of die casting?
9. What is palletizing?
10. Why are robots so well adapted to welding and spray painting?
11. List three types of welding that robots can do.
12. What are the objectives of CIM?

CHAPTER 8

Manufacturers' Equipment

Performance Objectives

After reading this chapter, you will be able to:

- Select a robot for a job from a list of manufacturers.
- Understand the language of the robotics industry.
- Describe various characteristics for manufacturers for a given job.
- Understand the terms used in the spec sheets for robots.
- Read a cross-comparison table for selecting the proper machine for a workstation.
- Answer the review questions at the end of the chapter.

A Little Robot History

It was in 1961 that the first industrial robot, named Unimate, joined the assembly line at one of General Motors' plants to work with heated die-casting machines. The robot took the die castings from the machines and performed welding on automobile bodies. It did some of the tasks that people find unpleasant. Unimate was able to obey step-by-step commands stored on a magnetic drum. These instructions allowed the 4,000-pound robot arm to show its versatility in performing a variety of tasks associated with an assembly line.

As more robots were built, they did a variety of other tasks. The robots were found to be good at loading and unloading machines. These first industrial robots now are among the most widely used industrial robots in the world. It took over 20 years of continued improvement for them to become highly reliable and easy to use.

Today's robots feature up to six fully programmable axes of motion and are designed for high-speed handling of parts, some weighing up to 500 pounds. The dedicated electronically controlled robot is regarded as one of the simplest controllers available in the industry today. It is used for teaching and operating industrial robots.

Unimate was conceived in 1956 at a meeting between inventors George Devol and Joseph Engelberger, when they were discussing science fiction writing. Together these two men made a serious commitment to develop a real, working robot.

Now that you have acquired some knowledge of the fundamentals of robots and robotics, the next questions to consider are, what do other systems look like, and what can they do?

To find the answers, we usually look at manufacturers' catalogs and check out the characteristics, capabilities, and technical qualities of robotic equipment manufactured in the United States, Europe, and Asia.

Robots are made by many companies the world over. A great many of them are made for the manufacture of automobiles. These robots do welding, painting, and assembly work very well in Italy, Sweden, Germany, Japan, Korea, and other countries. Robots are also used in the assembly of printed circuit boards for all types of electronic equipment. The electronics industry is one that can brag about lowering the price of its products rather than increasing them every year. In fact, you expect electronic equipment to decrease in price with time. One of the reasons for this is the elimination of costly hand labor. Robots have become the workhorses of the electronics industry as well as the automobile industry. You will find that many of the robots shown here are heavily involved and used in these two areas of manufacturing.

No one name brand of robots has captured the market. There are a number of manufacturers in the business. However, the number decreases all the time. Robot manufacturers are merging, being bought out, or declaring bankruptcy in great numbers. The shakeout in this industry is similar to that in the computer industry. Some of the more well-established firms and their literature have been selected for discussion here. This chapter will acquaint you with a broad cross section of what has been available. Computers and microprocessors are mentioned, but you should consult other sources and the Internet for more detailed information on them.

Selected Manufacturers and Equipment

A list of selected U.S. manufacturers and their locations follows. Keep in mind that the shakeout of manufacturers is constantly changing because the need for robots is constantly changing. Much of the world's manufacturing was once done in the United States. Today, most manufacturing is done in China and other Asian countries. Those countries are also realizing that the production of quality products at reasonable costs can only be accomplished by the use of robots.

- Binks Manufacturing Company (Franklin Park, IL)
- Cincinnati Milacron, Inc. (Lebanon, OH)
- Comau Productivity Systems, Inc. (Troy, MI)
- Cybotech Corporation (Indianapolis, IN)
- Elicon (Brea, CA)
- ESAB North America, Inc. (Fort Collins, CO)
- Fared Robot Systems, Inc. (Denver, CO)
- Feedback, Inc. (Berkeley Heights, NJ)
- GCA Corporation/Industrial Systems Group (Naperville, IL)
- Hobart Brothers Company (Troy, OH)
- International Business Machines (IBM) Corporation/Manufacturing Systems Products (Boca Raton, FL)
- International Robomation/Intelligence (Carlsbad, CA)
- I.S.I. Manufacturing, Inc. (Fraser, MI)
- Lamson Corporation (Syracuse, NY)
- Mack Corporation (Flagstaff, AZ)
- Microbot, Inc. (Mountain View, CA)
- Pick-O-Matic Systems, Inc. (Sterling Heights, MI)
- PRAB Robots, Inc. (Kalamazoo, MI)
- Schrader-Bellows, Division of Parker-Hannifin (Akron, OH)
- Seiko Instruments USA, Inc. (Torrance, CA)

- Transcom, Inc. (Mentor, OH)
- Thermwood Robotics (Dale, IN)
- Unimation, Inc., a Westinghouse Company (Danbury, CT)
- Yaskawa America, Inc. (Northbrook, IL)

Illustrative material provided by most of the listed manufacturers is presented in the following sections. These manufacturers were very enthusiastic and willing to supply data. We have tried to represent them properly, with all the information being taken from their brochures and other available instruction sheets. You will find specification sheets for some of the products mentioned here at the end of this chapter.

Binks Manufacturing Company (Franklin Park, IL)

88-800 Robot

The Binks 88-800 *is* the only robot designed and manufactured in the United States specifically for spray painting. It has received excellent acceptance. It is noted for its flexibility, ease of operation, and reasonable price (see Figure 1-8).

The advanced articulation of the 88-800 lets it apply uniform coatings to virtually any product that can be sprayed manually. Its 18-pound capacity permits the use of a wide range of spray equipment, including air, airless, electrostatic, and plural-component spray guns.

A flexible corrugated plastic shroud protects moving parts from dust, dirt, and overspray—a major advantage in the application of frits in porcelain enameling. The clean, unobstructed lines of the arm allow it to reach into recessed or enclosed areas to paint interior surfaces. Factory Mutual (FM) Approval has been obtained for use of the 88-800 in hazardous locations.

The Binks 88-800 is a servo-controlled, six-axis (additional axes are optional), continuous-path robot with features not previously available in spray-painting robots. Its solid-state computer control system can store up to eight programs. An optional disk memory can store hundreds of programs and 40 hours of total programming time.

A unique editing system makes it possible to correct a program by either gun motion or triggering without removing the robot from production. An exclusive adaptive correction system lets the machine perform any program sequence while automatically adjusting for changes in temperature, fluid viscosity, load, and other factors affecting its operation.

The 88-800 is easily programmed by manually leading it through the spraying operations to be performed. The lightweight counterbalanced arm allows free movement of the spray gun during programming sequences.

The completely solid-state control has no moving parts and is tolerant of humidity, dust, and temperature. At the push of a button, a diagnostic system quickly isolates electronic problems. The mechanical design of the 88-800 provides major economies. Its unique editing control permits the system to be fine-tuned to minimize material usage and rejects.

Manipulator Unit

The manipulator unit is a hydraulically driven, servo-controlled, six-axis unit that will accurately duplicate the motions of a skilled painter (Figure 8-1). The entire spraying sequence can be recorded by continually depressing the program trigger.

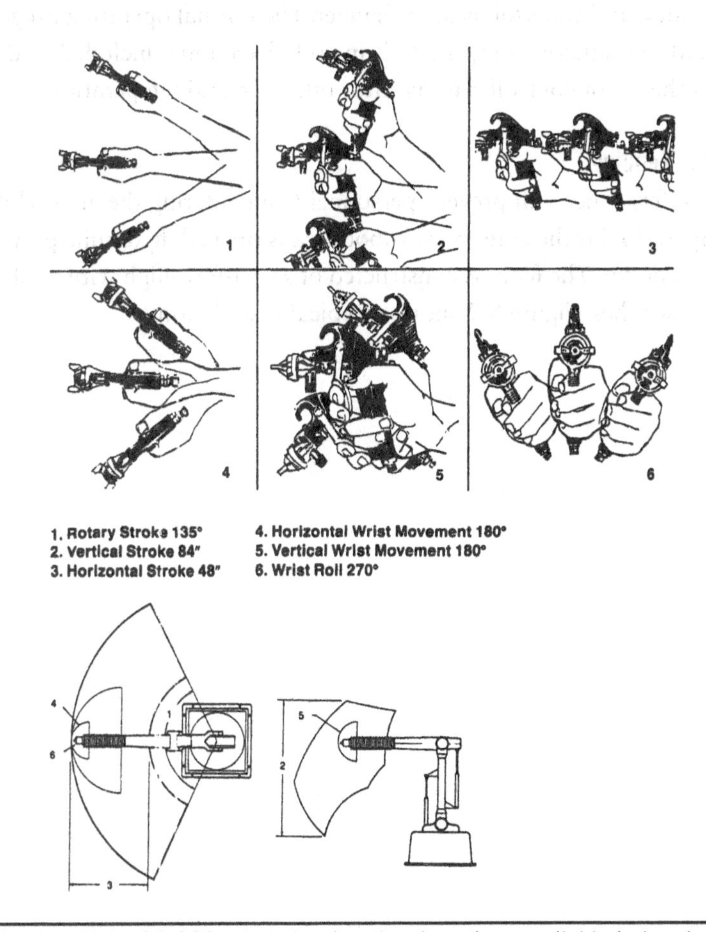

FIGURE 8-1 The Binks 88-800 robot showing the six actions available in hand spray gun operations. (*Courtesy of Binks Manufacturing Company.*)

A point-to-point program may be recorded by turning the switch on and off at each of the desired points. The spray gun trigger switch is located on the front handle. The manipulator can be used in Class I, II, and III, Division I, Groups D, F, and G hazardous locations. It is FM Approved.

Robot Controller

The solid-state control system eliminates the normal problems associated with moving-tape readers, card readers, magnetic-tape cartridges, and reels. The control system includes a complete electronic diagnostic system. The system will operate in temperatures between 40 and 120°F.

Hydraulic Power Supply

The hydraulic power supply is a standard-type unit complete with filter, pressure regulator, and water/oil heat exchanger. The normal operating range is 600 to 700 pounds per square foot (psi). Built-in fault detection is included, and it is interfaced with the robot controller to monitor oil, filter, and temperature.

Safety Fencing

The safety fence will prevent personnel from entering the area while the robot is in operation. If the gate to the robot area is opened, hydraulic power to the robot is shut down. The fence is constructed of seven-foot-high wire mesh with gate and limit switches. Figure 8-2 shows a typical installation.

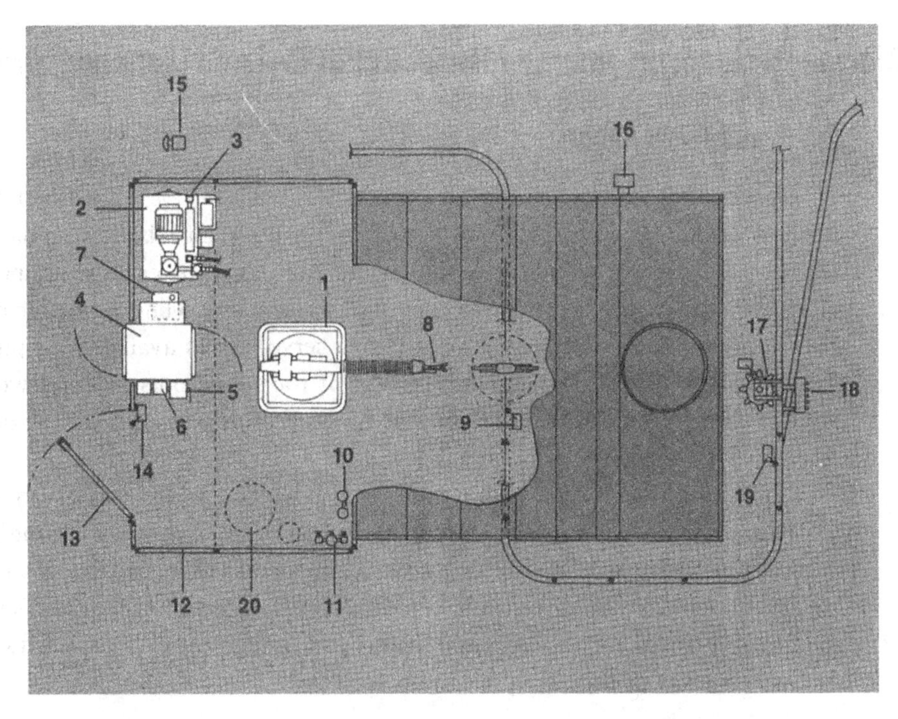

1. Manipulator
2. Hydraulic Power Supply
3. Water/Oil Heat Exchanger
4. Control Enclosure, Floor Standing
   Includes: a) Robot Controller, b) Hard Disk Memory, c) CRT Terminal, d) Air Conditioner for Enclosure.
5. System Disconnect
6. Transformers 220/110 or 440/110
7. Electrostatic Power Supply
8. Spray Gun
9. Program Start Limit Switch
10. Gun Trigger Solenoid Valves
11. Compressed Air Regulators and Filter Assembly
12. Wire Mesh Enclosure
13. Enclosure Gate
14. Gate Interlock Limit Switch
15. Remote Emergency Stop P.B. Station
16. Exhaust Flow Switch
17. Memory Parasitic Drive Assembly
18. Program Identification Manual Input System
19. Program "Enter" Limit Switch
20. Material Supply

FIGURE 8-2 Typical 88-800 robot installation. (*Courtesy of Binks Manufacturing Company.*)

Cincinnati Milacron, Inc., Industrial Robot Division (Lebanon, OH)

T³363 and T³746 Robots

A "simple robot for simple tasks" is the way the T³363 robot is described. It was designed and built to bring the productivity of off-the-shelf robotic automation to simple tasks, such as machine tending and medium-duty materials handling. This all-electric robot operates on three axes of movement—two linear and one rotary. For even more flexibility, an optional fourth servo axis is available for pitch or yaw (see Figure 2-9). The compact design of the robot puts its total working volume within 18 inches of the floor and allows it to deliver a full 300 degrees of rotation with a payload of up to 110 pounds.

Cincinnati Milacron makes a number of robots for a wide variety of jobs. The T³746 and the CR-35iA robots are typical types. The work envelope for the T³746 is shown in Figure 8-3. Spec Sheet 1 gives all the details of its capabilities. It is an electric-driven, computer-controlled, versatile industrial robot. Its direct-current (DC) motor drives and its 70-pound load capacity enable it to boost productivity in a wide variety of process applications such as:

- Arc welding
- Drilling
- Deburring
- Grinding
- Inspection
- Materials handling
- Polishing
- Routing
- Sealant application

Control

Cincinnati Milacron's microprocessor-based Acramatic Version 4 control and operator-friendly software make programming easy. There is no need for computer experience—just knowledge of the job the robot is to perform.

Simple use of the lightweight handheld teach pendant leads the robot through its required moves (Figure 8-4). The controlled-path-motion feature automatically coordinates all six axes to move the robot from one point to the next at the velocity selected in world coordinates. An average of 3,000 points can be programmed and stored within the control memory.

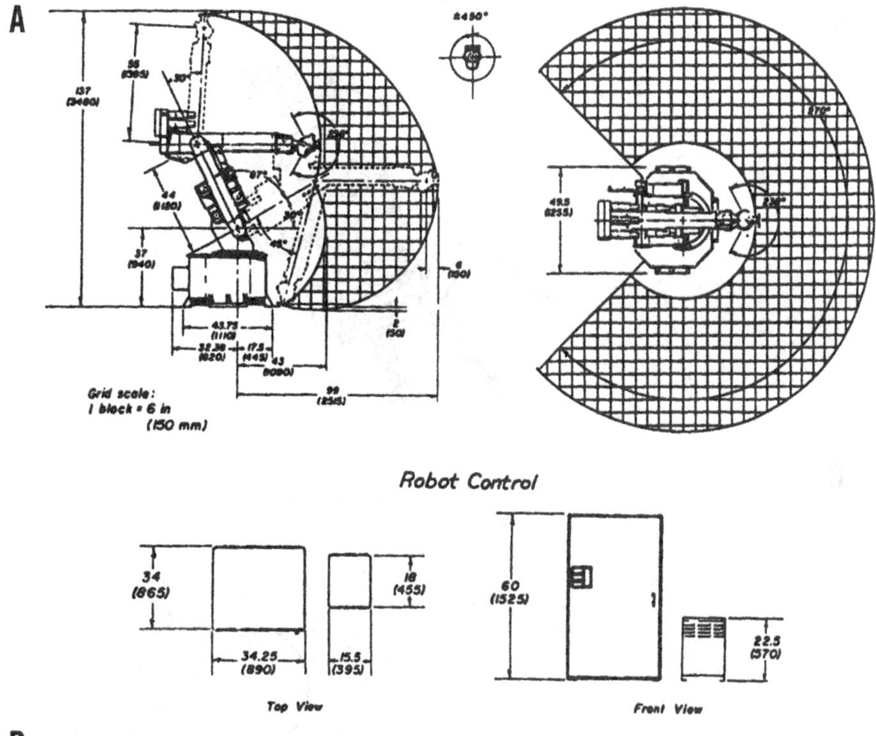

B

Model		CR-35iA
Type		Articulated Type
Controlled axes		6 axes (J1, J2, J3, J4, J5, J6)
Reach		1813 mm
Installation		Floor
Motion range	J1 axis rotation	340° / 370°(Option) 5.93 rad / 6.46 rad(Option)
	J2 axis rotation	165° 2.88 rad
	J3 axis rotation	312° 5.45 rad
	J4 axis wrist rotation	400° 6.98 rad
	J5 axis wrist swing	220° 3.84 rad
	J6 axis wrist rotation	900° 15.71 rad
Max. load capacity at wrist		35 kg
Max. load capacity on J3 casing		2 kg
Maximum speed (Note 1)		250 mm/s(Max. 750 mm/s) (Note 2)
Allowable load moment at wrist	J4 axis	110 N·m
	J5 axis	110 N·m
	J6 axis	60 N·m
Allowable load inertia at wrist	J4 axis	4.00 kg·m²
	J5 axis	4.00 kg·m²
	J6 axis	1.50 kg·m²
Drive method		Electric servo drive by AC servo motor
Repeatability		±0.08 mm
Mass (Note 3)		990 kg
Installation environment		Ambient temperature : 0 to 45°C Ambient humidity : Normally 75 %RH or less (No dew nor frost allowed) Short term Max.95 %RH or less (within one month) Vibration acceleration : 4.9 m/s² (0.5G) or less

Note 1) In case of short distance motion, the speed may not reach the maximum value stated.
Note 2) If the area is monitored by a safety sensor (located separately).
Note 3) Without controller.

Figure 8-3 (A) Work envelope and control cabinet for T³746 robot; (B) specs for CR-35iA. (*Courtesy of (A) Cincinnati Milacron and (B) FANUC.*)

A

B

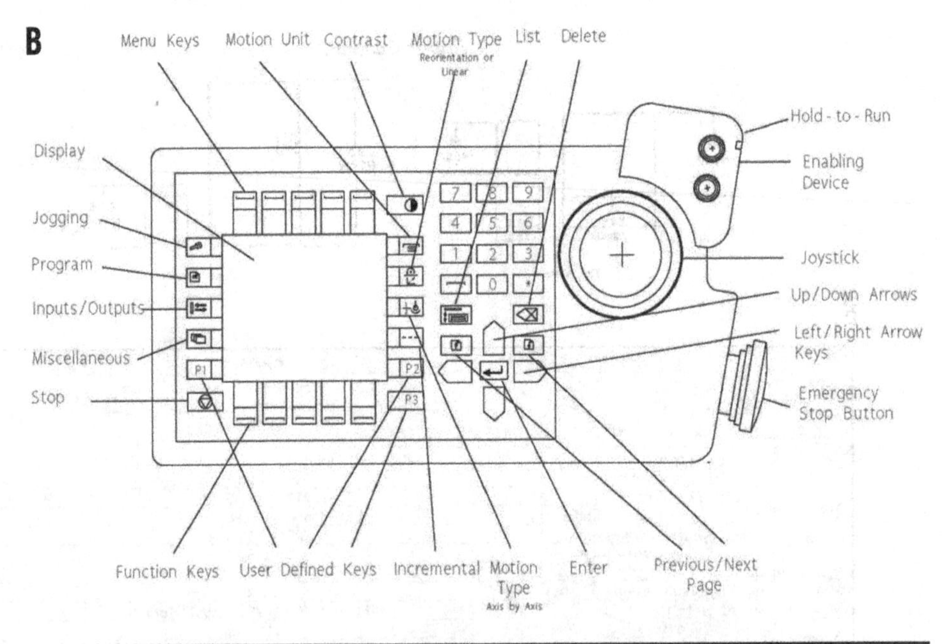

FIGURE 8-4 (A) Teach pendant used with the T³746 robot; (B) labeled parts of a teach pendant. (*(A) courtesy of Cincinnati Milacron.*)

The T³746 can be used as the central element in a robotic arc welding system. With this system, the power supply, wire feeder, positioning tables, and associated hardware are all interfaced with the robot through its own computer control.

Comau Productivity Systems, Inc. (Troy, MI) and Comau, S.p.A. (Torino, Italy)

SMART Robot

SMART is a six-axis, all-electric industrial robot designed by Comau of Torino, Italy. It is mounted on the floor, upside down, sideways, or in any position to best meet the application requirements. This robot is used extensively in the manufacture of automobiles.

The operator can program and operate the robot from either the operator's panel or a lightweight handheld microcomputer-based pendant unit (programming terminal). Robot programming is performed using the versatile Comau PDL (Process Description Language), a high-level language that permits the user to define motion sequences and logic sequences; activate, deactivate, and synchronize independent processes; and easily define complex cycles (e.g., palletizing or tasks in which the decision about the actual path of the arm is made at run time). Robot motion control during both programming and automatic operation is accomplished by point interpolation using one of the following methods: (1) points are generated for simultaneous coordinated motion of all axes traversing the tool center point (TCP) via the shortest path between programmed positions (point-to-point motion in the robot coordinate system); points are generated for coordinated motion of all axes, moving the TCP in a straight line from one programmed position to the next while performing programmed tool orientation (controlled path motion in the Cartesian world coordinate system X_B, Y_B, Z_B originating in the robot base center; Figure 8-5); and points are generated for coordinated motion of all axes, moving the TCP in a straight line from one programmed position to the next, always maintaining the taught tool orientation (controlled path motion in the Cartesian tool coordinate system X_T, Y_T, Z_T originating in the TCP itself). See Spec Sheet 2 for robot characteristics.

In controlled path motion, the ultimate servo commands are derived from point interpolation in the real world via the transformation from fixed world coordinates to robot coordinates. Advanced algorithms provide a higher-level feedback of TCP velocity and position, thus allowing full-speed positioning (or positioning at the selected speed) based on servo torque limits.

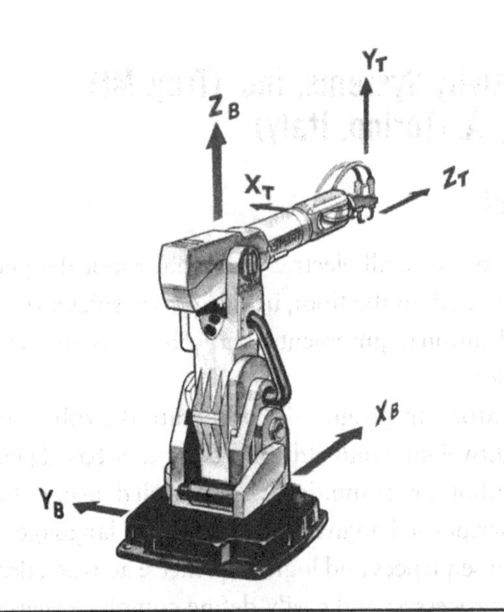

Figure 8-5 Six axes of the SMART robot. (*Courtesy of Comau Productivity Systems, Inc.*)

Controller

The computerized control system is housed in a single cabinet with cables that run 5 to 40 meters (Figure 8-6).

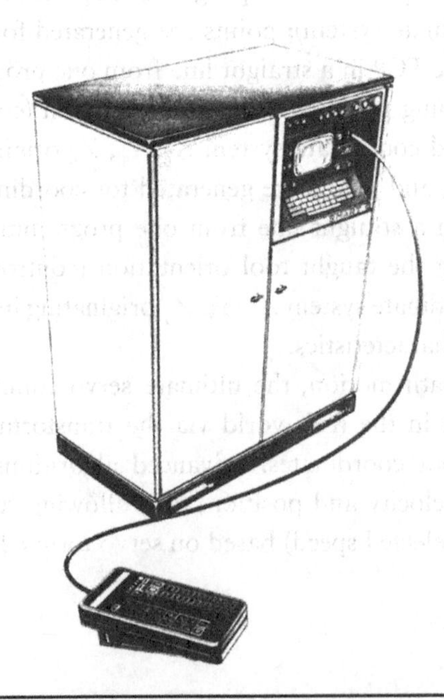

Figure 8-6 Control console for the SMART robot. (*Courtesy of Comau Productivity Systems, Inc.*)

The operator's panel has a 9-inch cathode-ray tube (CRT), an alphanumeric keyboard, and control panel. Optionally, a serial printer can be attached for printing source programs and/or machine parameters.

All robot axes are driven by DC servo motors. Figure 8-7 shows the work envelope for the SMART robot. Resolvers are mounted on the motor shaft and

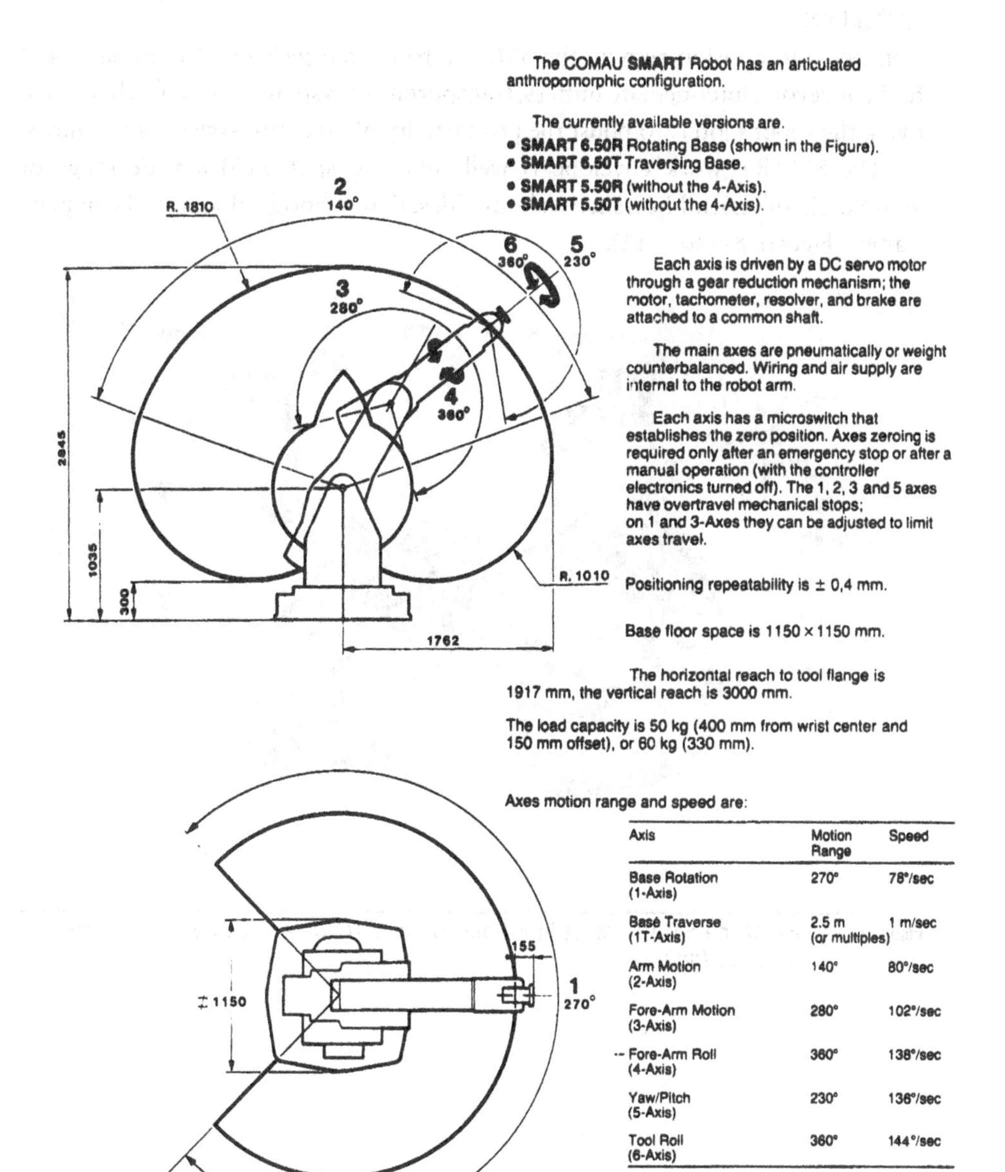
The COMAU SMART Robot has an articulated anthropomorphic configuration.
The currently available versions are.
• SMART 6.50R Rotating Base (shown in the Figure).
• SMART 6.50T Traversing Base.
• SMART 5.50R (without the 4-Axis).
• SMART 5.50T (without the 4-Axis).
R. 1810
2
140°
6
360°
5
230°
3
280°
4
360°
2945
1035
300
R. 1010
1762
Each axis is driven by a DC servo motor through a gear reduction mechanism; the motor, tachometer, resolver, and brake are attached to a common shaft.
The main axes are pneumatically or weight counterbalanced. Wiring and air supply are internal to the robot arm.
Each axis has a microswitch that establishes the zero position. Axes zeroing is required only after an emergency stop or after a manual operation (with the controller electronics turned off). The 1, 2, 3 and 5 axes have overtravel mechanical stops; on 1 and 3-Axes they can be adjusted to limit axes travel.
Positioning repeatability is ± 0,4 mm.
Base floor space is 1150 × 1150 mm.
The horizontal reach to tool flange is 1917 mm, the vertical reach is 3000 mm.
The load capacity is 50 kg (400 mm from wrist center and 150 mm offset), or 60 kg (330 mm).
1150
155
1
270°
Axes motion range and speed are:
The resulting velocity at a distance of 1.9 m from the base center is approximately 2.6 m/sec.

Axis	Motion Range	Speed
Base Rotation (1-Axis)	270°	78°/sec
Base Traverse (1T-Axis)	2.5 m (or multiples)	1 m/sec
Arm Motion (2-Axis)	140°	80°/sec
Fore-Arm Motion (3-Axis)	280°	102°/sec
Fore-Arm Roll (4-Axis)	360°	138°/sec
Yaw/Pitch (5-Axis)	230°	136°/sec
Tool Roll (6-Axis)	360°	144°/sec

FIGURE 8-7 Work envelope for the SMART robot. (*Courtesy of Comau Productivity Systems, Inc.*)

provide absolute cyclic position feedback. The axis zeroing procedure is necessary only in case of emergency manual operation. Compact modular, multiaxis pulse-width modulation (PWM) transistor DC motor drives are mounted in the controller cabinet.

Applications

Connected to a vision system, the SMART robot can pick semisorted parts and feed conveyors, intermediate buffers, transportation systems, and so forth. In such cases, the robot's job is to boost the productivity of an entire system of machines.

The SMART work envelope is well suited to spot weld a wide range of automobile subassemblies, such as body sides, doors, hoods, floors, and complete frames (Figures 8-8 to 8-11).

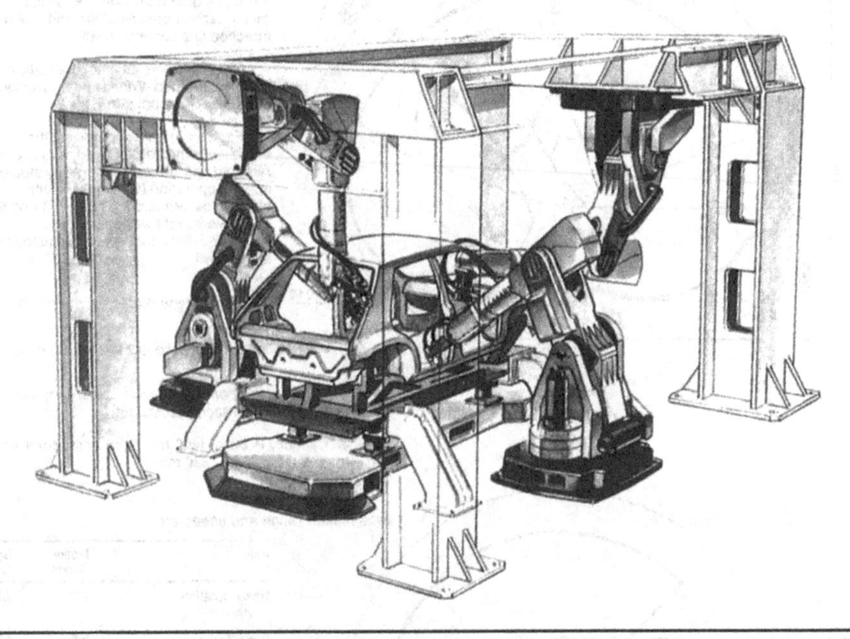

FIGURE 8-8 Use of the SMART robot in various mounts. (*Courtesy of Comau Productivity Systems, Inc.*)

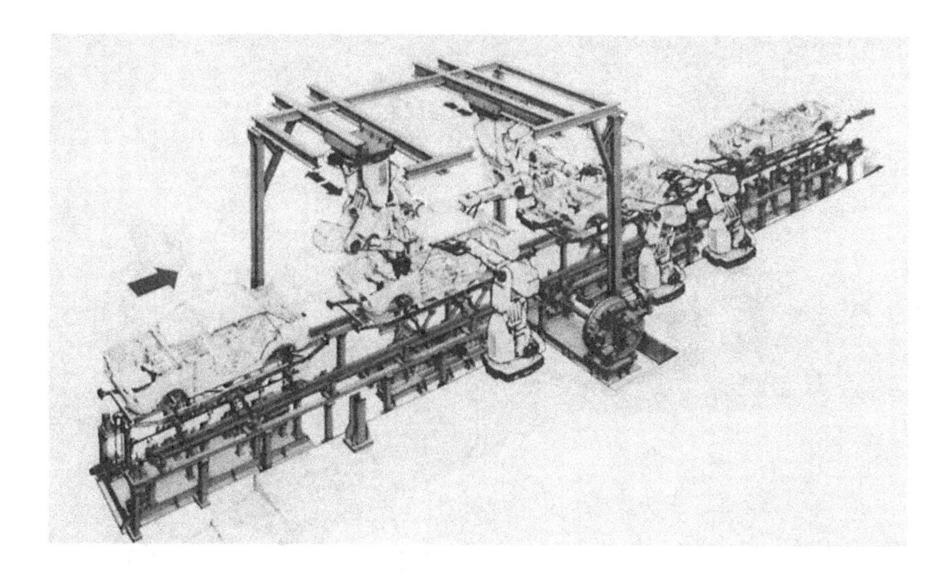

A. Using eight SMART robots to weld four different types of body shells

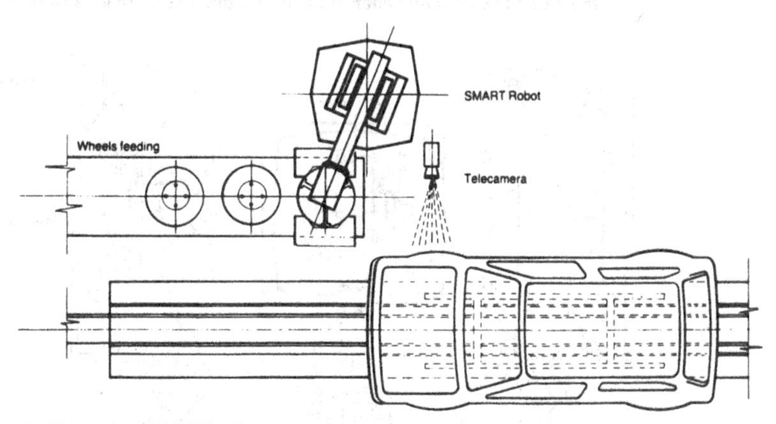

B. Using the SMART robot to mount wheels on a car as it moves on assembly line

Figure 8-9 Assembly-line applications of the SMART robot. (*Courtesy of Comau Productivity Systems, Inc.*)

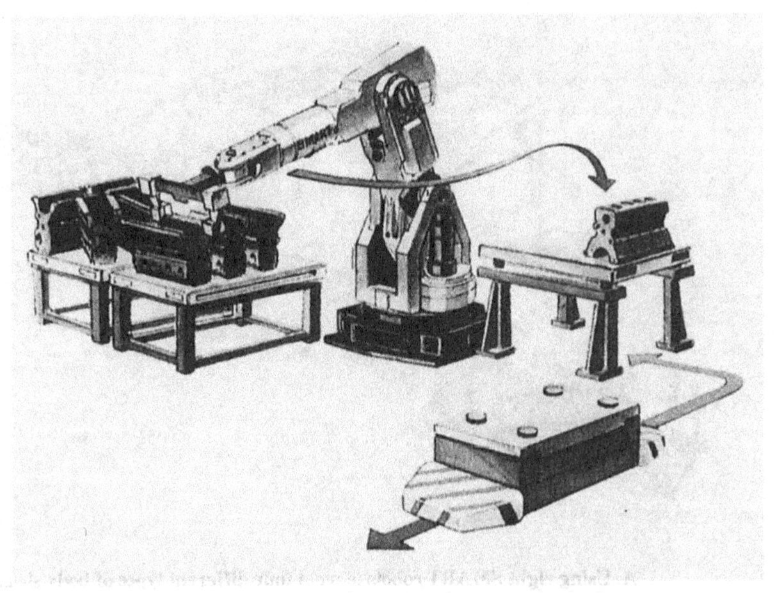

A. Robot picking up semi-ordered parts and placing them on a conveyor

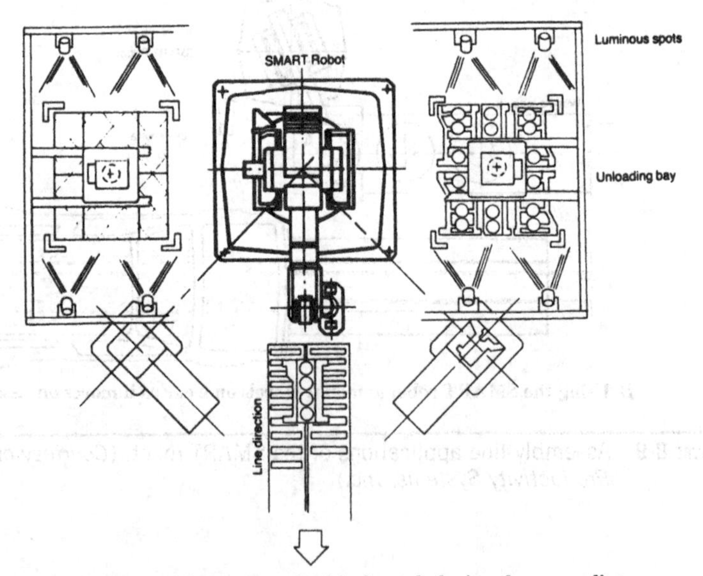

B. Robot picking up engine blocks and placing them on a line

Figure 8-10 Parts-handling applications of the SMART robot. (*Courtesy of Comau Productivity Systems, Inc.*)

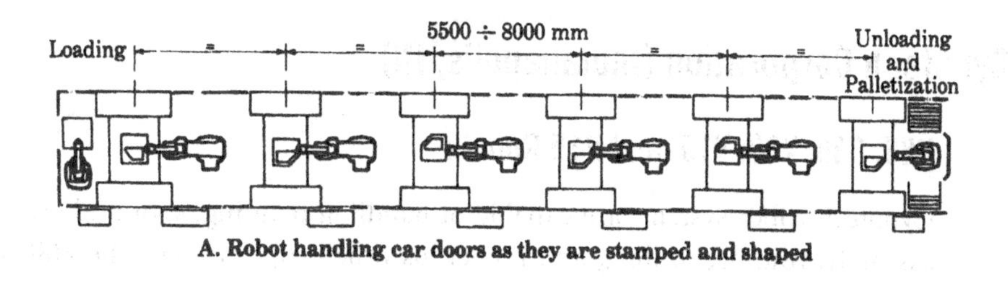

A. Robot handling car doors as they are stamped and shaped

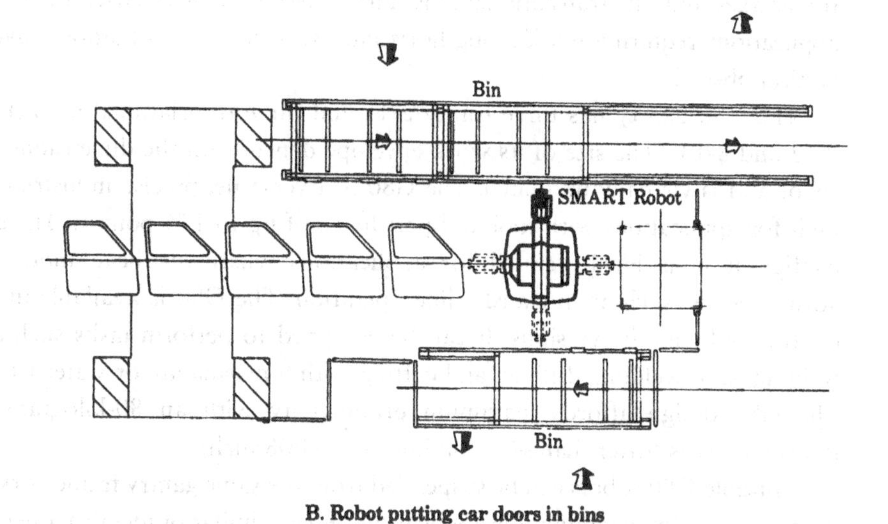

B. Robot putting car doors in bins

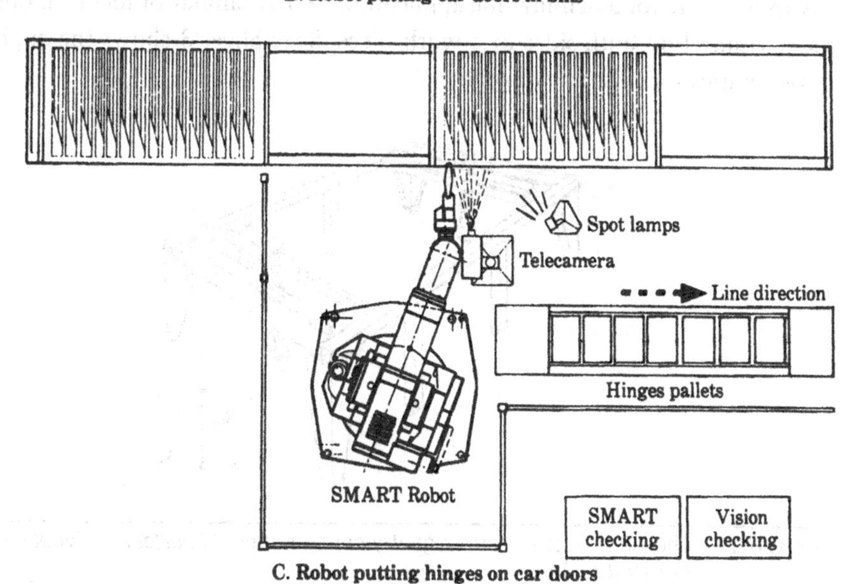

C. Robot putting hinges on car doors

Figure 8-11 Automatic application of the SMART robot. (*Courtesy of Comau Productivity Systems, Inc.*)

Cybotech Corporation (Indianapolis, IN)

H80, G80, V80, V15, and P15 Robots

Cybotech robot systems are capable of handling painting, spot and arc welding, assembly, routing, drilling, and numerous other applications. The H80 has five rotary axes and one translational axis. Either hydraulic or electric, it is ideal for applications requiring work along horizontal surfaces or in situations involving vertical obstacles.

The G80 gantry has three rotary axes and three translational axes (Figures 8-12 and 8-13). The size of its work envelope depends on the dimensions of the robot and is virtually unlimited. The G80 is a versatile, precise industrial robot built for applications with tool and part loads of up to 175 pounds. The gantry configuration is ideally suited for applications where overhead mounting is advantageous, such as in transfer-line operation. The G80 is available in either electric or hydraulic versions. It can be equipped to perform tasks such as arc welding, spot welding, drilling, and cutting with a plasma arc or water jet torch. The G80 design affords maximum productivity with an 80-kilogram load. Repeatability is better than ±0.2 millimeter (±0.008 inch).

Multiple G80 robots can be suspended from the same gantry framework. This is appropriate for assembly-line applications where similar or identical operations are required on both sides of a work piece. Spec Sheet 3 shows the applications possibilities for this type of robot.

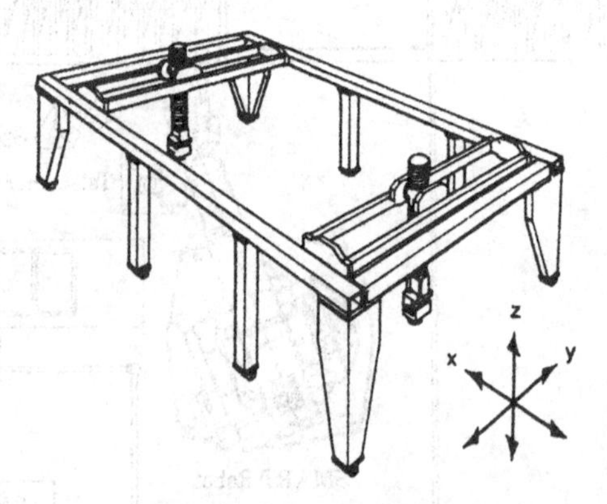

FIGURE 8-12 The G80 robot in double-gantry configuration. (*Courtesy of Cybotech Corporation.*)

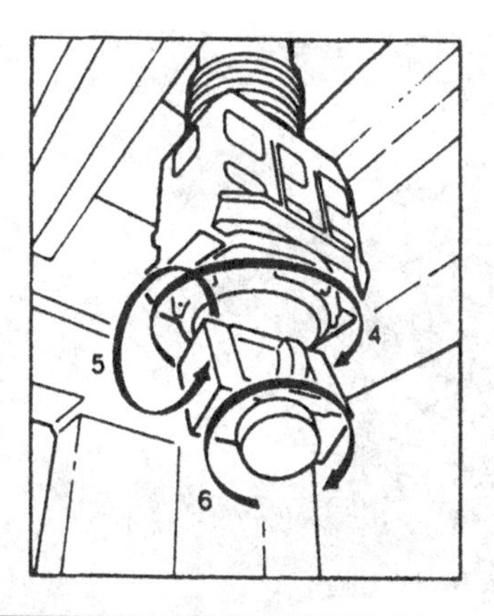

FIGURE 8-13 The G80 robot's three rotary axes. (*Courtesy of Cybotech Corporation.*)

The V80 is a hydraulic robot with six rotational axes. It is well adapted to applications where work must be reached from the top or where obstacles must be avoided (see Figure 3-15). The V15 has either five or six rotational axes. This compact, accurate electric robot is suited to numerous manufacturing operations such as welding, grinding, drilling, cutting, and parts handling. The P15 is a hydraulic robot with seven rotational axes. It is ideal for painting and coating because it is large and flexible enough to cover objects and volumes beyond the reach of other robots.

ESAB North America, Inc. (Fort Collins, CO)

MAC 500 Robotic Welding and Cutting System

MAC 500's easy-teach features hold training time to a minimum (Figure 8-14). The entire system is so simple to use that anyone can learn its programming and operating functions in about 15 minutes, according to the manufacturer. Complex welding applications can be mastered in an hour (see below).

The MAC 500 is a manual teach robot that uses a point-to-point teaching method. Its linear and circular interpolation capabilities cut programming time and increase productivity. With its remote teaching pendant, it allows the operator to program, fine-tune, and edit all movements of the robot.

The MAC 500 is not dedicated to just one function. It is specially designed for *gas metal arc welding* (GMAW) applications yet easily interfaces with flux-cored arc welding (FCAW), gas tungsten arc welding (GTAW), plasma arc welding (PAW), and plasma arc cutting (PAC) applications. See Spec Sheet 4.

FIGURE 8-14 The MAC 500 welding cell. (*Courtesy of Mack Corporation.*)

Repeatability and Accuracy

The MAC 500 has high-speed repeatability and accuracy from 2.4 to 2,362 inches per minute for a wide range of welding and cutting applications. Precision incremental pulse encoders and advanced design of the arm provide ±0.004 inch repeatability. Accuracy is maintained to ±0.008 inch. Quick operating speeds and programmable functions greatly increase versatility, productivity, and part quality. The MAC 500 robotic arc welding system is ideally suited for welding large batches (100 or more) of small, close-tolerance parts (Figure 8-15). The MAC 500 has experienced minimal downtime. A better than 97 percent reliability factor accounts for the MAC 500's excellent record.

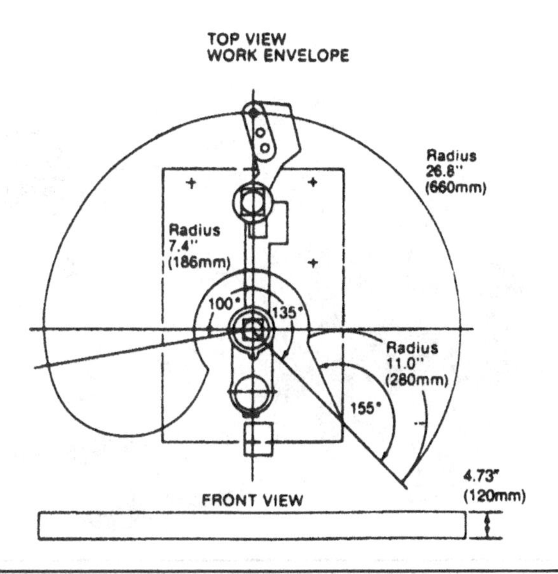

FIGURE 8-15 Work envelope for the MAC 500 robot. (*Courtesy of Mack Corporation.*)

The System

The system consists of a five-axis (four simultaneous axes and one independent axis) articulated selective compliance assembly *robot* arm (SCARA) robot including manipulator, controller, and power unit. Integrated welding and/or cutting equipment includes a power source, wire feeder, welding torch, and special interface for plasma cutting. Optional accessories include manual and pneumatic turntables, a cassette recorder, a printer, extra I/O boards, and a hand teach pendant.

The following systems are available from ESAB: LAH 500 MIG Welding System, LAK 350 Pulsed MIG Welding System, LAP 500 Pulsed MIG Welding System, and Sidewinder 55 Plasma Arc Cutting System.

Feedback, Inc. (Berkeley Heights, NJ)

Armdraulic

Feedback's Model EHA 1052A is an electrohydraulic robot designed and built for educational programs (Figure 8-16). It has closed-loop position control with inductive sensors. An onboard 6802 processer and a full-function teach pendant are also part of the package. Serial and parallel ports for connection to external computers are optional. Extensive documentation for service technician training is available.

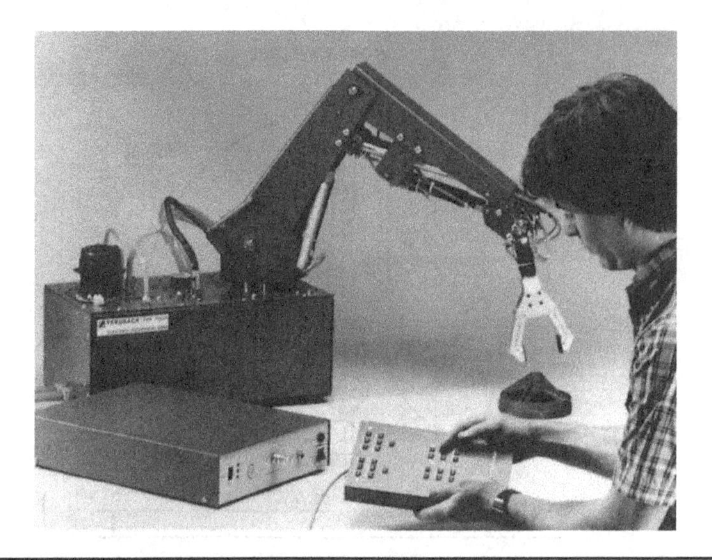

Figure 8-16 Feedback's EHA 1052A educational robot. (*Courtesy of Feedback, Inc.*)

Armatrol

Feedback's Model ESA 1010 is an electric servo, revolute robot designed for educational purposes (Figure 8-17). It uses closed-loop position sensing that is done using potentiometers. It is a low-cost introductory robot. The ESA 1010 requires an external computer. Software is available for most models to work with computers.

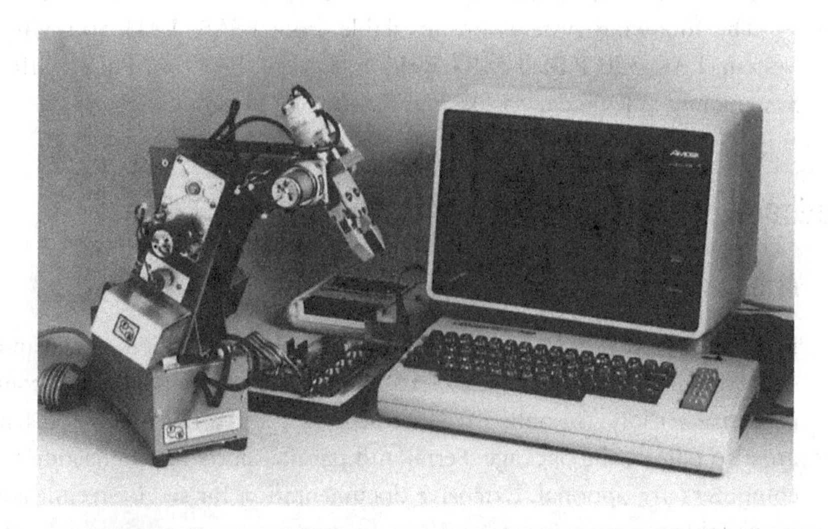

Figure 8-17 Feedback's ESA 1010 computer-controlled educational robot. (*Courtesy of Feedback, Inc.*)

GCA Corporation/Industrial Systems Group (Naperville, IL)

GCA/DKB3200 Robot

The GCA/DKB3200 robot is designed for heavy-duty industrial tasks such as materials handling, loading and unloading pallets, and long-reach spot welding (see Figure 2-8). The robot has an end-of-arm capacity of 220 pounds. It has a reach of more than 9 feet and a work envelope of about 650 cubic feet (Figure 8-18).

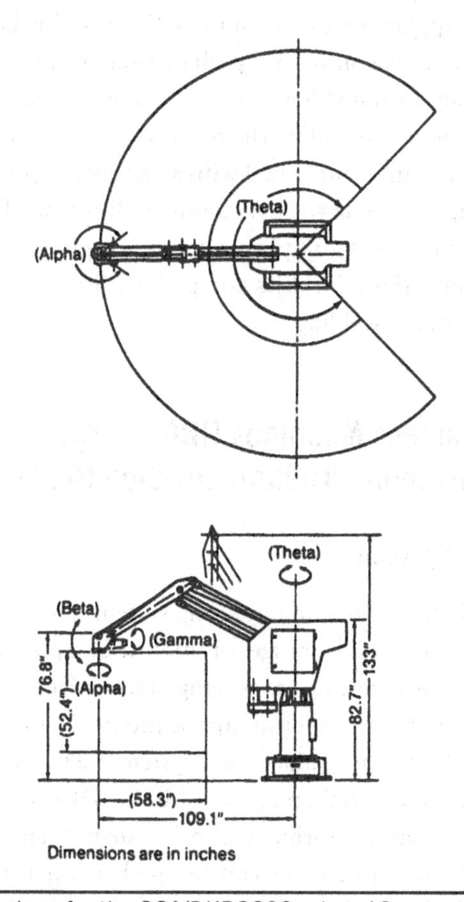

Figure 8-18 Work envelope for the GCA/DKB3200 robot. (*Courtesy of GCA Corporation/ Industrial systems Group.*)

All electric DC servo drives are on all axes, and a unique automatic counterbalancing system is located on the arm articulation axes. This ensures rapid, smooth, and responsive manipulation of maximum payloads at maximum speeds everywhere in the work envelope. The pneumatic counterbalancing facilitates the optional lead-through teaching and reduces power consumption to less than 10 percent of that required by some robots with similar payload capacities.

Up to six degrees of freedom are available with the B3200 robot. Four degrees are standard: horizontal travel (*X* axis), vertical travel (*Z* axis), horizontal sweep (θ axis), and wrist yaw (α axis). Wrist pitch (β axis) and wrist roll (γ axis) are optional. See Spec Sheet 5.

The B3200 robot uses a dual coordinate system (cylindrical and Cartesian) to maximize dexterity. The robot is offered with the standard robot controller for point-to-point applications or with the CIMROC 2 robot controller, where either six-axis simultaneous continuous path or similar integrations are required.

An externally mounted θ-axis motor makes maintenance easy. This motor can be changed without suspending the robot's body by a crane or other lifting device. The B3200 robot can be furnished with a variety of optional accessories, including mechanical gripper hands with vacuum-cup lifters for flat material such as glass, nonferrous metal sheets, or plastic items; magnetic lifters for iron, steel, and ferrous alloy materials; forklifts for pallets; lifters for cartons and boxes; and complete systems for welding.

International Business Machines (IBM) Corporation, Manufacturing Systems Products Division (Boca Raton, FL)

Manfacturing Systems

The IBM 7575 and 7576 manufacturing systems are electric-drive, programmable units suitable for a wide range of industrial applications such as electronic component insertion, testing, packaging, and surface-mount device placement, as well as many light mechanical applications. Figure 8-19 shows a typically configured IBM 7575 manufacturing system, including the manipulator, a rack-mountable 7532 industrial computer (Model 310), a 7572 servo power module, and an AML/2 manufacturing control system licensed program. An optional handheld push-button pendant can be used to teach the robotic arm assembly movements. The new systems perform tasks at speeds significantly faster than IBM's selective compliant assembly robot arm (SCARA) robotic product offerings.

Modularity and Flexibility

The rack-mountable 7532 industrial computer (Model 310) is one of the newest members of IBM's industrial computer family. The 7532 is designed to meet

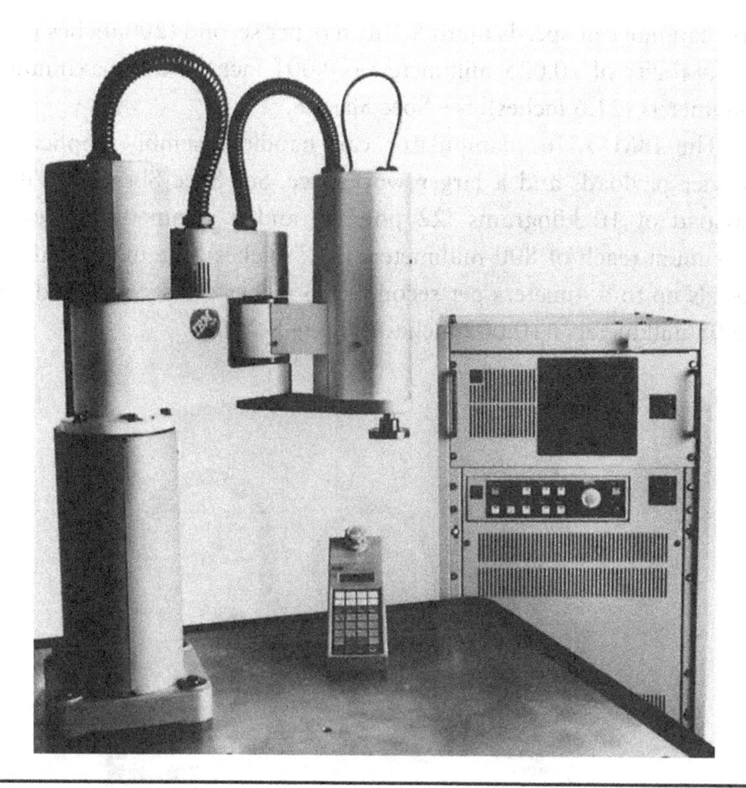

FIGURE 8-19 Typically configured IBM 7575 manufacturing system. (*Courtesy of IBM Corporation.*)

industrial environmental conditions such as temperature variations, vibration, and shock voltage surges and particulates in the air.

Model 310, based on the Intel 80286 microprocessor, offers the power and flexibility required for manipulator control. It has 512 kB of memory and an Intel 80287 math coprocessor. The computer is specially configured with two intelligent axis control cards based on the Motorola 68000 microprocessor, a 48-point digital input/digital output card, a four-port RS232 asynchronous communications adapter, and three expansion slots for optional features.

The 7572 servo power module, which can be mounted in a rack or on a panel, provides the power interface between the 7532 industrial computer and either the 7575 or 7576 manipulator. An easy-to-use operator control panel allows plant floor personnel to control the manufacturing system and the application program. A handheld push-button teach pendant is optional.

The 7575 manipulator is designed for a wide range of electronic and light mechanical assembly applications that require high speed, accuracy, and repeatability. With a maximum payload of 5 kilograms (11 pounds), the robotic arm can move at speeds up to 5.1 meters per second (200 inches per second) with repeatability of ±0.025 millimeters (±0.001 inch) and a maximum reach of 550 millimeters (21.6 inches). See Spec Sheet 6.

The IBM 7576 manipulator can handle assembly applications requiring heavier payloads and a larger workspace. See Spec Sheet 7. With a maximum payload of 10 kilograms (22 pounds) and a symmetrical workspace with a maximum reach of 800 millimeters (31.5 inches), the manipulator is capable of speeds up to 4.4 meters per second (173 inches per second) and repeatability of ±0.05 millimeters (±0.002 inches) (Figure 8-20).

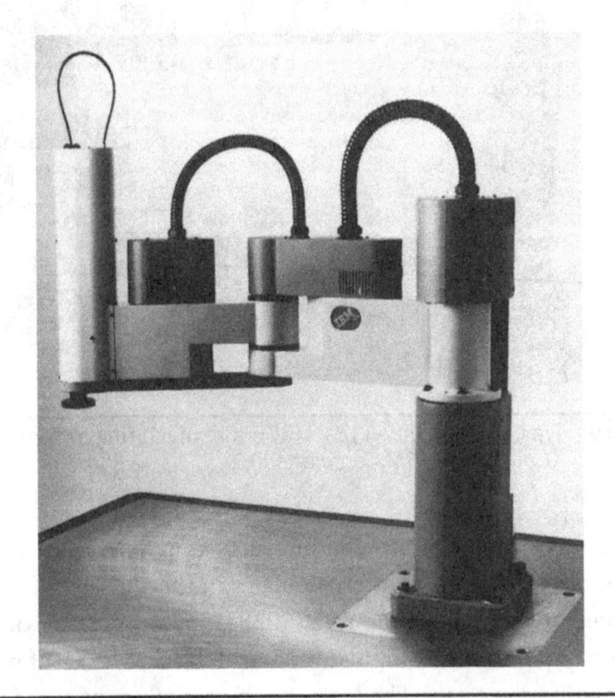

Figure 8-20 IBM 7576 manipulator. (*Courtesy of IBM Corporation.*)

New Programming Language

IBM has announced the following licensed programs for the system:

- **AML/2 Manufacturing Control System.** This is a prerequisite for each 7575 or 7576 manufacturing system. It contains a powerful and flexible, easy-to-use interpretive language in which robotic application programs are written.
- **AML/2 Application Development Environment.** This is recommended for each manufacturing site using a 7575 or 7576 manufacturing system. The

program is used to easily and efficiently develop and debug customized robotic application programs.

- **AML/2 Application Simulator.** This is a powerful off-line tool for simulating and debugging AML/2 application programs. Other uses include estimating an application's cycle time and familiarizing the user with AML/2.

International Robomation/Intelligence (Carlsbad, CA)

IRI M50E AC Servo Robot

Robot Command Language

The Robot Command Language (RCL) provides the user with an easy-to-use means of creating many applications for the robot. The RCL provides a small set of easy-to-use commands that allow the novice to create a complete applications program. It also provides an extended set of commands that will satisfy experts. The RCL is divided into the following command groups:

```
MOVE                        SYSTEM CONTROL
TEST AND BRANCH             INPUT/OUTPUT
SET VALUE                   DATA TRANSFER
DISPLAY                     EDITOR
GRIPPER CONTROL             LOGIC AND ARITHMETIC
```

In addition to the extremely versatile RCL, the firmware provides a complete online editor, system control commands, and application program load and save capability. See Spec Sheet 6.

Application of the M50 Robot

The work envelope is such that it allows a wide range of activities to be performed (Figure 8-21):

- Machine loading and unloading
  - Injection-molding
  - Chuckers and lathes
  - Compression-molding
  - Milling machines
  - Die casting
  - CNC machinery centers
  - Shear and punch

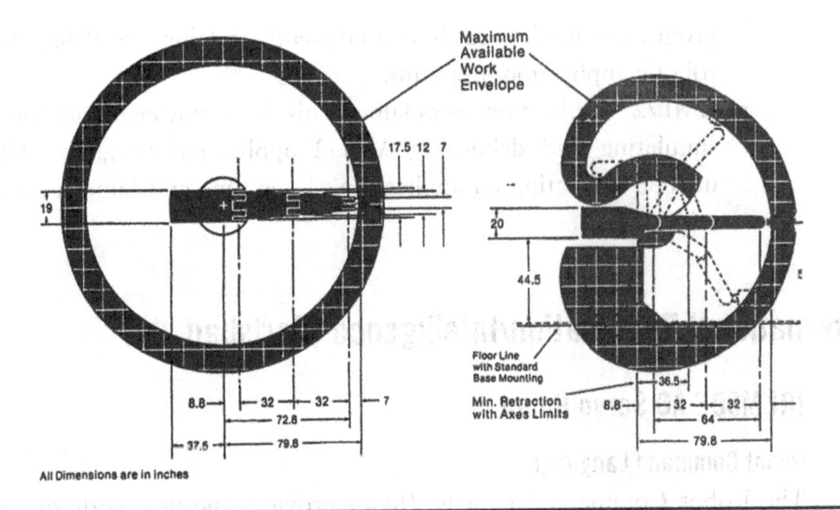

FIGURE 8-21 Work envelope for the IRI M50E AC servo robot. (*Courtesy of International Robomation/Intelligence.*)

- Induction furnace
- Broaching
- Automatic weld stations
- Palletizing and depalletizing
  - Product-to-carton
  - Machined parts
  - Carton-to-pallet
  - Wood products
  - Food products
  - Chemicals
  - Canned goods
  - Glass
  - Pharmaceuticals
- Spray applications
  - Mold release agents
  - Fogging
  - Sand blasting
  - Caustic washing
- Educational applications
  - Establishing a "hands-on" robot
  - Control laboratory and control theory development
- Pick-and-place applications
  - Part transfer for final assembly
  - Nondestructive testing

- Test fixture load and unload
- Finished parts packaging
- Conveyor-to-conveyor transfer
- Heat treating

Mack Corporation (Flagstaff, AZ)

Mack Products

Mack Corporation has a line of proprietary products for the automation and automatic equipment industry. Mack makes B·A·S·E grippers, gripper adaptors, mounting brackets, transporters, rotators, roll models, and pitch/yaw models. The company also manufactures actuators and various hydraulic and pneumatic cylinders for use in automatic equipment and robots.

Microbot, Inc. (Mountain View, CA)

Alpha II

The Alpha II is a proven low-cost robot system designed specifically to help manufacturing operations improve productivity by automating low-level tasks that humans find hazardous or difficult to repeat accurately for long periods of time (Figure 8-22).

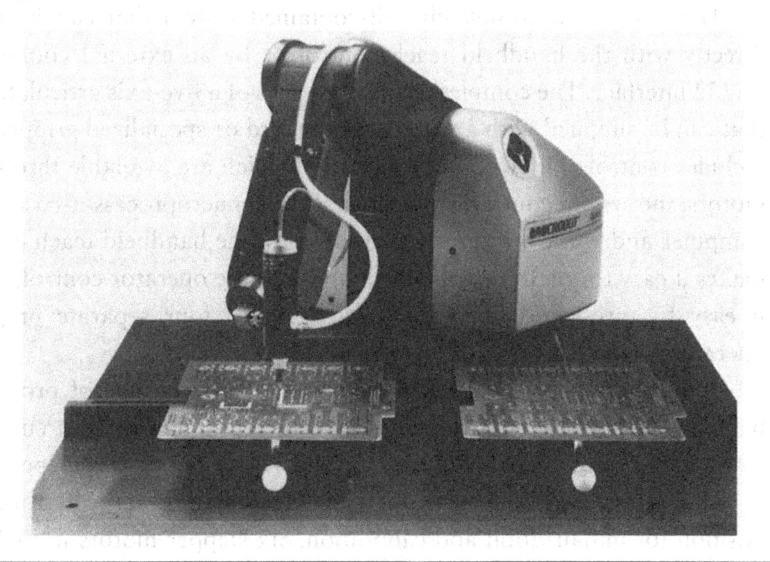

Figure 8-22 Alpha II robot system. (*Courtesy of Microbot, Inc.*)

Proven applications include machine loading and unloading, semiconductor wafer and cassette handling, solder masking, application of adhesive and coating materials, dip soldering, and packaging of pharmaceuticals and chemicals (Figure 8-23).

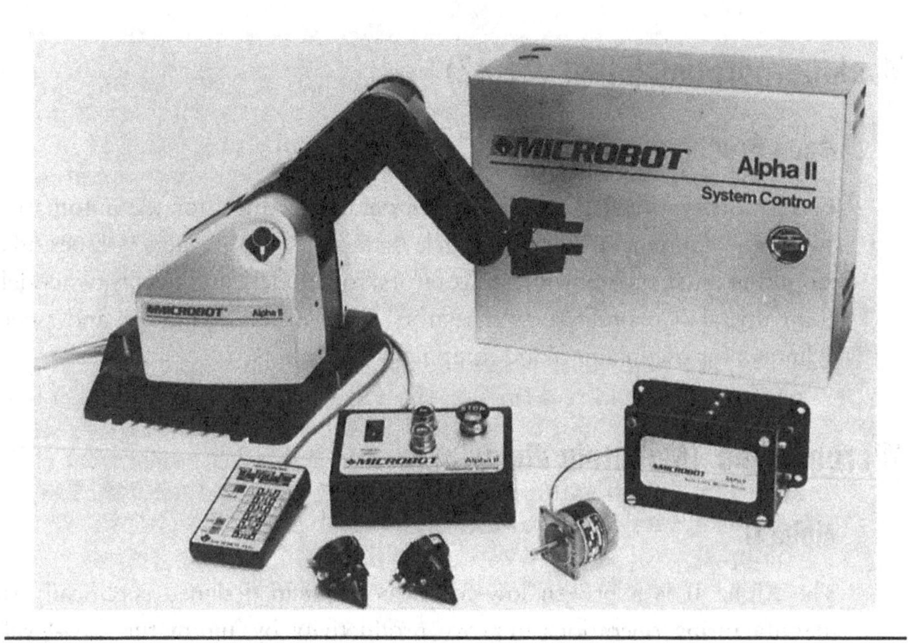

FIGURE 8-23 Alpha II dispensing system. (*Courtesy of Microbot, Inc.*)

This robot is a completely self-contained system that can be programmed directly with the handheld teach control or by an external computer with an RS232 interface. The complete system consists of a five-axis articulated robot arm that can be supplied with a variety of standard or specialized grippers. The robot includes controls for two additional axes, which are available through auxiliary motors; the system control, which contains the microprocessor-based control and computer and work-cell interface electronics; the handheld teach control, which makes it easy to teach the robot motions; and the operator control, which makes it easy for production people to select up to four separate preprogrammed operations.

This robot can operate continuously in a broad range of production tasks, handling payloads of up to 3 pounds with repeatability requirements of up to ±0.015 inch. Maximum arm speed is 51 inches per second. See Spec Sheet 9.

The semi-closed-loop control system references the arm location to a home position for initialization and calibration. Six stepper motors in its base operate the robot's six-jointed arm through a system of precision stainless steel cables.

Tests have shown that Alpha II has a mean time between failures (MTBF) of greater than 4,000 hours and an uptime of more than 98 percent.

The removable teach control allows you to program the robot's microprocessor in real time as the robot arm is moved through each of the motions required to execute its tasks. It has an easy-to-read LED display that shows the system's status and quickly tells what step the arm is on (which allows for easy programming and editing). The microprocessor retains up to 227 individual steps. The Alpha II is also equipped with other RS232-controlled devices through the computer. Two optional auxiliary motors may be used to operate application peripherals such as rotating tables, slides, and conveyor belts, all under robot control. Eighteen optically isolated I/O ports are available for interfacing the robot with work-cell sensors and device actuators at factory-level voltages.

The operator control is used by production personnel to run preprogrammed tasks. It contains an emergency shutoff switch, but the operator cannot in any way modify or erase any of the existing programs (Figure 8-24).

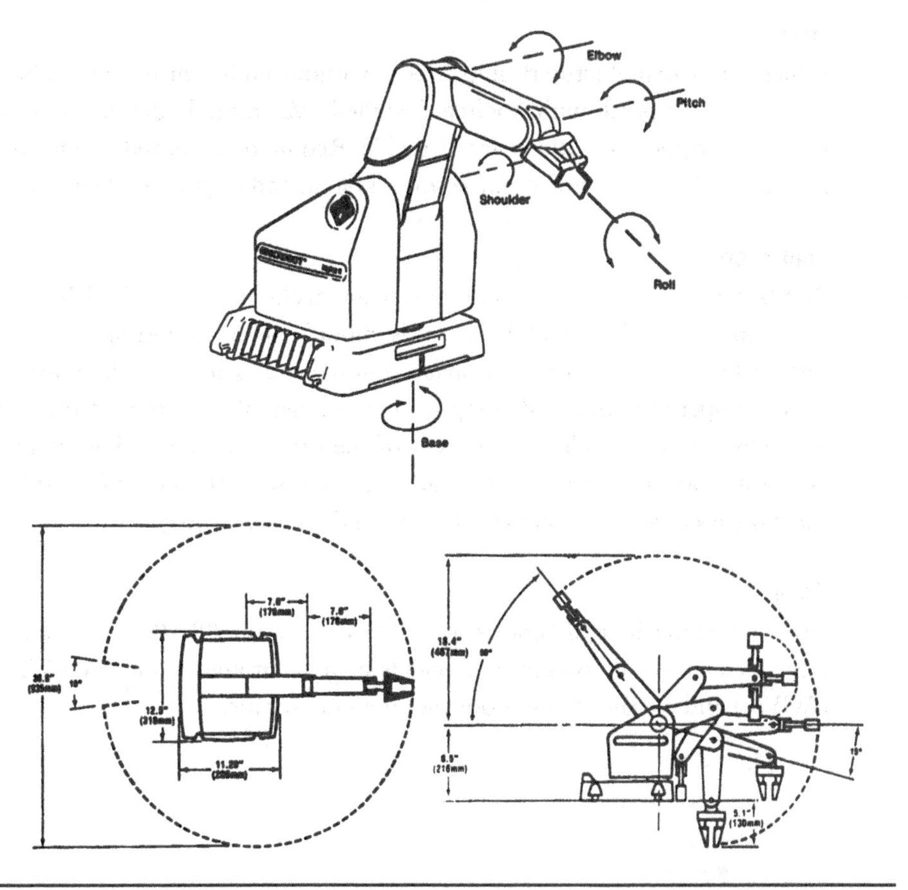

FIGURE 8-24 Operating envelope for the Alpha II robot. (*Courtesy of Microbot, Inc.*)

PRAB Robots, Inc. (Kalamazoo, MI)

PRAB Industrial Robots

Many of the industrial robots built by PRAB in the early 1960s are still in service today. As industry's acceptance of robots has increased, demand for additional models has likewise increased. Today, PRAB offers a complete line of industrial robots including electrohydraulic and totally electric robots. See Spec Sheet 8.

The line includes a full range of cylindrical coordinate, spherical coordinate, and articulated arm robots. These robots offer point-to-point, multipoint, continuous-path, and circular interpolation capabilities.

Although PRAB robots are capable of performing many different jobs, PRAB has established itself as a leader in materials-handling applications. Payloads of these robots range from a few ounces to 1 ton.

Tooling

Other robot manufacturers ship robots without end-of-arm tooling, but PRAB robots are shipped with tooling installed. Vacuum, hydraulic, electric, and magnetic gripper operations are available. Required repeatability and tolerances are factored into the design of the robot system and its tooling (Figure 8-25).

Controllers

PRAB controllers integrate state-of-the-art technology, which delivers greater robot uptime and control board axis module interchangeability, and ease of serviceability and minimal floor space requirements. The controller is designed for effective operation in a wide range of environmental conditions. Control boards are housed in the stable environment of the cabinet and have high immunity to electrical noise and temperature. Authorities consider this controller to be one of the most user-friendly robot controllers on the market today.

Vision

Vision tracking in real time is available on certain PRAB robots and can be interfaced with other system components such as automatic guided vehicle systems (AGVSs) and overhead- and floor-mounted monorails.

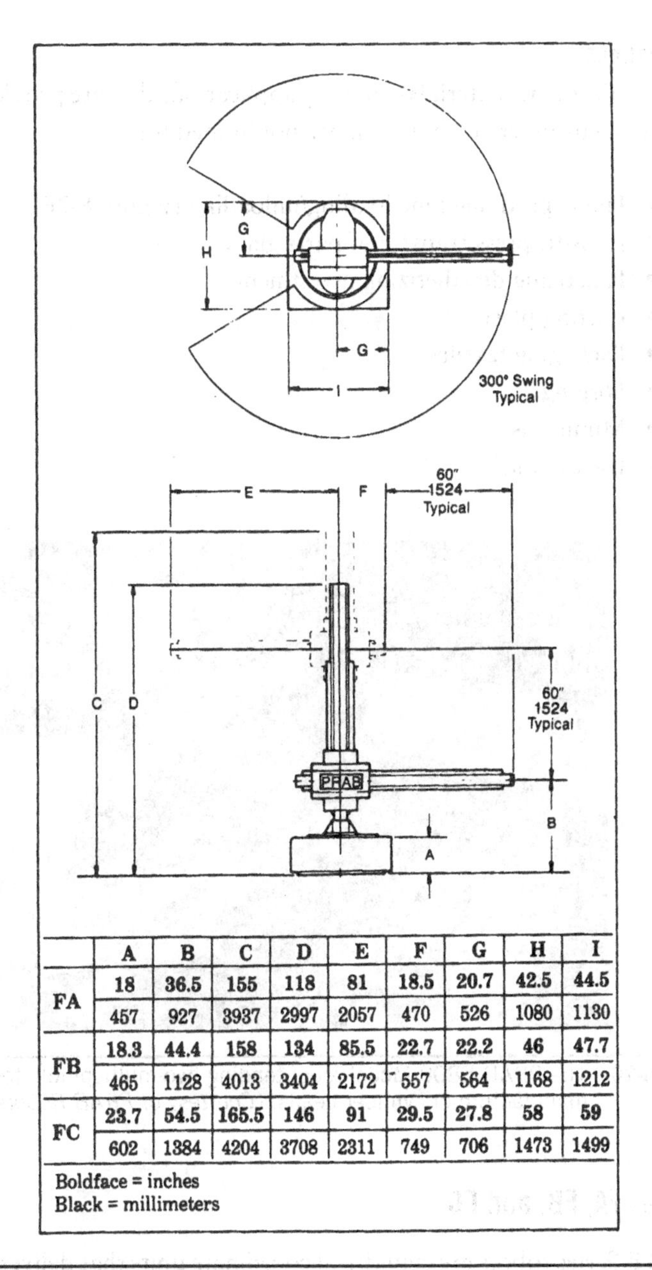

		A	B	C	D	E	F	G	H	I
FA	in	**18**	**36.5**	**155**	**118**	**81**	**18.5**	**20.7**	**42.5**	**44.5**
	mm	457	927	3937	2997	2057	470	526	1080	1130
FB	in	**18.3**	**44.4**	**158**	**134**	**85.5**	**22.7**	**22.2**	**46**	**47.7**
	mm	465	1128	4013	3404	2172	557	564	1168	1212
FC	in	**23.7**	**54.5**	**165.5**	**146**	**91**	**29.5**	**27.8**	**58**	**59**
	mm	602	1384	4204	3708	2311	749	706	1473	1499

Boldface = inches
Black = millimeters

Figure 8-25 Work envelope for PRAB F-Series robots. (*Courtesy of PRAB Robots.*)

Applications

Some of the many materials-handling applications that are providing cost-effective service to customers include, but are not limited to:

- High-speed machine loading/unloading (Figure 8-26)
- Press-to-press transfer of parts glass
- Palletizing/depalletizing investment
- Casting plastics
- Packaging textiles
- Forging
- Munitions
- Die casting

FIGURE 8-26 The PRAB 5800 Model robot tending two multispindle tools in the manufacture of cylinder heads. (*Courtesy of PRAB Robots.*)

Models FA, FB, and FC

PRAB F-Series robots are cylindrical coordinate units that deliver a smooth, steady motion for payloads of up to 2,000 pounds. They are ideally suited for heavy-duty, servo-controlled, high-speed materials handling. In terms of weight handling, function, control operations, and up to seven-axis motion, the ultimate in flexible automation is offered. These units can be equipped with real-time vision tracking and are easily interfaced with other system components such as AGVSs.

F-Series robots provide the rigidity needed for demanding applications. A servo feedback device on each axis is driven directly by the moving member rather than the actuator. This means accurate control and repeatability with greatly reduced backlash for long-term, close-tolerance work.

Simultaneous servo control of up to seven axes is possible with PRAB'S 700 controller, which features a 16-bit microcomputer. A free-standing, sound-enclosed, hydraulic power unit is connected to the robot by two 20-foot quick-disconnect hoses as standard equipment.

Applications

F-Series robots are being used effectively for the following applications:

- Stance (spot) welding
- Machine tool loading
- Material transfer
- Investment casting
- Forging
- Palletizing
- Glass handling
- Press loading
- Stacking/destacking

Schrader-Bellows, a Division of Parker-Hannifin (Akron, OH)

Robotics with Motion Mate

Motion Mate robots handle simple, repetitive materials-handling tasks with ease. Transferring, loading, and unloading are simple applications, but they are tasks that can free humans for more productive use (see Figure 1-3). Motion Mate robots are more than value-added devices; they are efficient links between different machines as well as between machines and material supply sources. They provide efficiency that pays for itself with increased production.

Typical Applications

A typical application for Motion Mate robots is as simple work cells. In this application, work is first processed by one machine and then by a second machine. A robot unloads the first machine and transfers the part to the second machine for the next operation (Figure 8-27). This linking of operations saves labor time and reduces in-process inventories. A complete work cell also must include a method for loading the first machine and unloading the second machine. An expanded work cell links three machines and uses two robots.

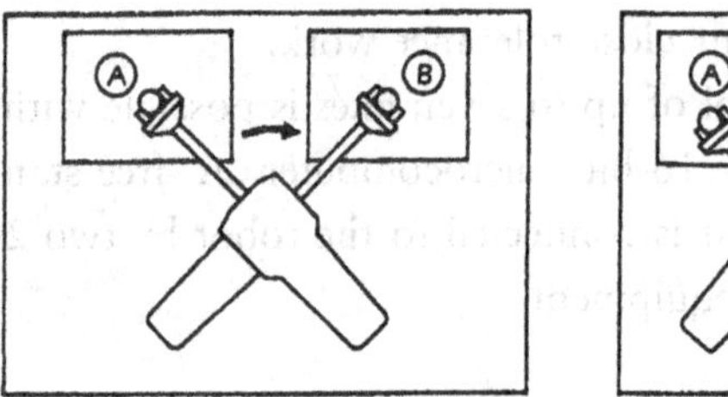

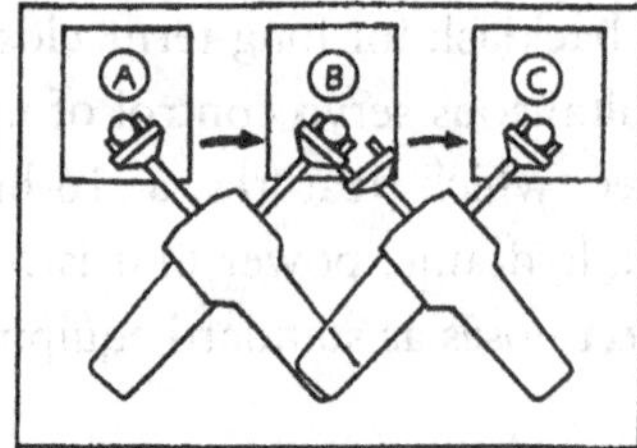

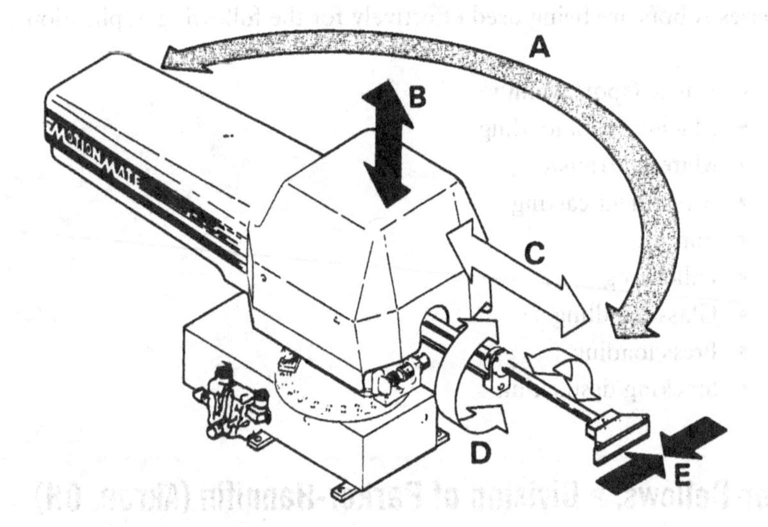

Figure 8-27 Simple work cells—a typical application for Motion Mate robots. (*Courtesy of Schrader-Bellows.*)

Motion Mate robots are also used in secondary operations. For example, a plastic part must not be damaged after molding due to cosmetic requirements. The robot removes the part from the opened mold and places it in a trim die for flash removal. The finished part with no scratches or mars is then placed on a conveyor (Figure 8-28). Another application for these versatile robots is punch-press loading. Figure 8-29 shows a punch press that is used to assemble two cup-shaped parts that form a wheel for office furniture. The robot arm automatically ejects the finished product as it moves the unfinished product into position. Even though, in this instance, the parts supply was fed from a hand-loaded magazine, the robot system tripled output because safety requirements for human punch press operators were eliminated.

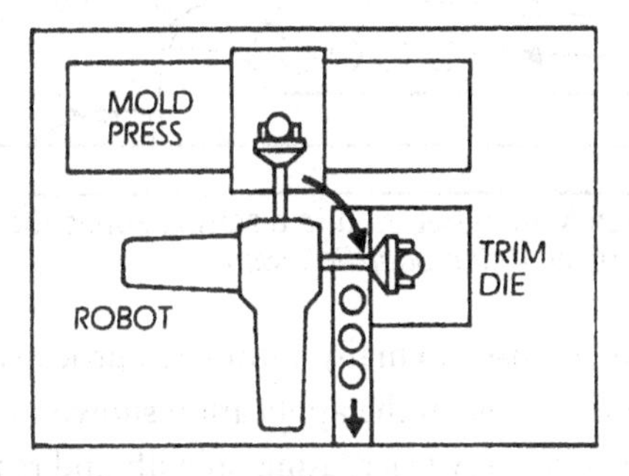

FIGURE 8-28 Motion Mate robot used in a secondary operation. (*Courtesy of Schrader-Bellows.*)

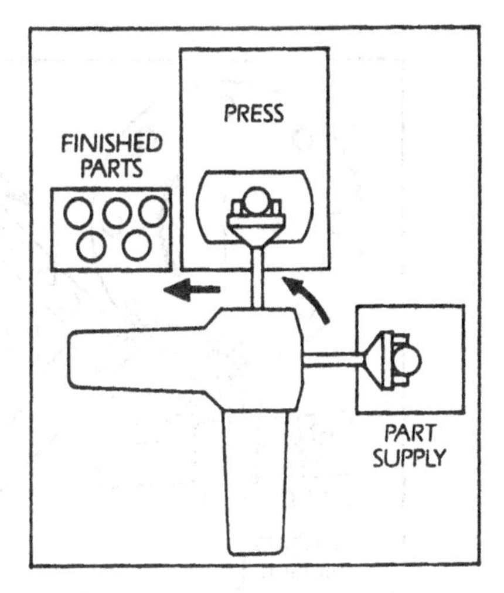

FIGURE 8-29 Motion Mate robot used in a press-loading application. (*Courtesy of Schrader-Bellows.*)

Motion Mate robots also can be used for sorting. For example, Figure 8-30 shows the robot sorting incoming parts in three categories, A, B, and C, based on the results of testing equipment. The multiple stops on the robot base and the communications linkage between the test equipment and robot are examples of custom system work that can significantly expand the capabilities of the basic robot unit.

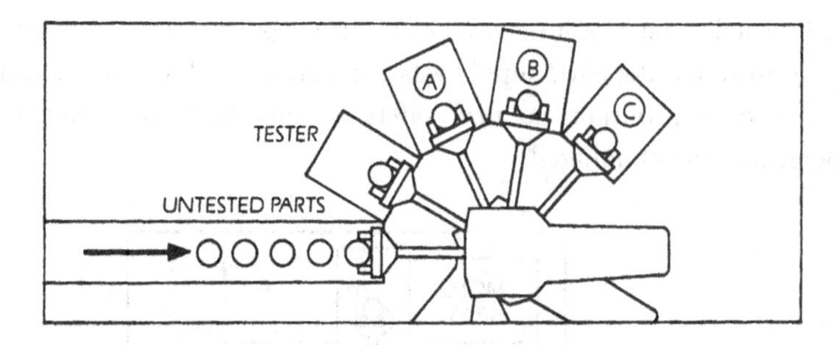

FIGURE 8-30 Motion Mate robot used in a sorting application. (*Courtesy of Schrader-Bellows.*)

Rotary tables are used in many automated processes to bring parts to a variety of different workstations. In the application shown in Figure 8-31, the load robot loads parts from a delivery tract, using only lift and reach movements. The unload robot, using information from the operations performed on the parts, places them in either good or reject bins.

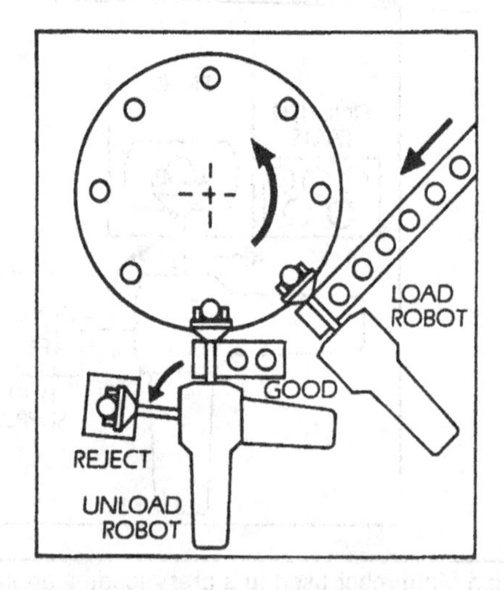

FIGURE 8-31 Motion Mate robots used in loading and unloading parts from a rotary table. (*Courtesy of Schrader-Bellows.*)

The System

This highly versatile, easily programmable robot is pneumatically powered and microprocessor controlled. The Motion Mate system consists of a manipulator, a robot controller, and a remote handheld teach module. The robot offers five axes of movement, including base rotation, lift, extension, wrist rotation, and grasp.

See Spec Sheet 11. The maximum payload is 5 pounds. One of the robot's major advantages is its ease of programming and resulting versatility. No special skills are required of the operator. The handheld teaching module programs the robot commander, which guides the robot through a sequence of operations for quick, precise transfer and placement of parts. This programming simplicity gives the robot the flexibility to be changed quickly for new production requirements.

Seiko Instruments USA, Inc., Robotics/Automation Division (Torrance, CA)

Cylindrical Coordinate Robots

Seiko's D-Tran RT3000-Series robots are used as the standard for precision comparisons (Figure 8-32). Accurate positioning with high speed allows the RT-Series robots to meet assembly and inspection problems with flexibility. The *A* axis has three possible mounting orientations: downward, upward, and outward. The *R* axis can be mounted in a 100-millimeter "extended" position. Figure 8-33 shows other details of the RT3000.

Figure 8-32 An early model of a precision robot. (*Courtesy of Seiko Instruments USA, Inc.*)

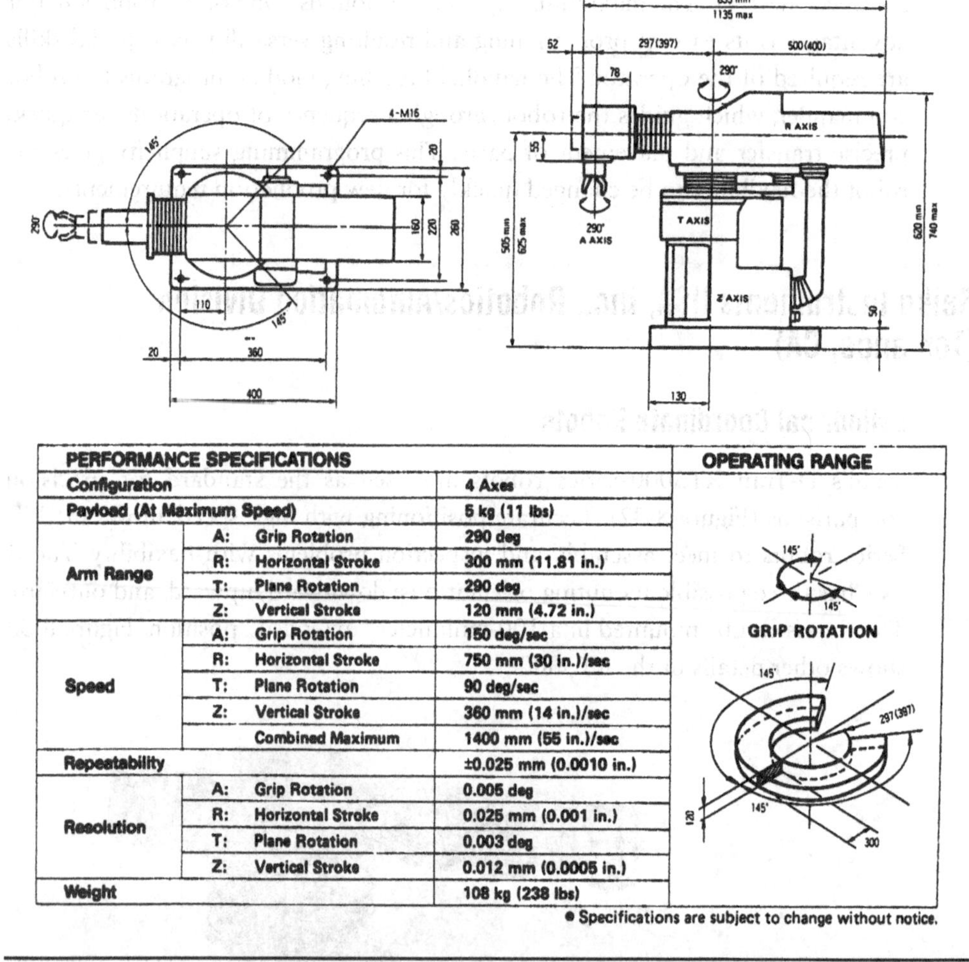

PERFORMANCE SPECIFICATIONS			
Configuration			4 Axes
Payload (At Maximum Speed)			5 kg (11 lbs)
Arm Range	A:	Grip Rotation	290 deg
	R:	Horizontal Stroke	300 mm (11.81 in.)
	T:	Plane Rotation	290 deg
	Z:	Vertical Stroke	120 mm (4.72 in.)
Speed	A:	Grip Rotation	150 deg/sec
	R:	Horizontal Stroke	750 mm (30 in.)/sec
	T:	Plane Rotation	90 deg/sec
	Z:	Vertical Stroke	360 mm (14 in.)/sec
		Combined Maximum	1400 mm (55 in.)/sec
Repeatability			±0.025 mm (0.0010 in.)
Resolution	A:	Grip Rotation	0.005 deg
	R:	Horizontal Stroke	0.025 mm (0.001 in.)
	T:	Plane Rotation	0.003 deg
	Z:	Vertical Stroke	0.012 mm (0.0005 in.)
Weight			108 kg (238 lbs)

• Specifications are subject to change without notice.

FIGURE 8-33 Details of the RT3000-Series robot. (*Courtesy of Seiko Instruments USA, Inc.*)

The Seiko D-Tran controller is common to all Seiko D-Tran robots, controlling four closed-loop DC servo axes and providing an optional single-phase or contamination-resistant controller. The teach terminal is a soft-touch, contamination-resistant, 70-key keyboard with 40 characters by four lines of alphanumeric liquid-crystal-display (LCD).

Thermwood Robotics (Dale, IN)

PR Series of Spray-Painting Robots

Thermwood has been providing industrial robots and computer-controlled automation for many years. More than 500 companies use Thermwood systems. The PR Series boasts an impressive array of today's advanced technological features from Path Perfect, an online editing system, to advanced electrical, electronic, and hydraulic diagnostic systems (Figure 8-34). Features include a menu-driven operating system, advanced digital servos, program queue, and fine synchronization. See the work envelopes for these robots in Figure 8-35.

FIGURE 8-34 Thermwood's Series PR 16, 24, and 36 spray-painting robots. (*Courtesy of Thermwood Corporation Robotics Division.*)

PR-Series robots are not delicate laboratory machines; they are designed to offer years of service in the hostile, rugged environment of a factory. Simple rugged construction borrows techniques from high-performance jet fighter aircraft to provide superb performance and extreme reliability. All-electric and hydraulic systems are located in the base, eliminating the problems caused by wires and hoses flexing in the arms. Oversized bearings, bushings, and push rods ensure a long, reliable operating life. See Spec Sheet 12 for more details on this series of spray-painting robots.

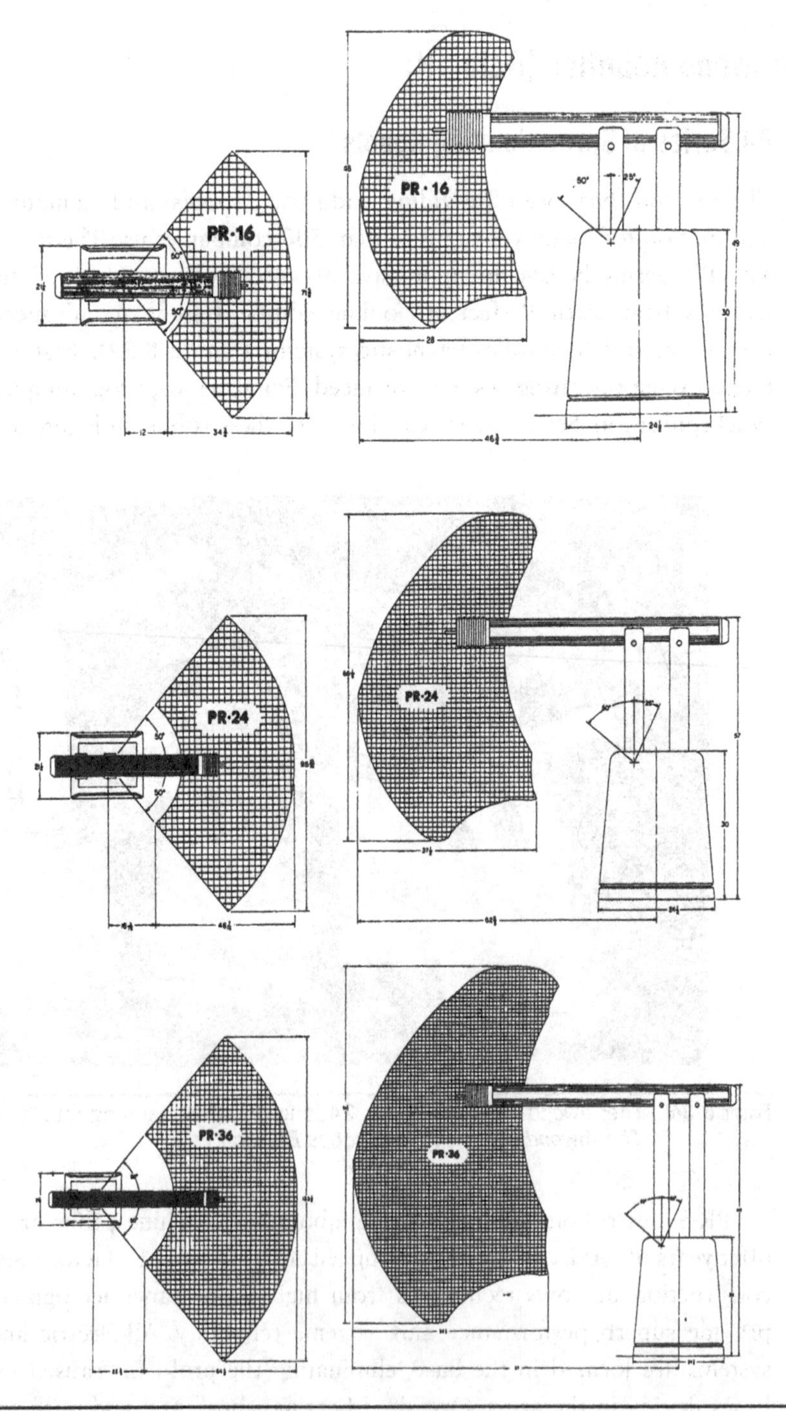

Figure 8-35 Work envelopes for Thermwood's Series PR 16, 24, and 36 robots. *(Courtesy of Thermwood Corporation Robotics Division.)*

Unimation, Inc., a Westinghouse Company (Danbury, CT)

Unimate Series 2000 and 4000

The Unimate Series 2000 and 4000 industrial robots are among the most widely used in the world (Figure 8-36). See Spec Sheet 13. With many years of continued improvement, they are highly reliable, easy-to-use robots. They have been used successfully for spot welding, die casting and investment casting, materials handling, and many other applications. See Spec Sheet 14.

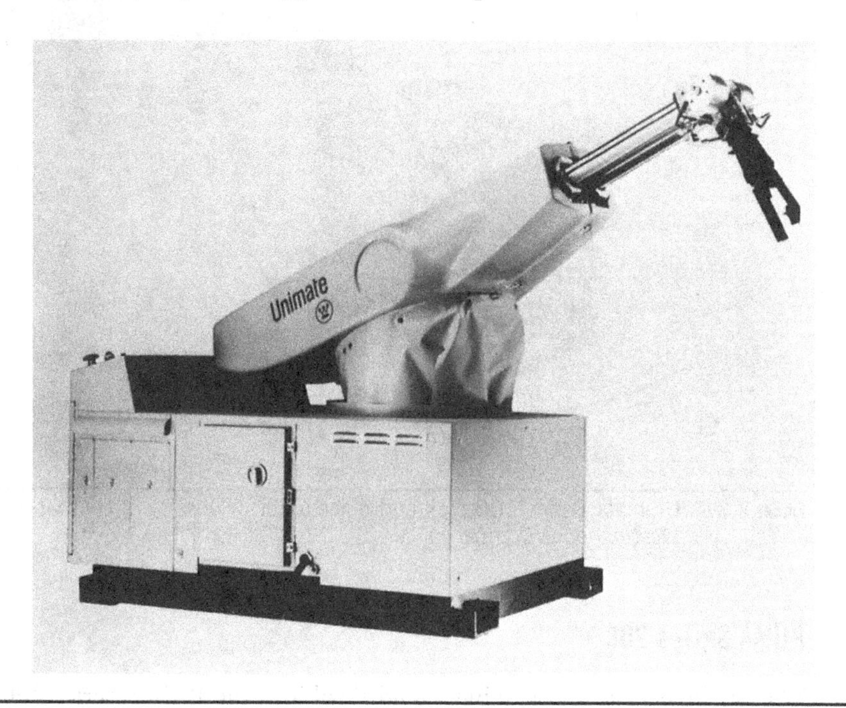

FIGURE 8-36 Unimate heavy-duty pick-and-place robot. (*Courtesy of Unimation, a Westinghouse Company.*)

The Unimate Series 2000 and 4000 robots feature six fully programmable axes of motion and are designed for high-speed handling of parts weighing up to 300 pounds. The dedicated electronic control is regarded as one of the simplest controllers available in the industry for teaching and operating industrial robots.

Unimate Series 100

The Unimate Series 100 robot is designed for high-speed small parts handling in assembly, inspection, packaging, material-transfer, and process applications in the automotive, appliance, electronics, aeronautical, and consumer products industries. Its high performance offers the advantages of improved production rates, consistent quality, low reject rates, and flexible operation (Figure 8-37). See Spec Sheet 15.

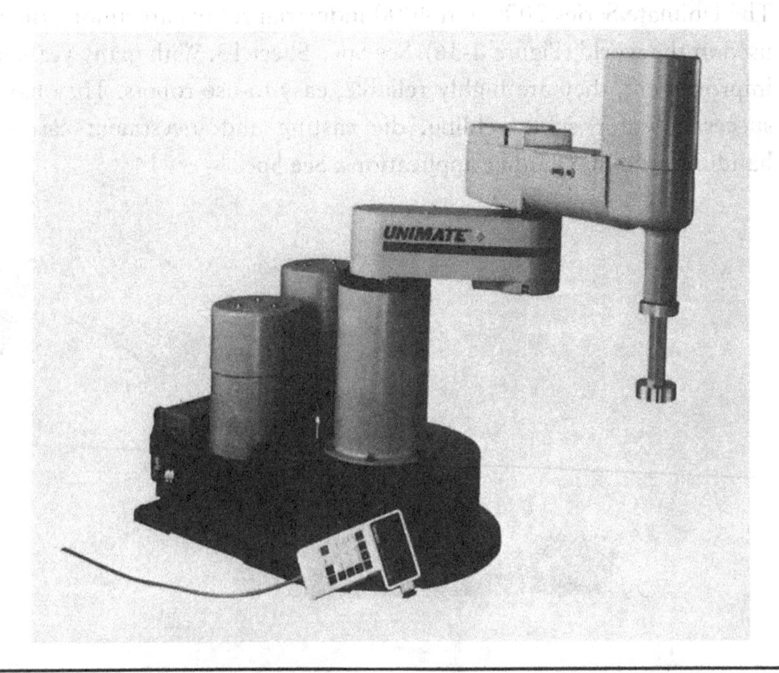

FIGURE 8-37 Unimate Series 100 pick-and-place robot. (*Courtesy of Unimation, a Westinghouse Company.*)

PUMA Series 200

With its high speed, repeatability, and flexibility, the PUMA Series 200 robot is suited to a wide range of small-parts-handling applications, and VAL control makes it easy to design application programs to carry out the most difficult robotic tasks (Figure 8-38).

Current assembly applications include automotive instrument panels, small electric motors, printed circuit boards, and subassemblies for radios, television sets, appliances, and more. Other applications include packaging functions in the pharmaceuticals, personal care, and food industries. Palletizing of small parts, inspection, and electronic parts handling in the computer, aerospace, and defense industries round out the present installed base. See Spec Sheet 16.

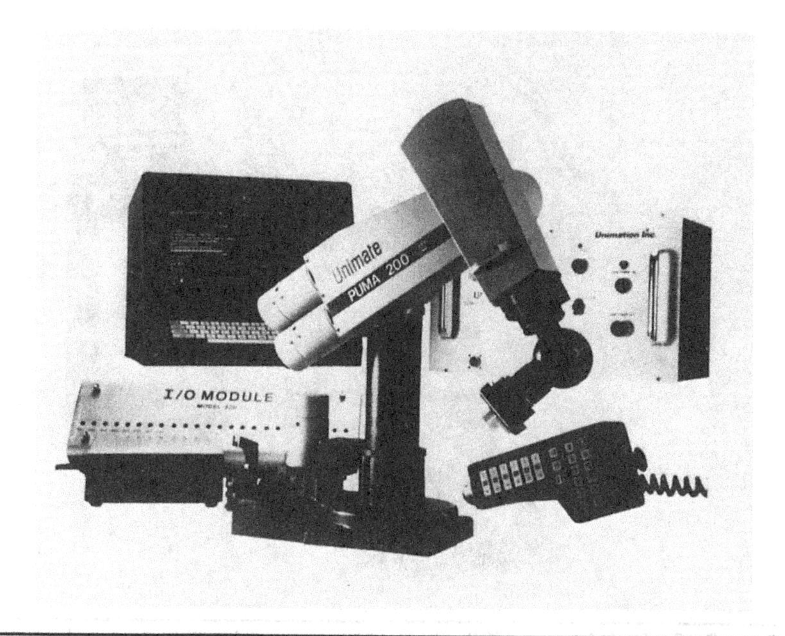

Figure 8-38 PUMA Series 200 with peripheral equipment. (*Courtesy of Unimation, a Westinghouse Company.*)

PUMA Series 700

PUMA Series 700 electric robots are designed with flexibility and durability to ensure long life and optimal performance in even the harshest, most demanding manufacturing environments (Figure 8-39). Specific customer needs for either higher payload or extended reach determine which model is suitable for a particular task (Figure 8-40). See Spec Sheet 17.

Program

A sample pick-and-place program follows:

```
000 SP   2   Sets arm speed to 1/6 of maximum speed.
001 MP   0   Moves arm to point 0 (system defined as
             the home position).
002 DE  10   Delays 1 second.
003 SP   4   Sets arm speed to 1/3 of maximum speed.
004 CS   1   Branches to subroutine 1.
005 MP   2   Moves to point 2 (first PLACE point).
006 CS   2   Branches to subroutine 2.
007 CS   1   Branches to subroutine 1.
```

FIGURE 8-39 PUMA Series 700 robot with control unit and teach pendant. (*Courtesy of Unimation, a Westinghouse Company.*)

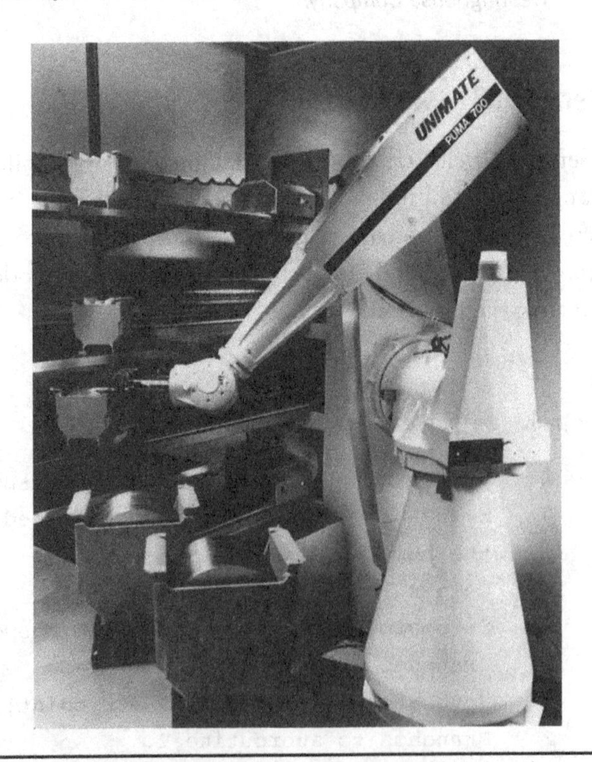

FIGURE 8-40 PUMA 700 series robot in action. (*Courtesy of Unimation, a Westinghouse Company.*)

```
008 MP   3    Moves to point 3 (second PLACE point).
009 CS   2    Branches to subroutine 2.
010 CS   1    Branches to subroutine 1.
011 MP   4    Moves to point 4 (third PLACE point).
012 CS   2    Branches to subroutine 2.
013 GO   0    Branches to label 0 (system defined as the start
              of the program).
014 LB   1    Defines the beginning of subroutine 1
              (PICK subroutine).
015 MP   1    Moves to point 1 (PICK point).
016 GR   0    Opens gripper.
017 GR   3    Lowers gripper.
018 WT   6    Waits for input signal to become true,
              then continues program.
019 GR   1    Closes gripper.
020 GR   2    Raises gripper.
021 WT   5    Waits for input signal to become true,
              then continues program
022 RE        Returns to original program and continues
023 LB   2    Defines the beginning of subroutine 2
              (PLACE subroutine).
024 GR   3    Lowers gripper.
025 WT   6    Waits for input signal to become true,
              then continues the program
026 GR   0    Opens gripper.
027 GR   2    Raises gripper.
028 RE        Returns to original program and continues.
```

Yaskawa America, Inc. (Northbrook, IL)

Motoman L-Series IA Robot

The Motoman L-Series IA robot has a maximum reach of 1,424 millimeters (56.06 inches). The manipulator provides 4 square meters (140 square feet) of volume of working range, the largest work area for this class of industrial robot (Figure 8-41).

The lightweight configuration of the manipulator, along with the sophisticated controller and its high-speed calculations, ensures accurate, smooth, and high-speed motions over the whole range of its operation. By using a cathode-ray tube

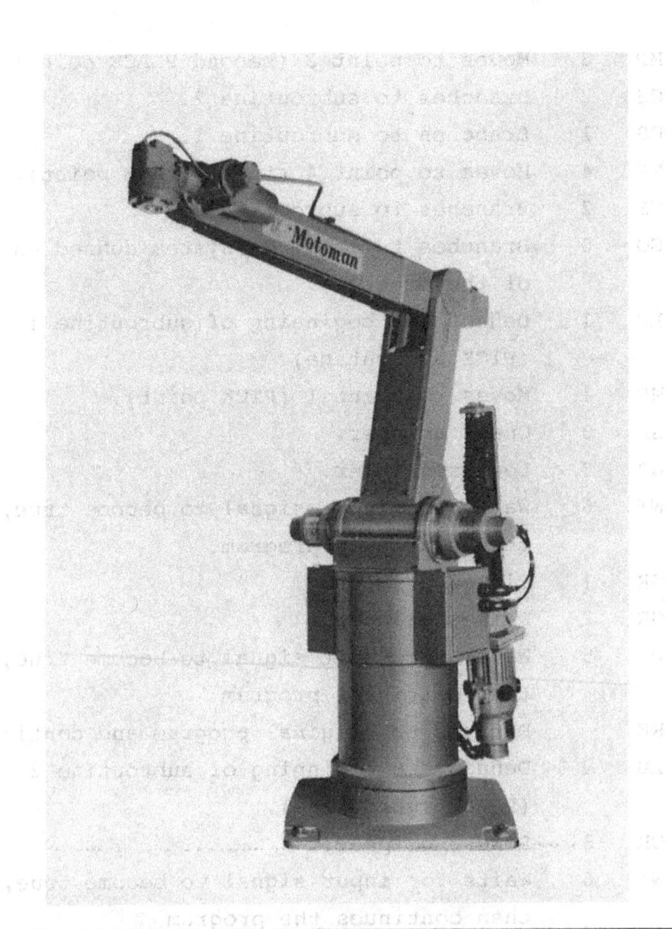

FIGURE 8-41 Motoman L-Series IA robot. (*Courtesy of Yaskawa America, Inc.*)

(CRT), the new RX controller displays various data such as job data, alarm contents, working time, position data, and diagnosis (Figure 8-42).

Advanced functions such as three-dimensional (3D) shift, mirror-image shift, arc sensor COM-ARC, and palletizing are available as options. By using 64 input and 31 output signals, 64 internal relays, and many instruction functions, sophisticated control of peripheral equipment is obtainable. By connecting the controller to a cassette tape recorder and to a printer, taught data can be easily presented and modified, if necessary. The robot can be mounted in the standard way, upside down, or on a wall. Manipulators are available for the various types of mounts. The main control unit can be separated from an operation panel by as much as 80 feet. A teach-lock model, a machine-lock function, an interference-free function, and an in-guard safety mode are built in. See Spec Sheets 18, 19, and 20 for details on the IA (which is used by Hobart Brothers Company and is shown in

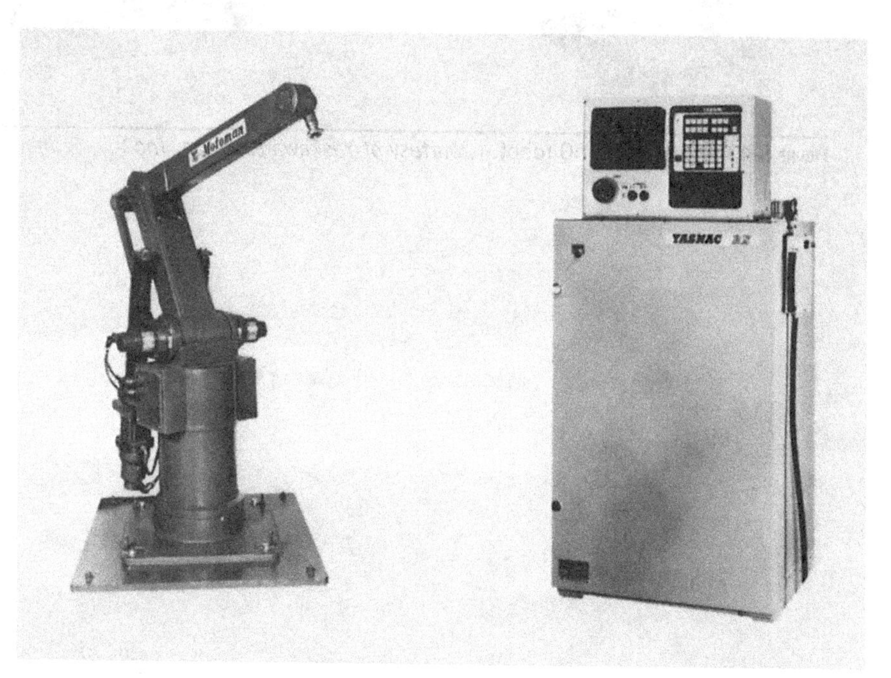

FIGURE 8-42 Motoman robots with YASNAC RX controller. (*Courtesy of Yaskawa America, Inc.*)

Figure 5-17 in a welding unit) and the S50 (which is shown in Figure 8-43). These are only two of the many Motoman robots available from one of Japan's most versatile robot manufacturers.

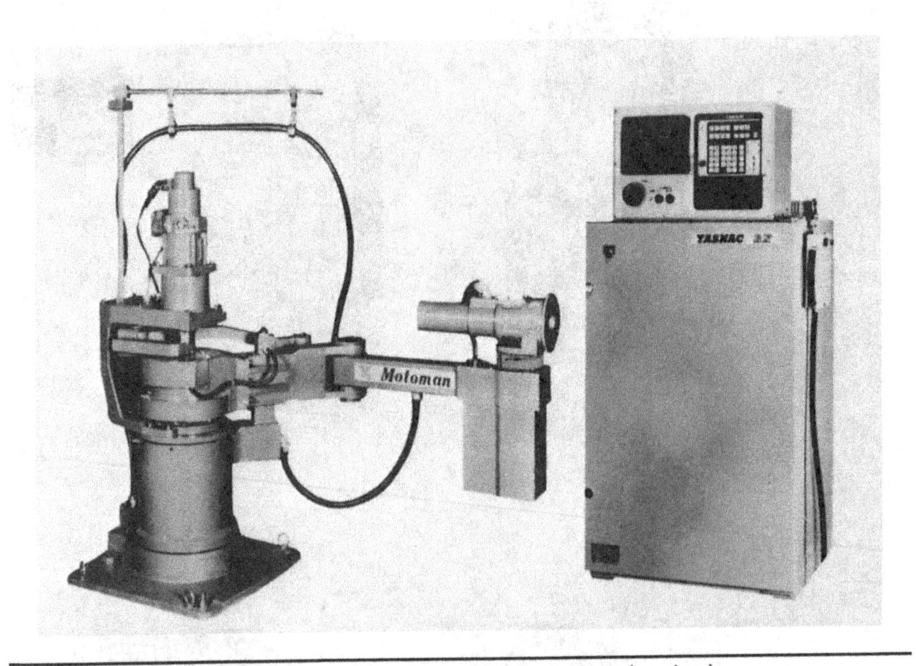

FIGURE 8-43 Motoman S50 robot. (*Courtesy of Yaskawa America, Inc.*)

Load Capacity
Load 10 in (250 mm) out and 5 in (125 mm) offset
from tool mounting plate . . . . 70 lb (32 kg)

Number of axes, configuration, servo system, control type
Number of servoed axes 6
Configuration . . . . Articulated
Drive system . . . . DC Motor
Position feedback . . . . Brushless resolver
Control type . . . . Controlled path at Tool Center Point

Positioning repeatability
Repeatability to any previously taught point . . . . ±0.010 in (±0.25 mm)

Range of motion, velocity
Base sweep . . . . 270 deg
Horizontal reach to tool mounting plate . . . . 99 in (2515 mm)
Vertical reach to tool mounting plate . . . . 137 in (3480 mm)
Pitch and Yaw relative to forearm . . . . 238 deg
Roll relative to wrist . . . . 900 deg
Nominal velocity at Tool Center Point (TCP) . . . . 25 ips (635 mm/sec)
Base slew rate . . . . 95 deg/sec

Memory capacity, I/O contacts
Data area size/Average number of data points . . . . 48 k byte/3000
Number of input contacts/Maximum optional 16/32
Number of output contacts/Maximum optional 16/32

Floor space and approximate net weight
Robot . . . . 15ft² (1.4 m²)/5,250 lb (2385 kg)
Robot control . . . . 8.1ft² (0.75 m²)/1,470 lb (670 kg)

Ambient temperature . . . . 40 to 105°F (5 to 40°C)
with air conditioner option . . . . 40 to 120°F (5 to 50°C)

Power requirements . . . . 460 volt, 3ϕ, 60 Hz*
Power rating/Power required for typical cycle 23kVA/4kW

*Other voltages and 50 Hz available

Spec Sheet 1 T³746 industrial robot specifications. (*Courtesy of Cincinnati Milacron.*)

SMART 6.50 GENERAL DATA			
Robot Type		All-electric, articulated anthropomorphic configuration	
Number of Axes		6 axes (1 or 1T, 2, 3, 4, 5, 6)	
Axes Motion Range and Speed	1-Axis	Base Rotation: 270° (78°/sec)	
	1T-Axis	Base Traverse: 2,5 m (or multiples) (1 m/sec)	Option
	2-Axis	Arm Motion: 140° (80°/sec)	
	3-Axis	Fore-Arm Motion: 280° (102°/sec)	
	4-Axis	Fore-Arm Roll: 360° (138°/sec)	
	5-Axis	Yaw/Pitch: 230° (136°/sec)	
	6-Axis	Tool Roll: 360° (144°/sec)	
Reach		Horizontal 1917 mm, vertical 3000 mm	
Repeatability		± 0.4 mm	
Base Floor Space		1150 × 1150 mm	
Load Capacity	Static Load	50 kg (400 mm from wrist center), or 60 kg (330 mm)	
	Moment of Inertia	5-Axis: 8 kgm² - 6-Axis: 1.125 kgm²	
	Static Torque	5-Axis: 200 Nm - 6-Axis: 75 Nm	
Drive		Electric DC servo motors with transistor PWM amplifiers	
Position Transducers		3 KHz resolvers mounted on motor shafts	
Axes Counterbalancing (1)		Pneumatic for the 2-Axis and weight for the 3-Axis (floor and roof mounting only)	
Axes Counterbalancing (2)		Weight for the 2 and 3 Axes (universal position mounting)	Option
Safety Flange		Available (for wrist overload protection)	Option
Axes Travel Limits		Programmable software limits	
		Adjustable limit switches for 1, 2 and 3 axes	Option
		Electrical limit switche for 4 axes (fixed)	
		Energy absorbing stops for 1, 2, 3, 4, 5 axes (1 and 3 may be adjusted)	
Robot Fine Calibration Device		Available	Option
Pins kit for Robot Lifting and Rotation		Available	Option
Provision for Fork-Lift Hoisting		Standard	
Computerized Control System		Multi-microcomputer configuration	
Serial Interface		Available for host computer and/or sensor subsystems connection (RS232 or RS422)	Option
Diagnostics		Wide range of controller diagnostic functions	
Controller Size		1250 × 1580 × 930 mm	
Controller Doors Interlocks		Available	Option
Program Storage		RAM CMOS with battery back-up (minimum 50-days retentivity)	
Program Storage Capacity		16, 32, 48, or 64K bytes. Number of programs limited only by memory capacity	16K standard
I/O Interface		14 Inputs, 9 Outputs user programmable standard	
I/O Interface Extension		5 additional I/O modules available (32 Inputs or 16 Outputs per module)	Option
Back-Up Recording Unit		Portable cassette tape unit	Option
Printer for Program Listing		Available	Option
Operator's Panel		9" CRT, Alphanumeric Keyboard, and Control Panel	
Control Panel		Electronics and Motor Drives ON/OFF, Emergency Stop, Feed Hold, Axes Zero, Feed Rate Selector, Program Execution Mode Selector, Cycle Start	
Programming Terminal (Pendant)		Microcomputer based. Includes key-pad, display, emergency stop, and dead-man switch	Option
Emergency Manual Control Box		Available	Option
Programming Method (1)		Teaching through the Programming Terminal (motion, interlocks, tool functions)	
Programming Method (2)		Manual Data Input (MDI) through the Operator's Panel using the PDL Programming System	
Programming Method (3)		Edit/Teach by MDI and teaching of key-positions	
Motion Mode (1)		Point-to-Point in the Robot Coordinate System	
Motion Mode (2)		Point-to-Point with path control (straight and circular line) of the tool centre point (TCP) in the Absolute Cartesian Coordinate System	Option
Motion Mode (3)		Point-to-Point with path control of tool centre point (TCP) in the Tool Cartesian System	Option
Selection of Motion Mode		From Program, Programming Terminal, and Operator's Panel	
Program Execution		Foreward and Backward Jogging, Automatic, Step-by-Step, Continuous Repeated	
Power Requirements		380 V three-phase + 10%, − 15%, (or others on request), 48-62 Hz, 14 KVA	
Air Supply		6 bar	
Environmental Operating Range		0-45°C	
Robot Weight		1600 kg (with pneumatic balancing)	

Spec Sheet 2 SMART 6.50 general data. (*Courtesy of Comau Productivity Systems, Inc.*)

Axis No./Degree of Freedom	Translation (T)/Rotation (R)	Maximum Velocity	Load Capacity
1 (x)[1]	T up to 7.5m (24.6 ft.)	1 m/sec. (3.3 ft./sec.)	80 kg (175 lb.)
2 (y)[1]	T up to 5.0m (16.4 ft.)	1 m/sec. (3.3 ft./sec.)	80 kg (175 lb.)
3 (z)[1]	T up to 1.5m (4.9 ft.)	0.5 m/sec. (1.65 ft./sec.)	80 kg (175 lb.)
4	R ± 167.5°	3.0 rad./sec.	20 m-kg (145 ft.-lb.)[2]
5	R ± 105°	3.0 rad./sec.	20 m-kg (145 ft.-lb.)[2]
6	R ± 172°	3.0 rad./sec.	10 m-kg (72.6 ft.-lb.)[2]

Specifications are subject to change without notice.

[1]These dimensions can be exceeded to fit a wide range of requirements.
[2]Allowable torque load perpendicular to axis of rotation

Drive System
Electric or hydraulic motors with with rack and pinion drives and gear reduction

Accuracy
With 80 kg load at maximum speed

Position Accuracy: ± 0.5mm (0.020″)
Position Repeatability: ± 0.2mm (0.008″)

Control System
The RC-6 is the standard Cybotech controller, specifically designed for ease of operation, flexibility, and simple maintenance. The RC-6 incorporates diagnostic circuits that greatly reduce troubleshooting time, while providing many safety features. Refer to RC-6 brochure for more details.

Power Requirements
20 kVA, 440 VAC, 3 phase (x2 for double G80) total for power unit and control system

Cooling Water (Hydraulic only)
11 gpm max. (x2 for double G80)

Weight
4,200 - 12,000 kg (9,000 - 26,000 lb.) depending on gantry dimensions and single or double configuration

Spec Sheet 3 G80 industrial robot engineering data. (*Courtesy of Cybotech Industries Corp.*)

MAC 500 MANIPULATOR

Structure	Horizontally Articulated
Axes of Motion	4 Simultaneous – 1 Independent
Load Capacity	6.6 lbs. (3 kg) Maximum
Position Repeatability	±0.004" (±.1mm) Maximum
Position Accuracy	±0.008" (±.2mm) Maximum
Weight–Manipulator	120 lbs. (54 kg)
Base	154 lbs. (70 kg)

OPERATING RANGE OF EACH AXIS

	Stroke	Maximum Speed
(1) First Arm	0-235°	143°/sec
(2) Second Arm	0-155°	204°/sec
(Z) Vertical Wrist	4.73" (120mm)	7.9"/sec (200mm)
(R) Turning Wrist	0-380°	150°/sec
(T) Torch Angle Wrist	0-180°	100°/sec

RANGE OF EFFECTIVE WORKING SPEED

2.4 IPM (0.06 m) to 2362 IPM (60 m)

MAC 500 CONTROLLER AND POWER UNIT

Weight–Control & Power Unit	319 lbs. (145 kg)
Teaching Method	"Direct" manual teaching MDI Teaching Box Operation (Optional)
Path Control	Continuous Path Control by PTP teaching
Interpolation	Linear or Circular
Number of Controlled Axes	4 Simultaneous Axes T Axis is independently controlled
Position Control	Digital Closed Loop
Position Detection Method	Pulse encoder, orgin position alignment
Speed Control	Constant Linear Speed Control
Memory/Memory Capacity	IC Memory/16K 500 steps/500 sequences
Battery Backup	Two Weeks (Rechargable)
Number of Programs	8
Program Size	255 Steps
Display	9 inch CRT

OPTIONAL EQUIPMENT

Teaching Box	16½' (5m) of Cable
Extended I/O Board	Input - 8 Terminals (for interlock) Output - 8 Terminals (for synchronizing)
Audio Cassette Tape Recorder Interface	External Memory for Data Saving and Data Loading
Printer Unit	Electrical Discharging Type Dot-Matrix

Spec Sheet 4 Mac 500 specifications. (*Courtesy of Mack Corporation.*)

Model		DKB3200 robot
Degrees of Freedom		6 (Beta and Gamma optional)
Payload, max including end effector weight		220 lb
Repeatability		±0.04 in.
Range (speed)	**Axes**	
	X (Horizontal, in and out)	58.3 in. (35.4 ips)
	Z (Vertical travel)	52.4 in. (35.4 ips)
	$Theta_1$ (Horizontal swing)	270° (75°/sec)
	$Theta_2$ (Main arm vertical rotation)	–
	$Theta_3$ (Forearm vertical rotation)	–
	Alpha (Wrist yaw)	300° (75°/sec)
	Beta (Wrist pitch)	120° (75°/sec)
	Gamma (Wrist rotation)	180° (75°/sec)
	Composite axes speed	–
Robot weight		3300 lb
Electric power requirement		AC 200/220v, 50/60 Hz, 2.5 KVA
Pneumatic counterbalancing (where applicable)		Pneumatic counterbalancing on X and Z axes

–All **GCA**/DK pedestal robots have a DC servomotor drive system

–Pneumatic power requirement (where applicable) is 85 psi.

Spec Sheet 5 GCA pedestal model robot specifications. (*Courtesy of GCA Corporation/Industrial Systems Group.*)

Technical Information

System Operating Specifications

Coordinate system	Cartesian, Joint, Region	
Degrees of freedom	4 (All servoed)	
Positioning control system	Point-to-point, circular, straight-line, via-point	
Digital I/O ports (Configurable)	48 Standard, 192 Maximum	
RS-232 Asynchronous Communications Port	4 Standard (1 for Pendant), 4 Optional	
RS-422 Asynchronous Communications Port	1 Optional (Via Combination Adapter II)	
Application program control	Operator Control Panel, Hand-Held Pendant, Host Communications	
Point teaching method	Hand-Held Pendant	
Maximum number of selectable programs	Seven via Operator Control Panel	
Primary storage	512KB Random Access Memory, expandable to 640KB	
Secondary storage	1.2MB Diskette (Standard), 1.2MB Diskette (Optional) 360KB Diskette (Optional), 20MB Fixed Disk (Optional)	
Power requirements	IBM 7575 receives DC power from 7572 IBM 7572 - 220 - 240 Vac at 47 - 63 Hz IBM 7532/310 - 220 - 240 Vac at 47 - 63 Hz (Switch selectable to 100 - 130 Vac at 47 - 63 Hz)	
System operating temperature	0°C - 40°C (32°F - 104°F)	
System relative humidity	8% - 80% Non-condensing	
Weight	IBM 7575	71 kg (156 lb)
	IBM 7572	48 kg (106 lb)
	IBM 7532/310	21.3 kg (47 lb)

All specifications are subject to change without notice. Performance is based on fixed operating conditions.

Manipulator Performance Specifications

X and Y axis repeatability	±0.025 mm (±0.001 in.)	
Maximum XY speed	5.1 m/sec (200 in./sec)	
Maximum Z speed	575 mm/sec (22.6 in./sec)	
Maximum roll speed	480 deg/sec	
Maximum payload (Reduced accelerations)	5 kg (11 lb)	
Symmetrical workspace:	Outside radius	550 mm (21.6 in.)
	Inside radius	222 mm (8.7 in.)
Theta 1 axis:	Arm length	325 mm (12.8 in.)
	Rotation	-120° - +120° (±1°)
Theta 2 axis:	Arm length	225 mm (8.9 in.)
	Rotation	-137° - +137° (±1°)
Z axis:	Stroke	150 mm (5.9 in.)
Roll axis:	Rotation	+3600° - -3600° (±2°) (±10 revolutions)

Spec Sheet 6 IBM 7575 manufacturing system specifications, features, and components. (*Courtesy of International Business Machines Corporation.*)

Workspace

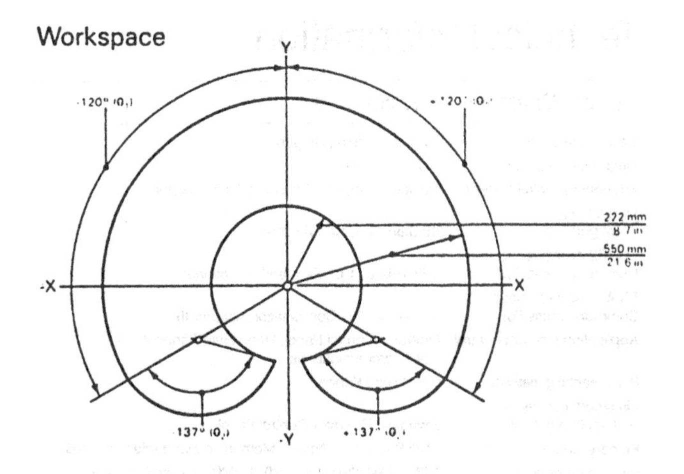

Optional Features

- 48-Point DI/DO Card (#6100)
- 48-Point DI/DO Cable (#6101)
- 4-Port RS-232 Asynchronous Communications Adapter (#4763)
- 4-Port RS-232 Asynchronous Communications Adapter Cable (#4704)
- 4-Port RS-232 Asynchronous Communications Adapter Rollup Cable (#6104)
- Hand-Held Pendant (#6105)
- 2-Meter Operator Panel Extension Kit (#6106)
- 3-Meter Manipulator Cable Set (#6108)
- 10-Meter Manipulator Cable Set (#6109)
- Color/Graphics Monitor Adapter (#4910)
- Serial/Parallel Adapter (#0215)
- Serial Adapter Cable (#0217)
- Serial Adapter Connector (#0242)
- Combination Adapter II (#6020)
- 128KB Memory Expansion for Combination Adapter II (#6040)
- Combination Adapter II Cable (#6001)
- 128KB Memory Expansion (#0209)
- 20MB Fixed Disk Drive (Customer-installed) (#6019)
- Second 1.2MB High-Capacity Diskette Drive (#6038)
- 360KB Dual-Sided Diskette Drive (Second drive) (#6039)

Other Supported Components

- IBM 5532 Industrial Color Display
- IBM 5533 Industrial Graphics Printer
- IBM 7534 Industrial Graphics Display

Spec Sheet 6 *(Continued)*

Technical Information

System Operating Specifications

Coordinate system	Cartesian, Joint, Region
Degrees of freedom	4 (All servoed)
Positioning control system	Point-to-point, circular, straight-line, via-point
Digital I/O ports (Configurable)	48 Standard, 192 Maximum
RS-232 Asynchronous Communications Port	4 Standard (1 for Pendant), 4 Optional
RS-422 Asynchronous Communications Port	1 Optional (Via Combination Adapter II)
Application program control	Operator Control Panel, Hand-Held Pendant, Host Communications
Point teaching method	Hand-Held Pendant
Maximum number of selectable programs	Seven via Operator Control Panel
Primary storage	512KB Random Access Memory, expandable to 640KB
Secondary storage	1.2MB Diskette (Standard), 1.2MB Diskette (Optional) 360KB Diskette (Optional), 20MB Fixed Disk (Optional)
Power requirements	IBM 7576 receives DC power from 7572 IBM 7572 - 220 - 240 Vac at 47-63 Hz IBM 7532/310 - 220 - 240 Vac at 47 - 63 Hz (Switch selectable to 100 - 130 Vac at 47 -63 Hz)
System operating temperature	0°C - 40°C (32°F - 104°F)
System relative humidity	8% - 80% Non-condensing
Weight	IBM 7576 76 kg (167 lb) IBM 7572 48 kg (106 lb) IBM 7532/310 21.3 kg (47 lb)

All specifications are subject to change without notice. Performance is based on fixed operating conditions.

Manipulator Performance Specifications

X and Y axis repeatability	±0.050 mm (±0.002 in.)	
Maximum XY speed	4.4 m/sec (173 in./sec)	
Maximum Z speed	575 mm/sec (22.6 in./sec)	
Maximum roll speed	480 deg/sec	
Maximum payload (Reduced accelerations)	10 kg (22 lb)	
Symmetrical workspace:	Outside radius	800 mm (31.5 in.)
	Inside radius	300 mm (11.8 in.)
Theta 1 axis:	Arm length	400 mm (15.7 in.)
	Rotation	-120° - +120° (±1°)
Theta 2 axis:	Arm length	400 mm (15.7 in.)
	Rotation	-136° - +136° (±1°)
Z axis:	Stroke	250 mm (9.8 in.)
Roll axis:	Rotation	+3600° - -3600° (±2°) (±10 revolutions)

Spec Sheet 7 IBM 7576 manufacturing system specifications, features, and components. (*Courtesy of International Business Machines Corporation.*)

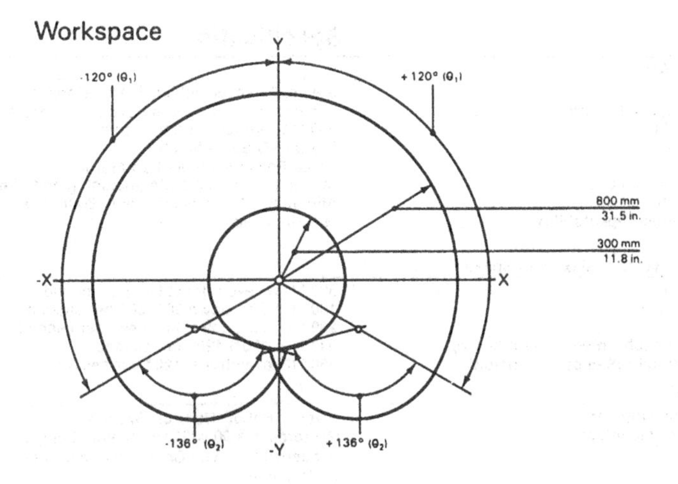

Optional Features

- 48-Point DI/DO Card (#6100)
- 48-Point DI/DO Cable (#6101)
- 4-Port RS-232 Asynchronous Communications Adapter (#4763)
- 4-Port RS-232 Asynchronous Communications Adapter Cable (#4704)
- 4-Port RS-232 Asynchronous Communications Adapter Rollup Cable (#6104)
- Hand-Held Pendant (#6105)
- 2-Meter Operator Panel Extension Kit (#6106)
- 3-Meter Manipulator Cable Set (#6108)
- 10-Meter Manipulator Cable Set (#6109)
- Color/Graphics Monitor Adapter (#4910)
- Serial/Parallel Adapter (#0215)
- Serial Adapter Cable (#0217)
- Serial Adapter Connector (#0242)
- Combination Adapter II (#6020)
- 128KB Memory Expansion for Combination Adapter II (#6040)
- Combination Adapter II Cable (#6001)
- 128KB Memory Expansion (#0209)
- 20MB Fixed Disk Drive (Customer-installed) (#6019)
- Second 1.2MB High-Capacity Diskette Drive (#6038)
- 360KB Dual-Sided Diskette Drive (Second drive) (#6039)

Other Supported Components

- IBM 5532 Industrial Color Display
- IBM 5533 Industrial Graphics Printer
- IBM 7534 Industrial Graphics Display

SPEC SHEET 7 *(Continued)*

Item	Specification
Robot Arm	
Axis	3 Axis, Torso, Shoulder, Elbow standard
Clearance Required	Sphere, 80-inch radius (excluding end-effector)
Weight	350 lbs. (without optional base)
Drive	Digital AC Servo Motor Three Phase AC Induction Motor
Maximum Load	50 lbs. at the face plate (including end-effector)
Reach	Minimum 20 inches/Maximum 80 inches
Position Repeatability	± .040 inches (± 1 mm)
Motion, Range & Maximum Speed	
Torso	360° (1½ revolutions) 60° per second
Shoulder	180° (up 90°, down 90°) 30° per second
Elbow	230° (up 90°, down 140°) 45° per second
Wrist-pitch * (See options below)	240° (± 120°) 120° per second
Wrist-roll * (See options below)	360° (2 revolutions) 120° per second
Control	
Microcomputer (Torso mounted)	Master Robot Control System (Motorola 68000 w/32KB Memory Computer) 3 Independent Axis Control System Computers (2 Optional) 3 AC Servo Output Drive Computers (2 Optional) Dedicated Safety Computer
Input/Output Ports	8 bit input/8 bit output (optional 16/16)
Communications	2 Independent RS232C Connections
Programming	
Language	IRI Robot Command Language (RCL)
Type	Basic-like language structure
Development	On Line Interpretive On line editor and teach-playback
Environment	
Ambient Temperature	50°C (122°F) Max.
Relative Humidity	90% Max. — non-condensing
Power Inputs	
Electrical	208 VAC, 3 phase, 60Hz-4 wire-30A
Air	80-100 psi, 20-130 scfm (depends on application) ½ inch supply line
Options	
*2 Axis Wrist-Pitch & Roll	
Base	48 inches high, 24 × 24 inch base
Universal End-effector	See Gripper Product Bulletins
Memory	Up to 128 KB
Move Pendant	
System Terminal	Keyboard, hard copy printer
Load Terminal	5¼" Floppy Disk
Utility I/O Module	8 digital in, 8 digital out (16/16 optional), E-Stop Switch
Base Installation Kit	
Pneumatic Installation Kit	FRL-Filter, Regulator, Lubricator
Back-up Battery	Supply to RAM memory
Power Input	220 VAC, 3 phase, 50Hz-4wire-30A

(Dimensions and specifications subject to change.)

Spec Sheet 8 IRI M50E AC servo robot specifications. (*Courtesy of International Robomation/Intelligence.*)

Description:
Robotic System with five-axis articulated arm plus controls for the cable-operated mechanical gripper and 2 additional axes (through optional auxiliary motors.)

Performance:
Maximum Payload: 3 lb. (1.362 kg) including gripper.

Maximum Speed at End of Arm: 51 inches/sec (1295mm/sec) with up to a 1.0 lb. (454 gm) payload. Maximum speed at 3 lb. (1.362kg) payload is 6 inches/sec.

Reach: 18.4" (467mm) spherical radius.

Repeatability: ± 0.015" (.38mm) with total payload up to 1.5 lb. (681 gm).

Axes: 5 axes plus controls for the mechanical gripper and 2 additional axes (through optional auxiliary motors.)

Motion Range & Speed:			
	Base rotation	345°	106°/sec
	Shoulder	145°	119°/sec
	Elbow	135°	202°/sec
	Pitch	180°	315°/sec
	Roll	540°	315°/sec

Control:
Teaching Methods: Teach Control with LED display & multifunction key pad or host computer.

Workcell Interface: Eighteen channels for optically isolated relay input/output at factory level voltages.

Computer Interface: Dual RS-232C asynchronous serial communications interface. Operates with any host computer language supporting ASCII I/O.

Control Method: Semi-closed loop control with automatic homing.

Drive: Electrical stepper motors—half step pulses.

Arm movement: Coordinated movement—all joints move simultaneously. Point-to-point or integrated path moves.

Programming Capacity: 227 program steps in non-volatile memory.

Controller: Real time 6502A microprocessor controls all functions.

Installation:
Operating Environment: Ambient temperature limits: 55°F to 104°F(13°C to 40°C).

Mounting: The robot may be mounted and operated right-side up or upside down with the base in a horizontal plane.

Power Requirements: 115VAC @ 1.5A; 220VAC @ .75A; 100VAC @ 1.7A; 47-400Hz.

Weight: Robot Arm: 30 lbs. (13.6 kg)
System Control: 30 lbs. (13.6 kg)
Total Shipping Weight: 75 lbs. (34.1 kg)

This data is based on laboratory tests conducted by Microbot. Specifications are subject to change without notice.

Spec Sheet 9 Alpha II system specifications. (*Courtesy of Microbot, Inc.*)

Technical Data	FA	FB	FC
Axes of motion (standard)	3	3	3
Payload capacity	250 lbs. (114 kg)	600 lbs. (273 kg)	2,000 lbs. (909 kg)
Repeatability	±.050" (±1.27 mm)	±.050" (±1.27 mm)	±.080" (±2.0 mm)
Motion			
Horizontal swing	300°	300°	300°
Vertical stroke			
• Standard	60" (1524 mm)	60" (1524 mm)	60" (1524 mm)
• Optional	48" (1219 mm)	48" (1219 mm)	48" (1219 mm)
Extend/retract stroke			
• Standard	60" (1524 mm)	60" (1524 mm)	60" (1524 mm)
• Optional	48" (1219 mm)	48" (1219 mm)	48" (1219 mm)
Optional wrist axes	Up to 3	Up to 3	Up to 3
Optional traverse base	Yes	Yes	Yes
Coordinate system	Cylindrical	Cylindrical	Cylindrical
Weight of robot	2,400 lbs. (1090 kg)	4,000 lbs. (1815 kg)	6,000 lbs. (2720 kg)
Hydraulic power unit (remote with 20' hose standard)	1,000 lbs. (454 kg)	1,350 lbs. (613 kg)	1,500 lbs. (680 kg)
Maximum horizontal reach (see chart 3 axis robot)	78.5" (1994 mm)	82" (2083 mm)	89.5" (2273 mm)
Servo/non-servo	Servo	Servo	Servo
Drive	Electro-hydraulic	Electro-hydraulic	Electro-hydraulic
Feedback	Absolute Resolver	Absolute Resolver	Absolute Resolver
Ambient temperatures for Robot and Hydraulic Power Unit			
• Standard water cooled	40°F-120°F (4°C-49°C)	40°F-120°F (4°C-49°C)	40°F-120°F (4°C-49°C)
• Optional air cooled	40°F-100°F (4°C-38°C)	40°F-120°F (4°C-38°C)	40°F-100°F (4°C-38°C)
• Optional refrigeration cooled	40°F-120°F (4°C-49°C)	40°F-120°F (4°C-49°C)	40°F-120°F (4°C-49°C)
Floor space required for mounting base	13 sq. feet (1.2 sq. m)	15 sq. feet (1.4 sq. m)	24 sq. feet (2.2 sq. m)

Note: Reach, speed, acceleration and deceleration all affect the above payloads and repeatability. Please consult Prab Robots, Inc., Kalamazoo, Michigan, for further information regarding your application. Payload calculation must include gripper, additional wrist axes (if required) and part weight.

Spec Sheet 10 PRAB models FA/FB/FC specifications. (*Courtesy of PRAB Robots, Inc.*)

AXIS OF MOTION					
AXIS	DESCRIPTION	FUNCTION	STROKE	MOTION/SPEED	AXIS DRIVE
A	Base rotate	Horizontal rotation	180°	180° per second	Rack and pinion, powered by dual air cylinders
B	Lift	Vertical linear	3"	12 to 24 i.p.s. depending upon load	Air cylinder
C	Extend	Horizontal linear	12"	12 to 24 i.p.s. depending upon load	Air cylinder
D	Wrist	Wrist rotation	90° 180°	¼ second ½ second	Cam actuated
E	Grasp	Open/close	±18°	⅛ second	Air cylinder actuated draw bar

Weight of robot	130 pounds
Max. load capacity at wrist	5 pounds
Repeatability	±.005
Axis drive method	Pneumatic 60 to 120 p.s.i.
Power source	115V/60Hz, 240V/50Hz optional
Axis motion detectors	Proximity switches on base rotate, lift and extend
Adjustable mechanical stops	On base rotate, lift and extend
Operating temperature	+20°F to +140°F
CONTROL	Electronic, microprocessor with remote portable teaching module
Memory capacity	Up to 300 steps
Program method	Push button via portable teach module
Maximum time interval between steps	10.79 minutes
Auxiliary equipment inputs/outputs	6 outputs and 8 inputs each rated at 115V/60Hz. Other voltages available

TORQUE DEVELOPED AT WRIST	90° WRIST ROTATION		180° WRIST ROTATION	
	P.S.I.G.	INCH LBS	P.S.I.G.	INCH LBS
	60	38	60	19
	80	54	80	27
	100	70	100	35

GRIPPER CLOSER FORCES	LINEAR		PIVOT (AT END OF 2" FINGER TOOLING)	
	P.S.I.G.	FORCE IN LBS.	P.S.I.G.	FORCE IN LBS.
	60	19	60	20
	80	26	80	27
	100	33	100	34

Spec Sheet 11 Motion Mate robot technical specifications. (*Courtesy of Schrader-Bellows.*)

Number of Axes: 6 (Additional Axes Optional)

Configuration: Jointed Arm

Mounting Position: Floor (Inverted Optional)

Coordinate System: Jointed Arm

Drive System: Hydraulic/Mechanical

Horizontal Stroke: PR16-28" PR24-37" PR36-52"

Horizontal Reach: PR16-46" PR24-62" PR36-87"

Vertical Stroke: PR16-48" PR24-66" PR36-91"

Vertical Reach: PR16-62" PR24-78" PR36-100"

Base Rotation: 100 Degrees

Wrist Roll: 450 Degrees

Wrist Movement: 120 Degree Cone

Control System: Closed loop digital servo microcomputer based solid state

Memory Type: Semiconductor/disc memory option

Memory Size: 5 minutes continuous path (Expandable to over 5 hr.) point to point dependent on program

Number of Programs: 9 (Expandable to 500)

Random Selection: Yes

Operating Modes: Continuous path/Point to point

Programming: Lead through teach

Editing: Yes

Adjustable Playback Speed: Yes

Outputs: 8

Inputs: Remote Program Select Lines

Interface Hardware: Terminal strip for external wiring RS232C Selectable baud rate asynchronous RS232C synchronous

Ambient Conditions: 40 to 120 F-5% to 95% humidity, non-condensing

Power Required: Maximum 10 kw

Spec Sheet 12 PR series robot specifications. (*Courtesy of Thermwood Corporation Robotics Division.*)

TYPICAL PRODUCTION ENVIRONMENT STRUCTURES

BASIC CONFIGURATION

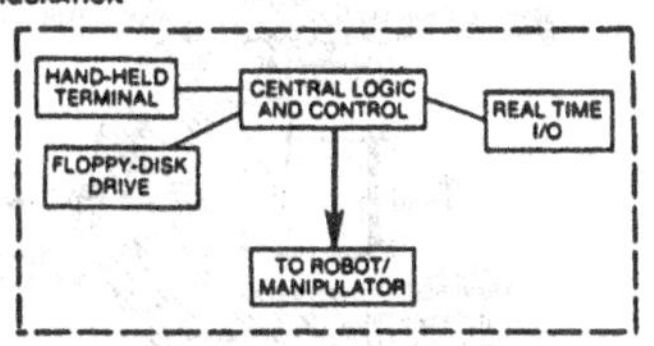

STANDARD CONFIGURATION

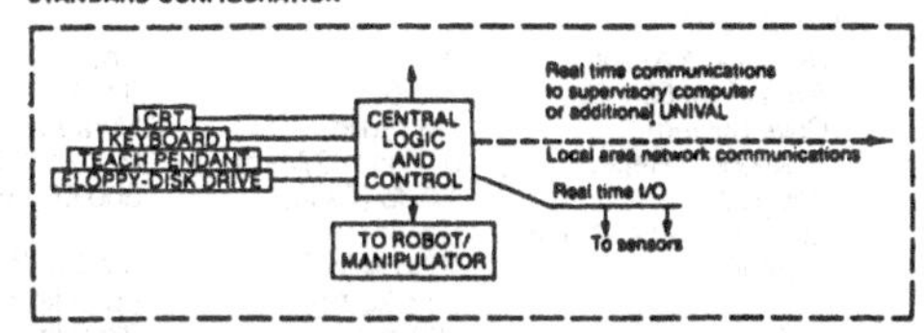

UNIVAL SYSTEM PERIPHERALS

System Terminal. Comprised of the CRT and keyboard. Program and diagnostic information are displayed via the CRT. Keyboard offers direct user interface for generating and modifying VAL III programs, as well as capabilities to initiate functions and recall stored data.

Teach Pendant. A microprocessor-based device used as a programming aid to teach location points and to move the robot arm. An alphanumeric display can show error messages and the names of previously stored locations. Individual push buttons provide manipulation of the robot arm in Joint, World, Tool, and Workpiece modes. While in the teach mode, the user gains total control over robot motion, coordinating all axes for movement along a straight line, and the ability to maintain a desired tool orientation relative to the workpiece.

Hand-Held Terminal. A computerized device used as a simplified operator interface for the BASIC configuration. It has facilities to: position the robot arm in Joint, World, or Tool modes: create and execute programs and subprograms; store and load programs on the floppy disk; and program I/O signals. The alphanumeric display informs the operator of program status and prompts the operator through the programming sequence. Simple push-button programming provides fast set-up without requiring programming experience.

Floppy Disk Drive. Provides a high-density, quick-access storage medium for programs and robot locations. It is a 1 megabyte, double-sided, double density diskette system which operates at 9600 baud and can store up to 10,000 program steps. It offers reliable program back-up, and allows fast program and data file referencing through VAL III commands.

SPECIFICATIONS

	BASIC	STANDARD
Central Processor	Motorola 68000	Motorola 68000
Language	(not applicable)	VAL III
Coordinate Frames	Joint, World, Tool	Joint, World, Tool, Workpiece
Input Lines	16 (24 VDC) or 16 (110 VAC)	32 (24 VDC) or 32 (110 VAC)
Output Lines	16 (contact closure)	32 (contact closure)
Memory Capacity	16K Bytes (CMOS) 256K Bytes (Optional)	256K Bytes (CMOS) 512K Bytes (Optional)
Memory Battery Life	30 days	30 days
Program Storage	High density floppy disk	High density floppy disk
Controller Size	49" × 31" × 26" (1245mm × 788mm × 660mm)	60" × 31" × 26" (1524mm × 788mm × 660mm) incl. CRT
Controller Weight	550 lbs. (250kg)	550 lbs. (250kg)
Number of Subroutines	Unlimited	Unlimited
Number of Nested/ Subroutine Calls	4	10
Number of Coordinate Transformations	4	Unlimited
Max. Number of Stored Points	300 (6000 Optional)	6000 (12000 Optional)
Number of Comm. Ports	1 (RS422)	1 (RS422; 4 or 8 Optional)

Spec Sheet 13 Typical production environment structures, peripherals, and specifications for Unival robot system. (*Courtesy of Unimation Incorporated, a Westinghouse Company.*)

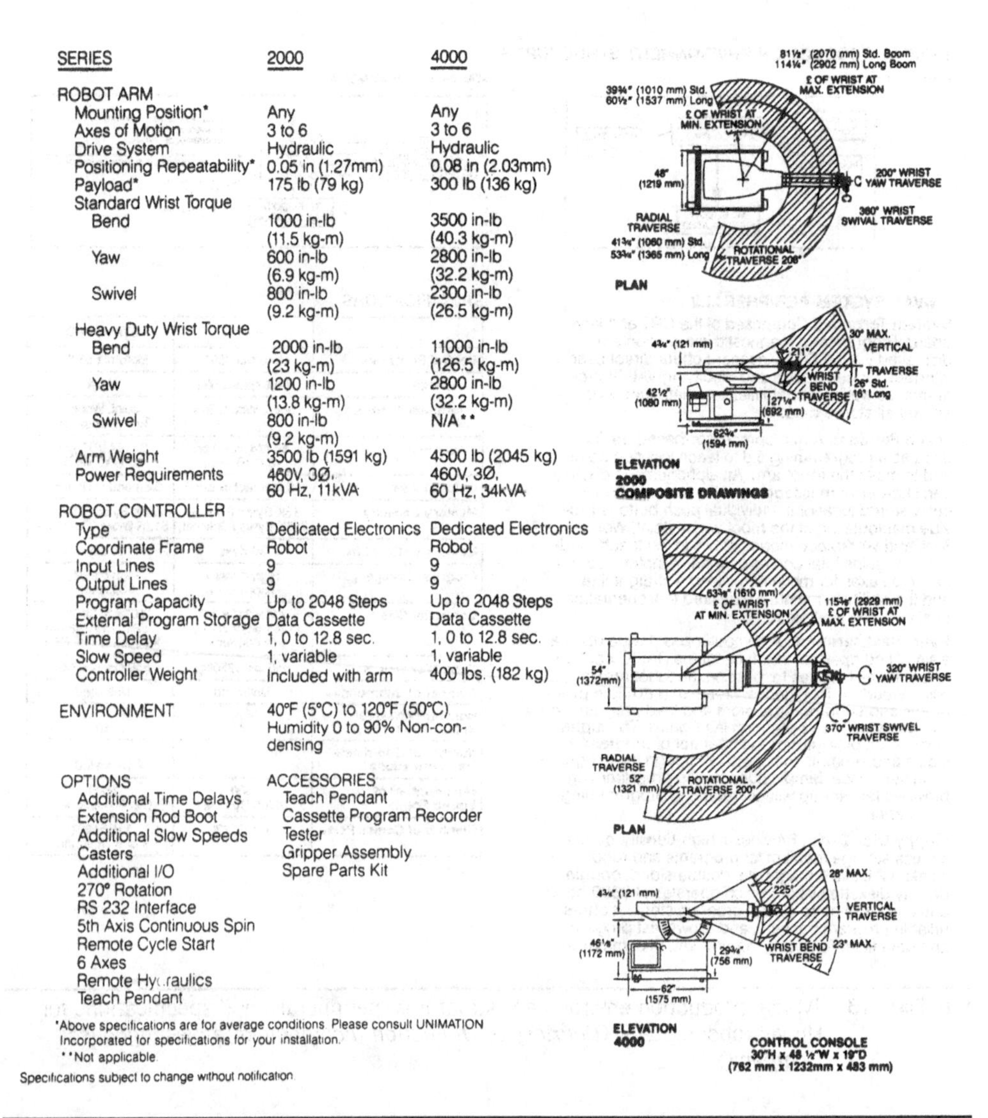

SERIES	2000	4000
ROBOT ARM		
Mounting Position*	Any	Any
Axes of Motion	3 to 6	3 to 6
Drive System	Hydraulic	Hydraulic
Positioning Repeatability*	0.05 in (1.27mm)	0.08 in (2.03mm)
Payload*	175 lb (79 kg)	300 lb (136 kg)
Standard Wrist Torque		
Bend	1000 in-lb (11.5 kg-m)	3500 in-lb (40.3 kg-m)
Yaw	600 in-lb (6.9 kg-m)	2800 in-lb (32.2 kg-m)
Swivel	800 in-lb (9.2 kg-m)	2300 in-lb (26.5 kg-m)
Heavy Duty Wrist Torque		
Bend	2000 in-lb (23 kg-m)	11000 in-lb (126.5 kg-m)
Yaw	1200 in-lb (13.8 kg-m)	2800 in-lb (32.2 kg-m)
Swivel	800 in-lb (9.2 kg-m)	N/A**
Arm Weight	3500 lb (1591 kg)	4500 lb (2045 kg)
Power Requirements	460V, 3Ø, 60 Hz, 11kVA	460V, 3Ø, 60 Hz, 34kVA
ROBOT CONTROLLER		
Type	Dedicated Electronics	Dedicated Electronics
Coordinate Frame	Robot	Robot
Input Lines	9	9
Output Lines	9	9
Program Capacity	Up to 2048 Steps	Up to 2048 Steps
External Program Storage	Data Cassette	Data Cassette
Time Delay	1, 0 to 12.8 sec.	1, 0 to 12.8 sec.
Slow Speed	1, variable	1, variable
Controller Weight	Included with arm	400 lbs. (182 kg)
ENVIRONMENT	40°F (5°C) to 120°F (50°C) Humidity 0 to 90% Non-condensing	

OPTIONS
- Additional Time Delays
- Extension Rod Boot
- Additional Slow Speeds
- Casters
- Additional I/O
- 270° Rotation
- RS 232 Interface
- 5th Axis Continuous Spin
- Remote Cycle Start
- 6 Axes
- Remote Hydraulics
- Teach Pendant

ACCESSORIES
- Teach Pendant
- Cassette Program Recorder
- Tester
- Gripper Assembly
- Spare Parts Kit

*Above specifications are for average conditions. Please consult UNIMATION Incorporated for specifications for your installation.
**Not applicable.

Specifications subject to change without notification.

Spec Sheet 14 Unimate Series 2000 and 4000 specifications. (*Courtesy of Unimation Incorporated, a Westinghouse Company.*)

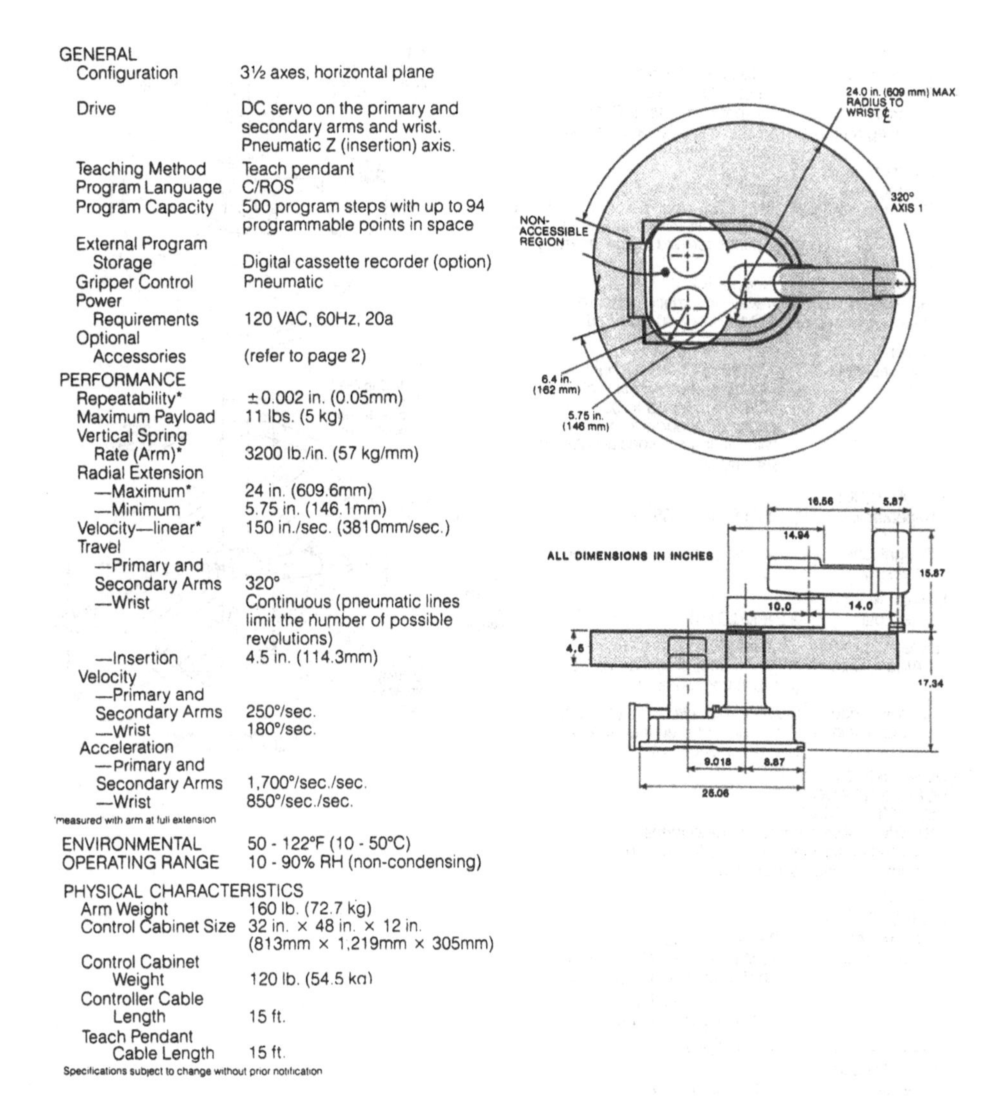

GENERAL

Configuration	3½ axes, horizontal plane
Drive	DC servo on the primary and secondary arms and wrist. Pneumatic Z (insertion) axis.
Teaching Method	Teach pendant
Program Language	C/ROS
Program Capacity	500 program steps with up to 94 programmable points in space
External Program Storage	Digital cassette recorder (option)
Gripper Control	Pneumatic
Power Requirements	120 VAC, 60Hz, 20a
Optional Accessories	(refer to page 2)

PERFORMANCE

Repeatability*	±0.002 in. (0.05mm)
Maximum Payload	11 lbs. (5 kg)
Vertical Spring Rate (Arm)*	3200 lb./in. (57 kg/mm)
Radial Extension	
—Maximum*	24 in. (609.6mm)
—Minimum	5.75 in. (146.1mm)
Velocity—linear*	150 in./sec. (3810mm/sec.)
Travel	
—Primary and Secondary Arms	320°
—Wrist	Continuous (pneumatic lines limit the number of possible revolutions)
—Insertion	4.5 in. (114.3mm)
Velocity	
—Primary and Secondary Arms	250°/sec.
—Wrist	180°/sec.
Acceleration	
—Primary and Secondary Arms	1,700°/sec./sec.
—Wrist	850°/sec./sec.

*measured with arm at full extension

ENVIRONMENTAL OPERATING RANGE	50 - 122°F (10 - 50°C) 10 - 90% RH (non-condensing)

PHYSICAL CHARACTERISTICS

Arm Weight	160 lb. (72.7 kg)
Control Cabinet Size	32 in. × 48 in. × 12 in. (813mm × 1,219mm × 305mm)
Control Cabinet Weight	120 lb. (54.5 kg)
Controller Cable Length	15 ft.
Teach Pendant Cable Length	15 ft.

Specifications subject to change without prior notification

Spec Sheet 15 Unimate Series 100 specifications. (*Courtesy of Unimation Incorporated, a Westinghouse Company.*)

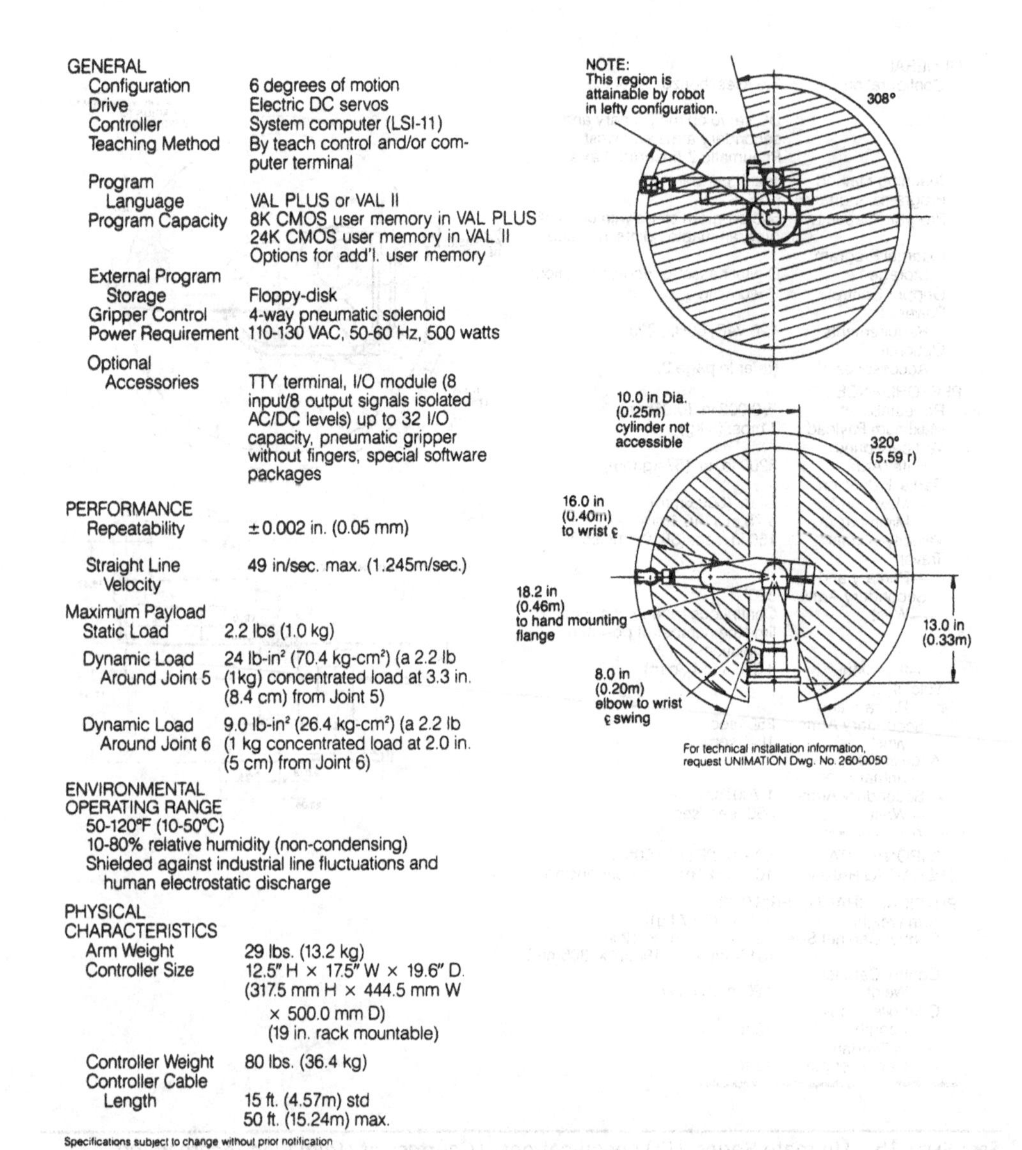

GENERAL

Configuration	6 degrees of motion
Drive	Electric DC servos
Controller	System computer (LSI-11)
Teaching Method	By teach control and/or computer terminal
Program Language	VAL PLUS or VAL II
Program Capacity	8K CMOS user memory in VAL PLUS 24K CMOS user memory in VAL II Options for add'l. user memory
External Program Storage	Floppy-disk
Gripper Control	4-way pneumatic solenoid
Power Requirement	110-130 VAC, 50-60 Hz, 500 watts
Optional Accessories	TTY terminal, I/O module (8 input/8 output signals isolated AC/DC levels) up to 32 I/O capacity, pneumatic gripper without fingers, special software packages

PERFORMANCE

Repeatability	±0.002 in. (0.05 mm)
Straight Line Velocity	49 in/sec. max. (1.245m/sec.)
Maximum Payload	
Static Load	2.2 lbs (1.0 kg)
Dynamic Load Around Joint 5	24 lb-in² (70.4 kg-cm²) (a 2.2 lb (1kg) concentrated load at 3.3 in. (8.4 cm) from Joint 5)
Dynamic Load Around Joint 6	9.0 lb-in² (26.4 kg-cm²) (a 2.2 lb (1 kg concentrated load at 2.0 in. (5 cm) from Joint 6)

ENVIRONMENTAL OPERATING RANGE
50-120°F (10-50°C)
10-80% relative humidity (non-condensing)
Shielded against industrial line fluctuations and human electrostatic discharge

PHYSICAL CHARACTERISTICS

Arm Weight	29 lbs. (13.2 kg)
Controller Size	12.5" H × 17.5" W × 19.6" D. (317.5 mm H × 444.5 mm W × 500.0 mm D) (19 in. rack mountable)
Controller Weight	80 lbs. (36.4 kg)
Controller Cable Length	15 ft. (4.57m) std 50 ft. (15.24m) max.

Specifications subject to change without prior notification

Spec Sheet 16 PUMA Series 200 specifications. (*Courtesy of Unimation Incorporated, a Westinghouse Company.*)

GENERAL

Configuration	6 degrees of motion
Drive	Electric DC Servos
Controller	System computer (LSI-11), with CRT
Teaching Method	By teach pendant and/or computer terminal
Program Language	VAL II
Program Capacity	24KW CMOS user memory in VAL II
External Program Storage	Floppy Disk (Double sided, double density)
I/O Module	32 input/32 output AC/DC user selectable voltage range, plus control I/O s
Gripper Control	4-way pneumatic solenoid
Power Requirement	240/380/415/480 Volts, 3-phase, 4500 watts (peak)

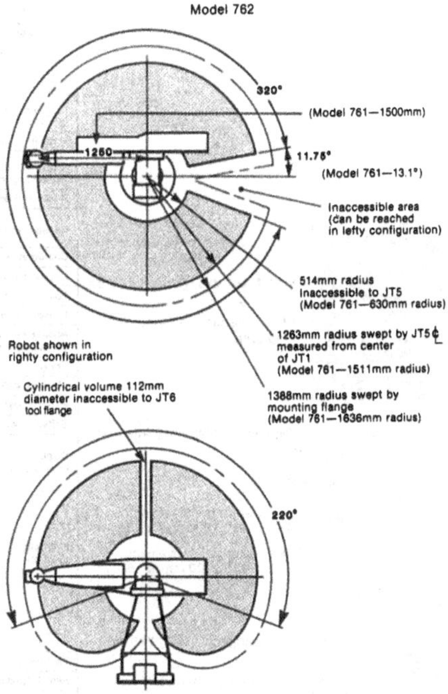

ENVIRONMENTAL OPERATING RANGE

50°-120°F (10°-50°C)
95% relative humidity (non-condensing)
Shielded against momentary industrial line fluctuations of -15%, +200% of nominal voltage and up to 10KV electrostatic discharge. All covers and rotary seals have gaskets to protect against water spray and dust.
Designed to comply with IP54 and NEMA 12 International Electrical Packaging Specifications.

PHYSICAL CHARACTERISTICS	Model 761	Model 762
Arm Weight	1276 lb. (580kg)	1298 lb. (590kg)
Base Diameter	23.6 in. (0.6m)	23.6 in. (0.6m)
Control Cabinet:		
Size	45.6 in.H x 23.6 in.W x 31.5 in.D (1160mm x 600mm x 800mm)	
Weight	440 lb. (200kg)	

PERFORMANCE	Model 761	Model 762
Repeatability	±0.008 in. (0.2mm)	±0.008 in. (0.2mm)
Maximum Payload		
Static Load	22 lb. (10kg)	44 lb. (20kg)
Dynamic Load Around Joint 5	1365 lb.-in.² (4000kg-cm²) 10kg load concentrated 20cm from JT5 (tool flange is 12.5cm from JT5)	4270 lb.-in.² (12500kg-cm²) 20kg load concentrated 25cm from JT5 (tool flange is 12.5cm from JT5)
Dynamic Load Around Joint 6	340 lb.-in.² (1000kg-cm²) 10kg concentrated at 10cm from JT6.	683 lb.-in.² (2000kg-cm²) 20kg concentrated at 10cm from JT6.
Straight Line Velocity	40 in./sec. (1.0m/sec.) max.	40 in./sec. (1.0m/sec.) max.

Model 762
320°
(Model 761—1500mm)
1260
11.75°
(Model 761—13.1°)
Inaccessible area (can be reached in lefty configuration)
514mm radius inaccessible to JT5 (Model 761—630mm radius)
Robot shown in righty configuration
1263mm radius swept by JT5 measured from center of JT1 (Model 761—1511mm radius)
Cylindrical volume 112mm diameter inaccessible to JT6 tool flange
1388mm radius swept by mounting flange (Model 761—1636mm radius)
220°

Specifications subject to change without prior notification.

SPEC SHEET 17 PUMA Series 700 specifications. (*Courtesy of Unimation Incorporated, a Westinghouse Company.*)

Specifications		Motoman-L10W
Features		• Most suitable robot for arc welding. • Accurate tasks in wide working area.
Controlled Axes		5 degrees of freedom, jointed-arm type
Motion Range and Maximum Speed	Arm	S-axis turning: 300°, 90°/s
		L-axis lower arm movement: +45°, −40°, 1000 mm/s 39.4"/s
		U-axis upper arm movement: +20°, −45°, 1400 mm/s 55.1"/s
	Wrist	–
		B-axis swinging: 180°, 240°/s
		T-axis twisting: 370°, 360°/s
Repetitive Positioning Accuracy		±0.2 mm ±0.008"
Payload		10 kg 22 lbs max
Weight		280 kg 616 lbs
Applications	Arc Welding	●
	Spot Welding	
	Cutting	●
	Laser Beam Cutting	●
	Gluing	●
	Material Handling	●
	Deburring, Finishing	
	Assembling	
	Inspection	●

■ **Control Panel *YASNAC RX***

Controlled axes:		5 axes, possible to add 1 axis optionally
Position teaching method:		Teaching play-back, Geometric position data input
Positioning method:		Point-to-point (P.T.P.) and Linear/circular interpolation (C.P.)
Position memory:		IC memory with battery back-up
Programming capacity (for 5 axes):		2,200 positions, 1,200 instructions/or 5,000 positions, 2,500 instructions (Option in total)
Number of jobs:		249 max.
Drive units:		Transistor PWM control
Positioning method:		Incremental digital positioning
Accel. and decel. control:		Software type
Speed setting:	Teaching:	Up to 8 steps (20% of play-back speed)
	Play-back:	Setting by traverse time in 0.01 sec. unit; Speed setting in 1cm/min. unit
Coordinate system:		Rectangular/Cylindrical/Jointed arm coordinates
Interpolation function:		Linear/Circular interpolation
Tool center point control:		Possible
Forward/reverse of step:		Possible by teach box, Jog return possible
Adding, deleting and correcting teaching points:		Possible by teach box
Dry run:		Possible
Machine lock:		Possible
Waiting position teaching:		Possible
Software weaving:		Superposed on straight or circular line
Display information	Display unit: Characters: No. of characters: No. of display:	9" CRT green-colored character display Alphabets and numerals 32 characters × 16 lines 8 categories, more than 128 displays in total
Instruction function:	M codes:	Linear/circular interpolation; Acceleration and deceleration
	A codes:	Timer: 0.1 ~ 25.5 sec. (0.1 sec. unit); 0.01 ~ 2.55 sec. (0.01 sec. unit) Input signal, setting/resetting/scanning Internal relay, setting/resetting/scanning
	H codes:	Counters/Robot halt/Index for jump function
	J codes:	Jump/Call/Skip jump functions
	F codes:	Welder control/Weaving function/Arc sensor
Files:	Welding condition:	Possible to program up to 16 files
	Weaving condition:	Possible to program up to 16 files
Parameters:		100 kinds
Search function:		Program, job, step, position, instruction
Job copy:		Possible
Input signals:		48 for general use/16 for specified use
Output signals:		24 for general use/7 for specified use
Internal relays:		64, built-in
Analog output channels:		2 channels (0V to +/− 14V/DC), 2 channels for option
Changing analog output voltage during play-back:		Possible for 2 channels simultaneously (welding voltage/current)
Cassette interface:		for TCM-5000
Printer interface:		for EPSON RP-80
Computer or sensor interface:		RS-232 C
Alarm indication:		48 kinds
Error indication:		120 kinds
Ambient temperature:		0 ~ 45°C (32 ~ 113°F)
Construction:		Free-stand/completely closed-type
External dimensions:	Box:	700 (W) × 1100 (H) × 580 (D) mm (27.56" × 43.31" × 22.83")
	Panel:	550 (W) × 400 (H) × 250 (D) mm (21.65" × 15.75" × 9.84")
Weight:		200 kg (440 lbs)
Finish color:		Muncell 7.5 BG 6/1.5 (Blue Green)
Power supply:		AC 200/220/230V +10%, −15%; 50/60Hz ±1Hz, 3-phase 3KVA

Spec Sheet 18 Motoman L10W specifications. (*Courtesy of Yaskawa America, Inc.*)

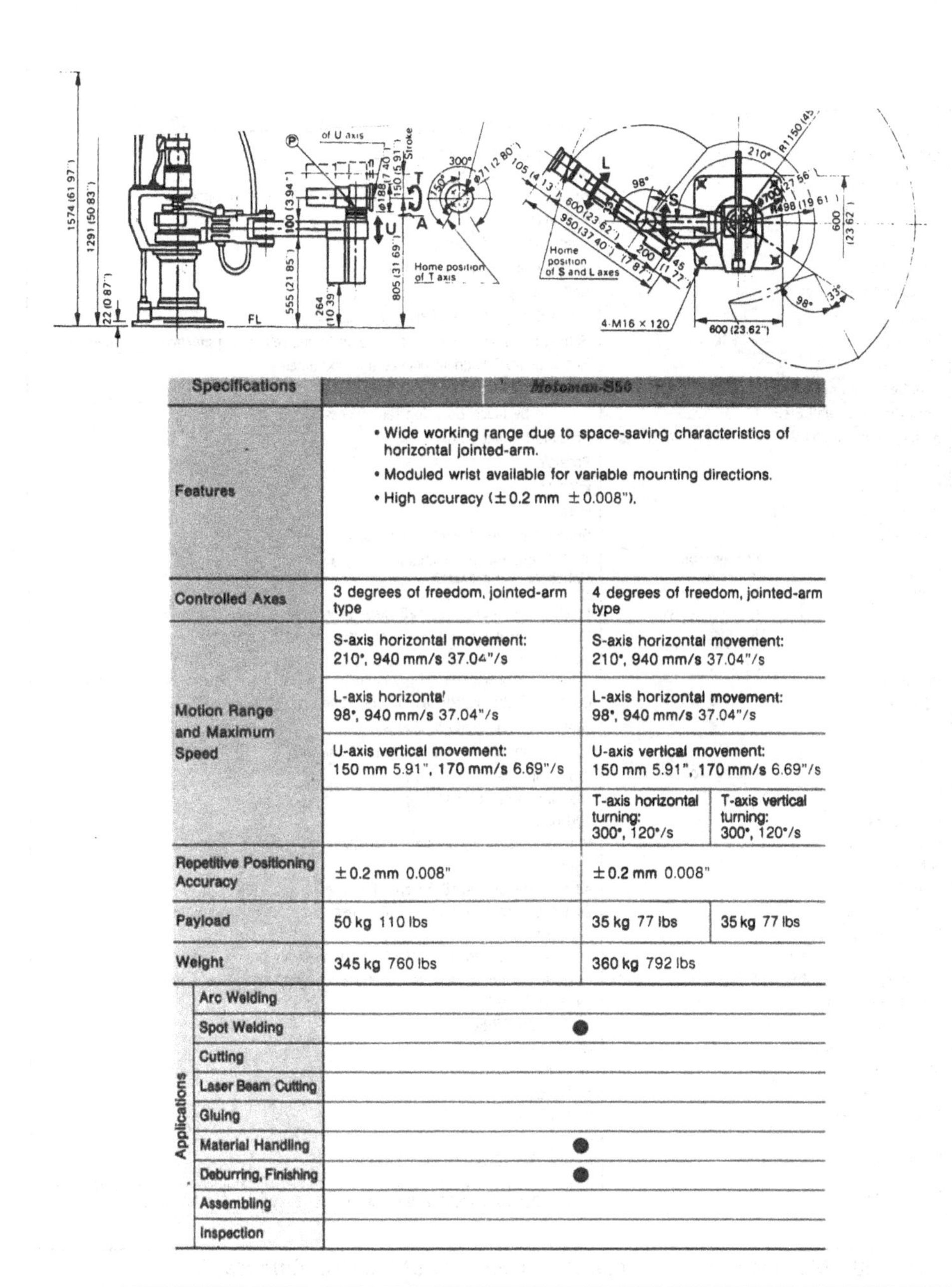

Specifications	Motoman-S50		
Features	• Wide working range due to space-saving characteristics of horizontal jointed-arm. • Moduled wrist available for variable mounting directions. • High accuracy (±0.2 mm ±0.008").		
Controlled Axes	3 degrees of freedom, jointed-arm type	4 degrees of freedom, jointed-arm type	
Motion Range and Maximum Speed	S-axis horizontal movement: 210°, 940 mm/s 37.04"/s	S-axis horizontal movement: 210°, 940 mm/s 37.04"/s	
	L-axis horizonta' 98°, 940 mm/s 37.04"/s	L-axis horizontal movement: 98°, 940 mm/s 37.04"/s	
	U-axis vertical movement: 150 mm 5.91", 170 mm/s 6.69"/s	U-axis vertical movement: 150 mm 5.91", 170 mm/s 6.69"/s	
		T-axis horizontal turning: 300°, 120°/s	T-axis vertical turning: 300°, 120°/s
Repetitive Positioning Accuracy	±0.2 mm 0.008"	±0.2 mm 0.008"	
Payload	50 kg 110 lbs	35 kg 77 lbs	35 kg 77 lbs
Weight	345 kg 760 lbs	360 kg 792 lbs	
Applications: Arc Welding			
Applications: Spot Welding	●		
Applications: Cutting			
Applications: Laser Beam Cutting			
Applications: Gluing			
Applications: Material Handling	●		
Applications: Deburring, Finishing	●		
Applications: Assembling			
Applications: Inspection			

Spec Sheet 19 Motoman S50 specifications. (*Courtesy of Yaskawa America, Inc.*)

Controlled Axes		6 axes, possible to add 2 axes optionally
Position Teaching Method		Teaching play-back, Geometric position data input
Interpolation		Point-to-point (P.T.P.) and Linear/circular interpolation (C.P.)
Position Memory		IC memory with battery back-up
Programming Capacity		2,200 positions, 1,200 instructions. Up to 5,000 positions, 2,500 instructions available as an option.
Number of Jobs		249 max
Drive Units		Transistor PWM control
Positioning		Incremental digital positioning
Accel. and Decel. Control		Software type
Speed Setting	Teaching	Up to 8 steps (20 % of play-back speed)
	Play-back	Setting by traverse time in 0.01 s. unit; Speed setting in 1 cm/min. /min. unit.
Coordinate System		Rectangular/Cylindrical/Jointed arm coordinates
Tool Center Point Control		Programmable up to 8 files
Forward/Reverse of Step		Possible by teach box, Jog return possible
Adding, Deleting and Correcting Teaching Points		Possible by teach box
Dry run		Possible
Machine Lock		Possible
Waiting Position Teaching		Possible
Software Weaving		Superposed on straight or circular line
Display Information	Display Unit	9" CRT green-colored character display
	Characters	Alphabets and numerals
	No. of Characters	32 characters × 16 lines
	No. of Display	8 categories, more than 128 displays in total
Instruction Function	M Codes	Linear/circular interpolation; Acceleration and deceleration
	A Codes	Timer: 0.1 to 25.5 s (0.1 s unit); 0.01 to 2.55 s (0.01 s unit). Input signal setting/resetting/scanning. Internal relay, setting/resetting/scanning
	H Codes	Counters/Robot halt/Index for jump function
	J Codes	Jump/Call/Skip jump functions
	F Codes	Welder control/Weaving function/Arc sensor
Files	Welding Condition	Programmable up to 16 files
	Weaving Condition	Programmable up to 16 files
Parameters		100 kinds
Search Function		Program, job, step, position, instruction
Job Copy		Possible
Input Signals		48 for general use/16 for specified use
Output Signals		24 for general use/7 for specified use
Internal Relays		64, built-in
Analog Output Channels		2 channels (0 V to ± 14 V/DC), 2 channels for option
Changing Analog Output Voltage During Play-back		Possible for 2 channels simultaneously (welding voltage/current)
Cassette Interface		for TCM-5000 EV
Printer Interface		for EPSON RP-80
Computer or Sensor Interface		RS-232C
Alarm Indication		48 kinds
Error Indication		120 kinds
Ambient Temperature		0 to 45°C (32 to 113°F)
Construction		Free standing
Finish Color		Munsell 7.5 BG 6/1.5 (Blue Green)
Power Supply		200/220/230VAC +10%, −15%; 50/60Hz ±1Hz, 3-phase 3kVA (V6, S50, L10W), 5kVA (V12, L106), 13kVA (L30, L100), 15kVA (L15), 20kVA (L60/L60W), 50kVA (L120)

Spec Sheet 20 Motoman S50 specifications. (*Courtesy of Yaskawa America, Inc.*)

Review Questions

1. Where is a robot's manipulator located?
2. How is a robot controlled?
3. Why is a safety fence needed when installing a robot?
4. Who made the 99-800 robot?
5. What is a SMART robot?
6. Where are SMART robots used?
7. What is a work envelope?
8. What are four standard degrees of freedom for a robot?
9. What is a teach control?
10. What is on a spec sheet?

CHAPTER 9

Troubleshooting and Maintenance

Performance Objectives

After studying this chapter, you will be able to:

- Explain how preventive maintenance can prolong trouble-free operation of a system.
- Explain how to prevent accidental shock.
- Troubleshoot AC and DC motors.
- Troubleshoot power-supply disturbances.
- Troubleshoot circuits using a volt-ohm meter (VOM).
- Troubleshoot circuits using an oscilloscope.
- Troubleshoot relays.
- Troubleshoot solid-state motor control equipment.
- Answer the review questions at the end of the chapter.

Troubleshooting and the Robotics Technician

Troubleshooting is another of the tasks performed by robotics technicians. It tests your ability to observe everything around you and your ability to understand how things work. One of the best ways to prevent trouble is to check off certain items as a routine procedure to catch trouble before it becomes a major item and causes fire, damage, and/or death. Electrical problems are many, and every connection and every device is a potential problem. Each device and service, as well as circuits, should have been wired properly, but that is not always the case.

For instance, one of the biggest problems is troubleshooting electric motors. A troubleshooting chart will aid in this task as well as some of the more obvious observations made by the person on the scene. All textbook troubleshooting can do is identify the logical problems. On-scene facts are not always detailed in textbooks, so a good observer must also be able to uncover the facts needed to make a diagnosis. Once the problem is found, it is usually easily corrected.

Preventive Maintenance

Damp and Wet Areas

One thing that can cause problems in any home or shop wiring system is dampness and wetness. Watertight equipment should be installed wherever there is a danger of water coming into contact with live wires. One of the largest problems is condensation of moisture inside a panel board (Figure 9-1). Moisture condenses where warm, moist air in a basement, say, moves up to come into contact with the cold air outside, making the riser cold. In areas where this is a problem, either an underground entrance can be built or an outside riser can be mounted alongside the building, which will make its entrance only when it reaches the panel board. An entrance as low as possible is preferred so that any moisture that does condense will easily drain out the bottom of the panel board without contacting the hot side of the distribution panel. Another problem is rust and corrosion. Anywhere there is moisture, there is the possibility of rust and corrosion. Both rust and corrosion can cause contact problems with metals and remove or place high resistance in the path of a ground system. Removing a ground produces a situation that can be very hazardous (Figure 9-2). In a single-phase system, the current on the neutral side of a properly installed 120/240-volt system carries the difference between the current flowing on the hot lines. If the ground is removed by corrosion or rust, preventing contact with the proper grounding lugs, it is the same as having an open ground.

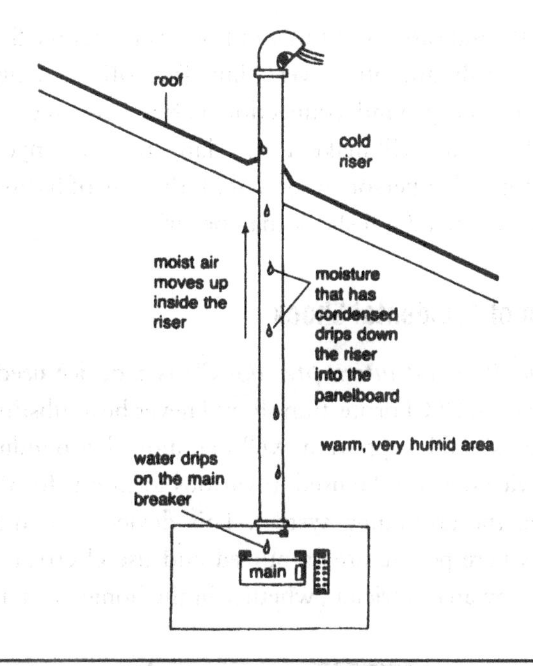

FIGURE 9-1 Moisture in a panel board can cause problems.

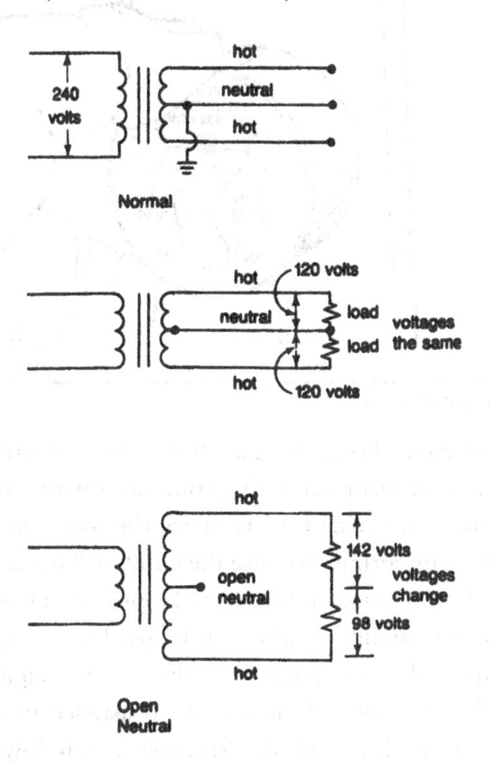

FIGURE 9-2 Removing a ground.

One of the indications of this condition is that some lights in the building will appear very bright and others very dim. Turn off the main switch and locate the open or corroded ground connection before allowing continued operation. A situation of this sort will make it very dangerous for anyone who touches any of the conductors. That person or animal (in the case of barnyards) will complete the ground circuit, and a fatal shock may occur.

Prevention of Accidental Shock

A *ground-fault circuit interrupter* (GFCI) is a device used to prevent accidental shock. However, GFCI protection should never be a substitute for good grounding practice but should support a well-maintained grounding system. Figure 9-3 shows a device that can be used in various locations for the prevention of shock by checking the grounding system. This device is used in homes, plants, and businesses where people are employed and use electrical equipment. It is often encountered by an electrician, whether in the home or on the job.

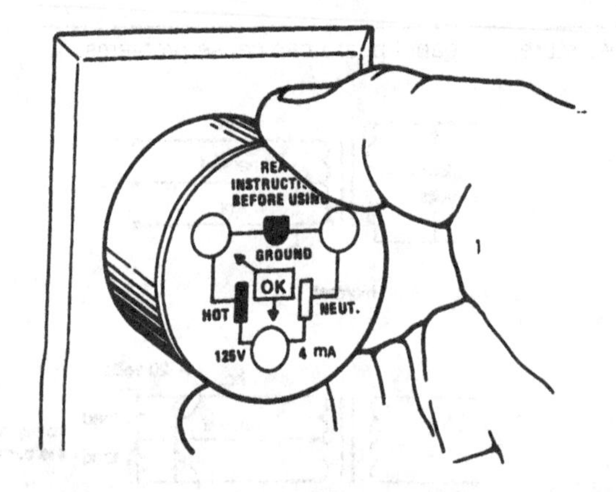

FIGURE 9-3 Ground monitor.

The arrangement shown in Figure 9-4A checks polarity and grounding. It also diagnoses five other incorrect wiring conditions with a plug-in tester. Figure 9-4B shows the same tester being used to check the continuity of the ground path of a tool. This is very important because the hand drill has a metal handle.

In Figure 9-4C, the meter is a ground loop tester; it measures the ground loop impedance of live circuits. It also can be used to check for grounding of tools, piping systems, and other equipment. The meter in Figure 9-4D is used to check the 500- and 1,000-volt DC insulation resistance of deenergized circuits and electrical equipment. It also checks for continuity in low-resistance circuits.

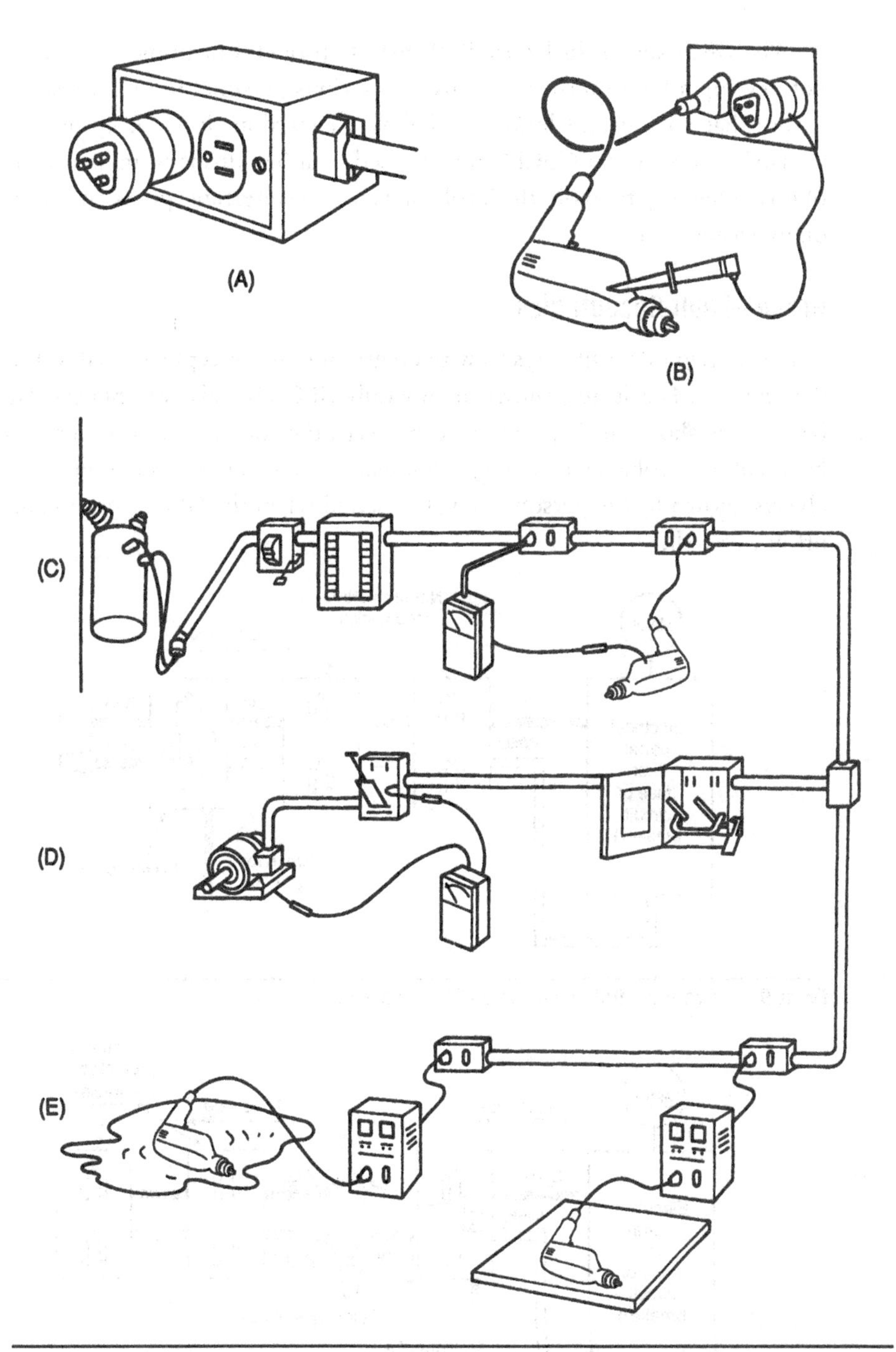

FIGURE 9-4 (A) Ground monitor used to check polarity and grounding. (B) Checking the grounding path for a tool. (C) Ground loop tester. (D) Checking the impedance of the live circuit. (E) Testing tools to ensure that leakage is below a hazardous level.

The GFCI shown in Figure 9-4E mainly provides insurance that a tool will not develop a fault on the job, causing a serious personal injury. It tests tools to ensure that any current leakage is below a hazardous level. Keep in mind that nuisance tripping of a GFCI can be caused by a few drops of moisture or flecks of dust. One way to avoid this problem is to use watertight plugs and connectors on extension cords.

Ground-Fault Receptacles

There are two different ways to wire up ground-fault receptacles (GFRs) (Figures 9-5 and 9-6). The devices shown are not only GFCIs but also receptacles. They can be used, as shown in Figure 9-6, to protect other downstream receptacles. This brings about problems in some places, inasmuch as the protected outlets are not always known by the persons using them, and when the GFR trips, it takes them off-line as well.

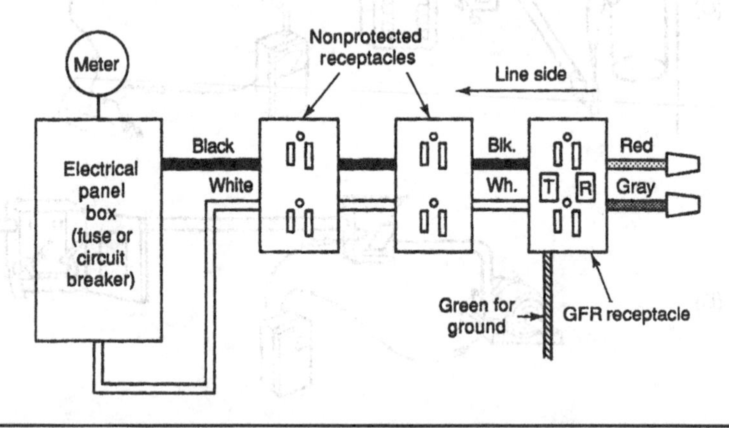

FIGURE 9-5 Ground-fault receptacle (GFR) installed in a box.

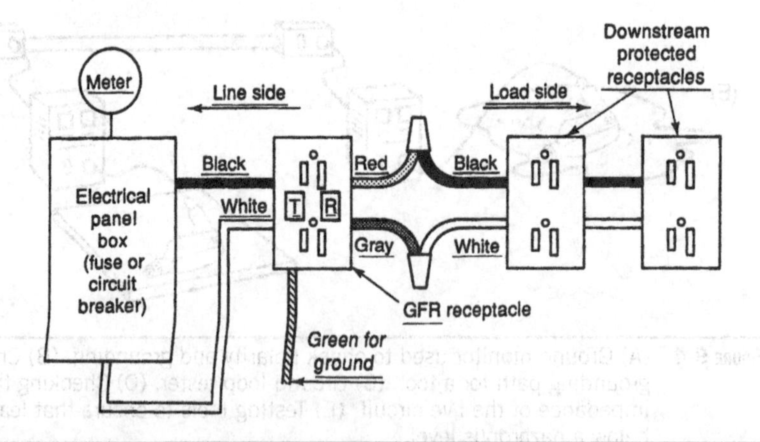

FIGURE 9-6 GFR wired to protect downstream devices.

One way to check for a terminal installation is to check the red and gray wires. If they are capped with a wire nut, you know that the GFR does not service any other outlets.

Wiring Devices

Using the proper wiring devices is a form of preventive maintenance inasmuch as it prevents problems later. Shock hazards are minimized by the dielectric strength of the material used for the molded interior walls and the individual wire pocket areas. Nylon seems to be best for this job. Nylon devices withstand high impact in heavy-duty industrial and commercial applications. Each molded piece has to support adjacent molded pieces to result in good resiliency and strength. Devices made of vinyl, neoprene, urea, or phenolic materials can crack or be damaged under pressure. Damage can be invisible and cause direct shorts and other hazards. Nylon also has the ability to withstand high voltages without breaking down.

Maintenance of Small Electric Motors

Small motors usually operate with so little trouble that they are apt to be neglected. They should be thoroughly inspected twice yearly to detect wear and to remove any conditions that might lead to further wear. Special care must be taken to inspect motor bearings, cutouts, and other wearing parts. Make sure that dirt and dust are not interfering with ventilation or clogging moving parts.

Adequate Wiring

When installing a new motor or transferring a motor from one installation to another, it is well to check the wiring. Be sure that adequate wire sizes are used to feed electrical power to the motor; in many cases, replacement of wires will prevent future breakdown. Adequate wiring assists in preventing overheating of motors and reduces electrical power costs.

Check Internal Switches

Start winding switches usually give little trouble, but regular attention makes them last even longer. Use fine sandpaper to clean contacts. Make sure that the sliding member on the shaft that operates the start winding switch moves freely. Check for loose screws.

Check Load Condition

Check the driven load regularly. Sometimes additional friction develops gradually within the machine and imposes an overload on the motor, so watch the motor temperature. Protect motors with properly rated fuses or overload cutouts.

Take Extra Care in Lubrication

A motor running three times as much as usual will need three times as much attention to lubrication. Motors should be lubricated according to the manufacturer's recommendation. Provide enough oil, but do not *overdo* it.

Keep Commutators Clean

Do not allow a commutator on a direct-current (DC) motor to become covered with dust or oil. It should be wiped occasionally with a clean, dry cloth or one moistened with a solvent that does not leave a film. If it is necessary to use sandpaper, use No. 0000 paper or finer. Sandpaper or abrasive papers are available with ratings as high as 1,500 grit.

Motors Must Have a Proper Service Rating

Sometimes it is necessary to move a motor from one job to another or to operate a machine continuously when it has previously been running for short periods of time. Whenever a motor is operated under different conditions or on a new application, make sure that it is rated properly. A motor is rated for intermittent duty because the temperature rise within the motor will not be excessive when it is operated for short periods. Putting such a motor on a continuous-duty application will result in an excessive temperature rise, which will cause the insulation to deteriorate or may even cause burnout.

Replace Worn Brushes

Brushes should be inspected at regular intervals so that replacements can be made if necessary. Whenever a brush is removed for inspection, be sure to replace it in the same axial position; that is, it must not be turned around in the brush holder when it is put back in the motor. If the contact surface, which has been "worn in" to fit the commutator, is not replaced in the same position, excessive sparking and loss of power will result. Brushes naturally wear down and should be replaced before they are less than ¼ inch in length. The commutator also should be inspected when brushes are removed. See the section "DC Motor Problems."

Motor Problems

Certain danger signals are presented before a motor overheats or burns out.

Ball Bearing Motors

Danger Signals

- A sudden increase in the temperature differential between the motor and bearing temperatures is an indication of malfunction of the bearing lubricant.
- A temperature higher than that recommended for the lubricant warns of a reduction in bearing life. The rule of thumb is that grease life is halved for each 25°F increase in operating temperature.
- An increase in bearing noise, accompanied by a bearing temperature rise, is an indication of a serious malfunction of a bearing.

Major Duties of Ball Bearing Lubricant

- To dissipate heat caused by friction of bearing members under load.
- To protect bearing members from rust or corrosion.
- To offer maximum protection against the entrance of foreign matter into the bearings.

Causes of Bearing Failure

- Foreign matter in bearing from dirty grease or ineffective seals.
- Deterioration of grease because of excessive temperature or contamination.
- Overheated bearings as a result of too much grease.

Sleeve Bearing Motors

The lubricant used with sleeve bearings actually must provide an oil film that completely separates the bearing surface from the rotating shaft member and, ideally, eliminates metal-to-metal contact.

Lubricant

Oil, because of its adhesive properties and because of its viscosity or resistance to flow, is dragged along by the rotating shaft of the motor and forms a wedge-shaped film between the shaft and the bearing. The oil film forms automatically when the shaft begins to turn and is maintained by the motion. The forward motion sets up a pressure in the oil film, which, in turn, supports the load. This wedge-shaped film of oil is an absolutely essential feature of effective hydrodynamic sleeve bearing lubrication. Without it, no great load can be carried, except with high friction loss and resulting destruction of the bearing. When lubrication is effective and an adequate oil film is maintained, the sleeve bearing serves chiefly as a guide to ensure alignment. In the event of failure of the oil film, the bearing functions as a safeguard to prevent actual damage to the motor shaft.

Selection of Oil

The selection of the oil that will provide the most effective bearing lubrication and not require frequent renewal merits careful consideration. Good lubricants are essential to low maintenance costs. Top-grade oils are recommended because they are refined from pure petroleum; are substantially noncorrosive as far as metal surfaces to be lubricated are concerned; are free from sediment, dirt, or other foreign materials; and are stable with respect to heat and moisture encountered in the motor. In performance terms, the higher-priced oils prove to be cheaper in the long run.

An oil film is built up of many layers or laminations that slide on one another as the shaft rotates. The internal friction of the oil, which is due to the sliding action of the many oil layers, is measured as the *viscosity*. The viscosity of the oil chosen for a particular application should provide ample oiliness to prevent wear and seizure at ambient temperature, low speeds, and heavy loads before the oil film is established and operating temperature is reached. Low-viscosity oils are recommended for use with fractional-horsepower motors because they offer low internal friction, permit fuller realization of the motor's efficiency, and minimize the operating temperature of the bearing.

Standard Oils

High ambient temperatures and high motor operating temperatures will have a destructive effect on sleeve bearings lubricated with standard-temperature-range oil by increasing the bearing operating temperature beyond the oil's capabilities. Such destructive effects include reduction in oil viscosity, an increase in corrosive oxidation products in the lubricant, and usually a reduction in the quantity of the lubricant in contact with the bearing. Special oils are available, however, for motor applications at high temperatures as well as for motor applications at low temperatures. The care exercised in selecting the proper lubricant for expected extremes in bearing operating temperatures will have a decided influence on motor performance and bearing life.

Wear

Although sleeve bearings are less sensitive to a limited amount of abrasive or foreign materials than are ball bearings, owing to the ability of the relatively soft surface of a sleeve bearing to absorb hard particles of foreign materials, good maintenance practice recommends that the oil and bearing be kept clean. Frequency of oil changing will depend on local conditions, such as the severity and continuity of service and operating temperature. A conservative lubrication maintenance program should call for periodic inspections of the oil level and cleaning and refilling with new oil every six months.

WARNING: *Overlubrication should be avoided. Insulation damage by excess motor lubricant represents one of the most common causes of motor winding insulation failure in both sleeve and ball bearing motors.*

Common Motor Problems and Their Causes

Easy-to-detect symptoms, in many cases, indicate exactly what is wrong with a fractional-horsepower motor. However, where general types of trouble have similar symptoms, it becomes necessary to check each possible cause separately. Table 9-1 lists some of the more common ailments of small motors, together with suggestions as to possible causes. Most common motor problems can be checked by some test or inspection. The order of making these tests rests with the troubleshooter, but it is natural to do the simplest tests first. For instance, when a motor fails to start, you first inspect the motor connections because that is an easy and simple thing to do.

TABLE 9-1 Squirrel-Cage Motor Problems

Symptom and Possible Cause	Possible Remedy
Motor Will not Start	
(a) Overload control tripped	(a) Wait for overload to cool. Try starting again. If motor still does not start, check all the causes as outlined in the following.
(b) Power not connected	(b) Connect power to control and control to motor. Check clip contacts.
(c) Faulty (open) fuses	(c) Test fuses.
(d) Low voltage	(d) Check motor name-plate values with power supply. Also check voltage at motor terminals with motor under load to be sure wire size is adequate.
(e) Wrong control connections	(e) Check connections with control wiring diagram.
(f) Loose terminal lead connection	(f) Tighten connections.
(g) Driven machine locked	(g) Disconnect motor from load. If motor starts satisfactorily, check driven machine.
(h) Open circuit in stator or rotor winding	(h) Check for open circuits.
(i) Short circuit in stator winding	(i) Check for shorted coil.
(j) Winding grounded	(j) Test for grounded winding.
(k) Bearing stiff	(k) Free bearings or replace.
(l) Grease too stiff	(l) Use special lubricant for special conditions.
(m) Faulty control	(m) Check control wiring.
(n) Overload	(n) Reduce load.
Motor Noisy	
(a) Motor running single phase	(a) Stop motor, then try to start. (It will not start on single phase.) Check for "open" in one of the lines or circuits.

Problem Diagnosis

In diagnosing problems, a combination of symptoms will often give a definite clue to the source of the trouble and hence eliminate other possibilities. For instance, in the case just cited of a motor that will not start, if heating occurs, it offers the suggestion that a short or ground exists in one of the windings and eliminates the likelihood of an open circuit, poor line connection, or defective starter switch.

Centrifugal Switches

Centrifugal starting switches, found in many types of single-phase fractional-horsepower motors, occasionally are a source of trouble. If the mechanism sticks in the running position, the motor will not start. However, if the switch is stuck in the closed position, the motor will not attain speed, and the starting winding heats up quickly. The motor also may fail to start if the contact points of the switch are out of adjustment or coated with oxide. It is important to remember, however, that any adjustment of the switch or contacts should be made only at the factory or an authorized service station.

Commutator-Type Motors

More maintenance is required by motors with commutators. High-speed series-wound motors should not be used on long, continuous-duty-cycle applications because the commutator and brushes are a potential source of trouble. Gummy commutators and oil-soaked brushes can cause sluggish action and severe sparking. The commutator can be cleaned with fine sandpaper. However, if pitted spots still appear, the commutator should be reground.

Troubleshooting Aids

Connection Diagrams

Figure 9-7 shows motor connection diagrams as an aid in troubleshooting. Knowing the arrangement of coils aids in checking out the shorts and grounds as well as opens.

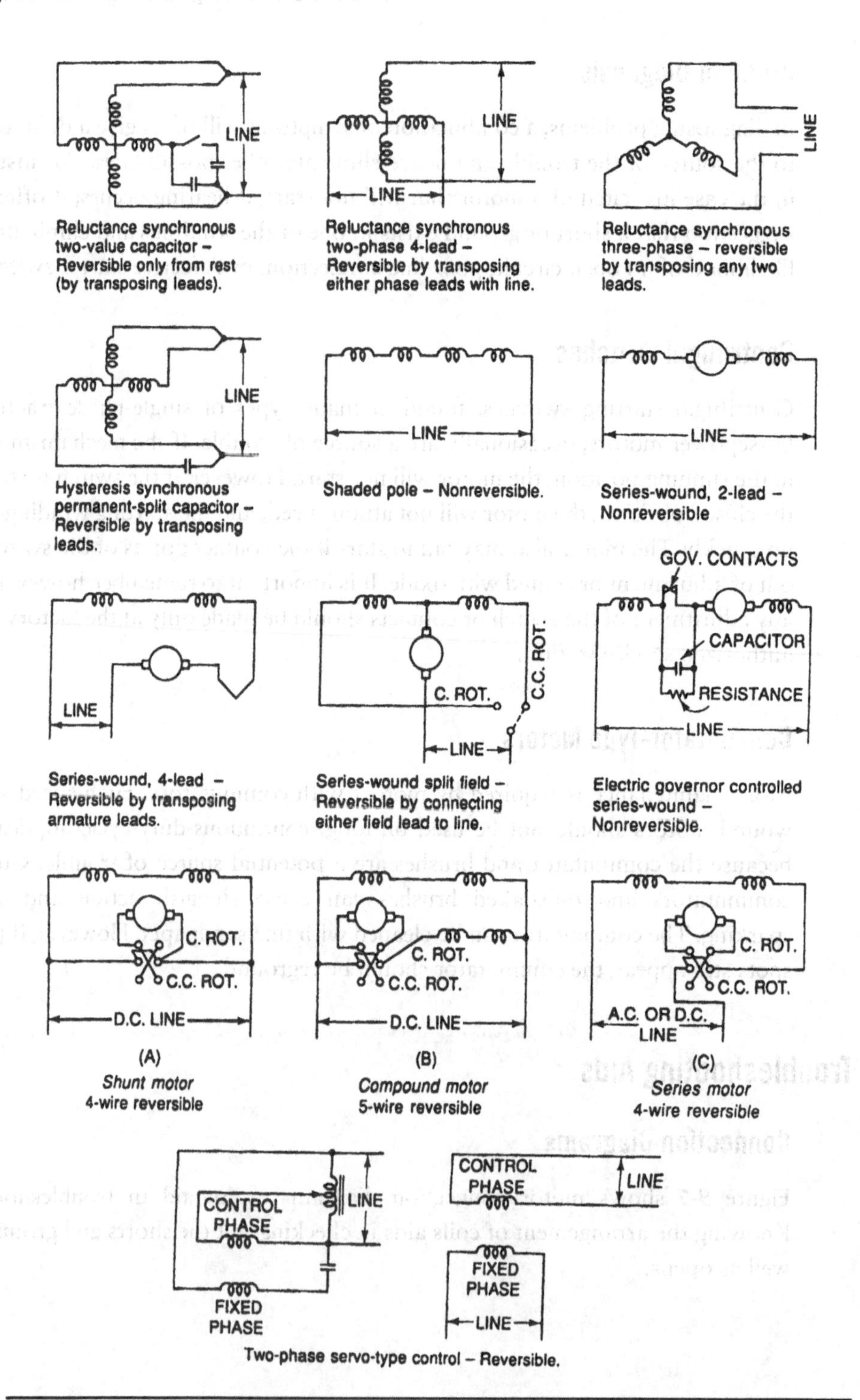
LINE
Reluctance synchronous two-value capacitor – Reversible only from rest (by transposing leads).
LINE
LINE
Reluctance synchronous two-phase 4-lead – Reversible by transposing either phase leads with line.
LINE
Reluctance synchronous three-phase – reversible by transposing any two leads.
LINE
Hysteresis synchronous permanent-split capacitor – Reversible by transposing leads.
LINE
Shaded pole – Nonreversible.
LINE
Series-wound, 2-lead – Nonreversible
LINE
Series-wound, 4-lead – Reversible by transposing armature leads.
C.C. ROT.
C. ROT.
LINE
Series-wound split field – Reversible by connecting either field lead to line.
GOV. CONTACTS
CAPACITOR
RESISTANCE
LINE
Electric governor controlled series-wound – Nonreversible.
C. ROT.
C.C. ROT.
D.C. LINE
(A)
Shunt motor
4-wire reversible
C. ROT.
C.C. ROT.
D.C. LINE
(B)
Compound motor
5-wire reversible
C. ROT.
C.C. ROT.
A.C. OR D.C. LINE
(C)
Series motor
4-wire reversible
CONTROL PHASE
LINE
FIXED PHASE
CONTROL PHASE
LINE
FIXED PHASE
LINE
Two-phase servo-type control – Reversible.

FIGURE 9-7 Motor connection diagrams.

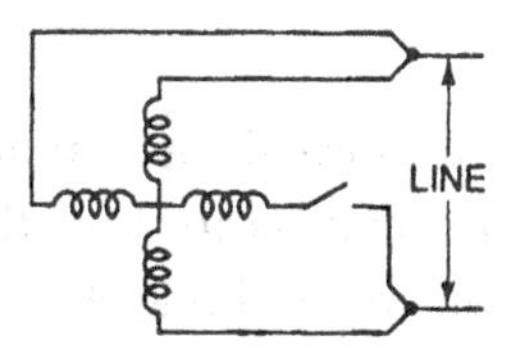

Split-phase – Reversible only from rest (by transposing leads).

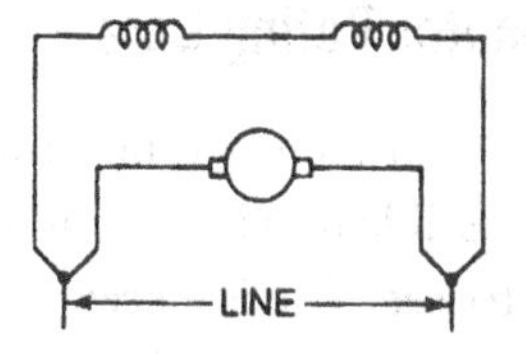

Shunt-wound – Reversible by transposing leads.

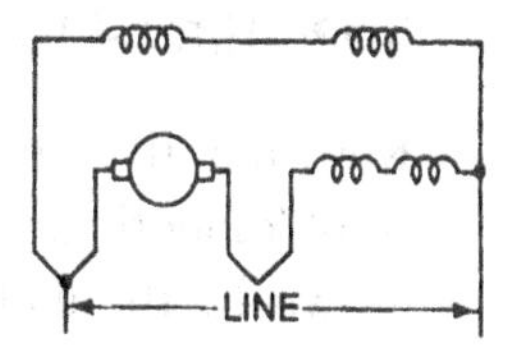

Compound-wound – Reversible by transposing armature leads.

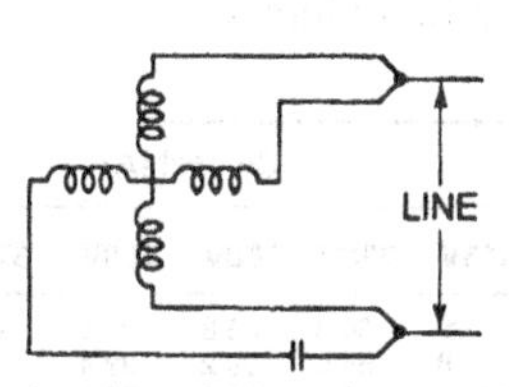

Permanent-split capacitor, 4-lead – Reversible by transposing leads.

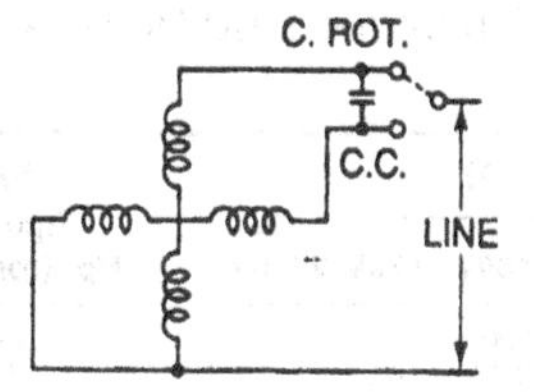

Permanent-split capacitor 3-lead – Reversible by connecting either side of capacitor to line.

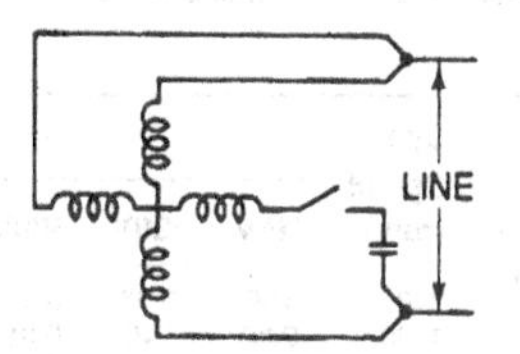

Capacitor-start – Reversible only from rest (by transposing leads).

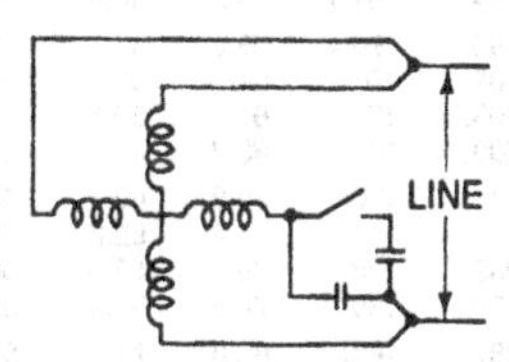

Two-value capacitor – Reversible only from rest (by transposing leads).

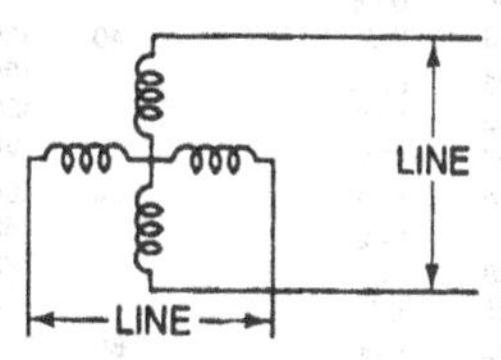

Two-phase 4-lead – Reversible by transposing either phase leads with line.

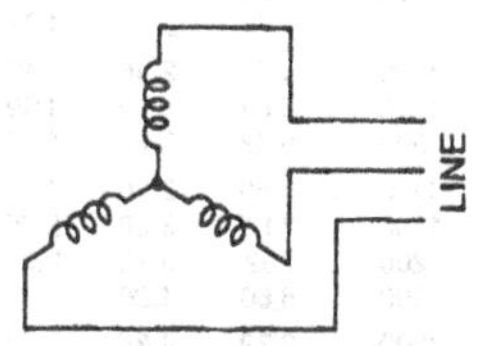

Three-phase, single voltage – Reversible by transposing any two leads.

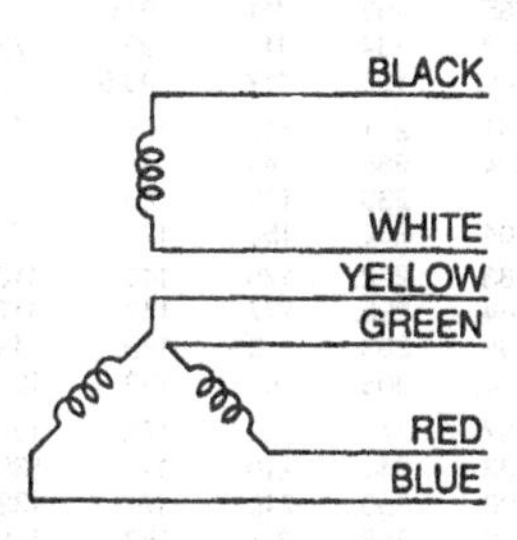

3-phase, star-delta 6-lead reversible – For 440 volts connect together white, yellow, and green: Connect to line black, red, and blue. To reverse rotation, transpose any two line leads. For 220 volts connect white to blue, black to green, and yellow to red: Then connect each junction point to line.

To reverse rotation, transpose any two junction points with line.

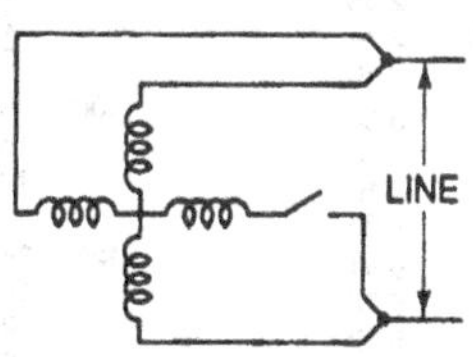

Reluctance synchronous split-phase – Reversible only from rest (by transposing leads).

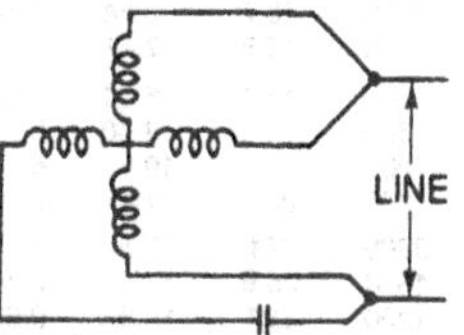

Reluctance synchronous permanent split-capacitor, 4-lead – Reversible by transposing lead.

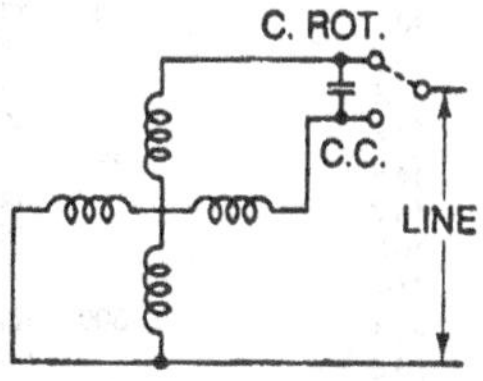

Reluctance synchronous permanent-split capacitor, 3-lead – Reversible by connecting either side of capacitor to line.

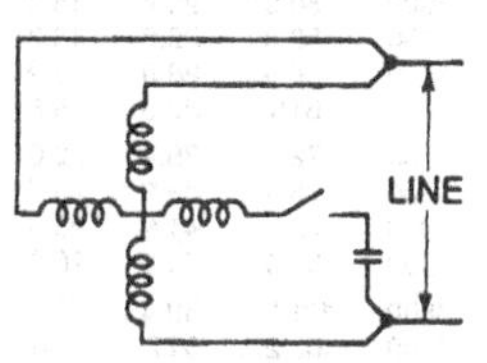

Reluctance synchronous capacitor-start – Reversible only from rest (by transposing leads).

Figure 9-7 *(Continued)*

Small Three-Phase Motor Rating Data

Knowing the current expected to be drawn in normal operation aids in troubleshooting. It is possible to use a clamp-on type of meter to check the current drawn by the motor to see if it is excessive or incorrect. Table 9-2 shows the ampere ratings of AC motors that operate on three-phase power.

Table 9-2 Ampere Ratings of Three-Phase, 60-Hertz, AC Induction Motors

Hp	Syn. Speed (rpm)	Current (A)					
		115V	230V	380V	460V	575V	2200V
$\frac{1}{4}$	1800	1.90	0.95	0.55	0.48	0.38	
	1200	2.80	1.40	0.81	0.70	0.56	
	900	3.20	1.60	0.93	0.80	0.64	
$\frac{1}{3}$	1800	2.38	1.19	0.69	0.60	0.48	
	1200	3.60	1.80	1.04	0.90	0.72	
	900	3.60	1.80	1.04	0.90	0.72	
$\frac{1}{2}$	1800	3.44	1.72	0.99	0.86	0.69	
	1200	4.30	2.15	1.24	1.08	0.86	
	900	4.76	2.38	1.38	1.19	0.95	
$\frac{3}{4}$	1800	4.92	2.46	1.42	1.23	0.98	
	1200	5.84	2.92	1.69	1.46	1.17	
	900	6.52	3.26	1.88	1.63	1.30	
1	3600	5.60	2.80	1.70	1.40	1.12	
	1800	7.12	3.56	2.06	1.78	1.42	
	1200	7.52	3.76	2.28	1.88	1.50	
	900	8.60	4.30	2.60	2.15	1.72	
$1\frac{1}{2}$	3600	8.72	4.36	2.64	2.18	1.74	
	1800	9.71	4.86	2.94	2.43	1.94	
	1200	10.5	5.28	3.20	2.64	2.11	
	900	11.2	5.60	3.39	2.80	2.24	
2	3600	11.2	5.60	3.39	2.80	2.24	
	1800	12.8	6.40	3.87	3.20	2.56	
	1200	13.7	6.84	4.14	3.42	2.74	
	900	15.8	7.90	4.77	3.95	3.16	
3	3600	16.7	8.34	5.02	4.17	3.34	
	1800	18.8	9.40	5.70	4.70	3.76	
	1200	20.5	10.2	6.20	5.12	4.10	
	900	22.8	11.4	6.90	5.70	4.55	
5	3600	27.1	13.5	8.20	6.76	5.41	
	1800	28.9	14.4	8.74	7.21	5.78	
	1200	31.7	15.8	9.59	7.91	6.32	
	900	31.0	15.5	9.38	7.75	6.20	
$7\frac{1}{2}$	3600	39.1	19.5	11.8	9.79	7.81	
	1800	43.0	21.5	13.0	10.7	8.55	
	1200	43.7	21.8	13.2	10.9	8.70	
	900	46.0	23.0	13.9	11.5	9.19	
10	3600	50.8	25.4	15.4	12.7	10.1	
	1800	53.8	26.8	16.3	13.4	10.7	
	1200	56.0	28.0	16.9	14.0	11.2	
	900	61.0	30.5	18.5	15.2	12.2	
15	3600	72.7	36.4	22.0	18.2	14.5	
	1800	78.4	39.2	23.7	19.6	15.7	
	1200	82.7	41.4	25.0	20.7	16.5	
	900	89.0	44.5	26.9	22.2	17.8	
20	3600	101.1	50.4	30.5	25.2	20.1	
	1800	102.2	51.2	31.0	25.6	20.5	
	1200	105.7	52.8	31.9	26.4	21.1	
	900	109.5	54.9	33.2	27.4	21.9	
25	3600	121.5	60.8	36.8	30.4	24.3	
	1800	129.8	64.8	39.2	32.4	25.9	
	1200	131.2	65.6	39.6	32.8	26.2	
	900	134.5	67.3	40.7	33.7	27.0	
30	3600	147.	73.7	44.4	36.8	29.4	
	1800	151.	75.6	45.7	37.8	30.2	
	1200	158.	78.8	47.6	39.4	31.5	
	900	164.	81.8	49.5	40.9	32.7	
40	3600	193.	96.4	58.2	48.2	38.5	
	1800	202.	101.	61.0	50.4	40.3	
	1200	203.	102.	61.2	50.6	40.4	
	900	209.	105.	63.2	52.2	41.7	
50	3600	241.	120.	72.9	60.1	48.2	
	1800	249.	124.	75.2	62.2	49.7	
	1200	252.	126.	76.2	63.0	50.4	
	900	260.	130.	78.5	65.0	52.0	
60	3600	287.	143.	86.8	71.7	57.3	
	1800	298.	149.	90.0	74.5	59.4	
	1200	300.	150.	91.0	75.0	60.0	
	900	308.	154.	93.1	77.0	61.5	
75	3600	359.	179.	108.	89.6	71.7	
	1800	365.	183.	111.	91.6	73.2	
	1200	368.	184.	112.	92.0	73.5	
	900	386.	193.	117.	96.5	77.5	
100	3600	461.	231.	140.	115.	92.2	
	1800	474.	236.	144.	118.	94.8	23.6
	1200	478.	239.	145.	120.	95.6	24.2
	900	504.	252.	153.	126.	101.	24.8
125	3600	583.	292.	176.	146.	116.	
	1800	584.	293.	177.	147.	117.	29.2
	1200	596.	298.	180.	149.	119.	29.9
	900	610.	305.	186.	153.	122.	30.9
150	3600	687.	343.	208.	171.	137.	
	1800	693.	348.	210.	174.	139.	34.8
	1200	700.	350.	210.	174.	139.	35.8
	900	730.	365.	211.	183.	146.	37.0
200	3600	904.	452.	274.	226.	181.	
	1800	915.	458.	277.	229.	184.	46.7
	1200	920.	460.	266.	230.	184.	47.0
	900	964.	482.	279.	241.	193.	49.4
250	3600	1118.	559.	338.	279.	223.	
	1800	1136.	568.	343.	284.	227.	57.5
	1200	1146.	573.	345.	287.	229.	58.5
	900	1200	600.	347.	300.	240.	60.5
300	1800	1356.	678.	392.	339.	274.	69.0
	1200	1368.	684.	395.	342.	274.	70.0
400	1800	1792.	896.	518.	448.	358.	91.8
500	1800	2220.	1110.	642.	555.	444.	116.

Source: Courtesy of Bodine Electric Company, Chicago.
*Ampere ratings of motors vary somewhat. The values given here are for drip-proof class B insulated (T frame) where available, 1.15 service factor, NEMA design B motors. The values represent an average full-load motor current that was calculated from the motor performance data published by several motor manufacturers. In the case of high-torque squirrel-cage motors, the ampere ratings will be at least 10% greater than the values shown.

Ampere ratings of motors vary somewhat depending on the type of motor. The values given in Table 9-2 are for drip-proof Class B insulated (T-frame) 1.15 service factor NEMA Design B motors. These values represent an average full-load motor current that was calculated from the motor performance data published by several motor manufacturers. In the case of high-torque squirrel-cage motors, the ampere ratings will be at least 10 percent greater than the values given below.

Ampere ratings of motors vary somewhat. The values given in the table are for drip-proof Class B insulated (T-frame), where available, 1.15 service factor NEMA Design B motors. The values represent an average full-load motor current that was calculated from the motor performance data published by several motor manufacturers. In the case of high-torque squirrel-cage motors, the ampere ratings will be at least 10 percent greater than the values shown.

Power-Supply Disturbances

Maintenance of equipment is affected by the quality of power supplied to it. There are a number of problems associated with various line power disturbances. Three types of irregularity that affect power supply are voltage fluctuations, transients, and power outages.

Voltage Fluctuations

In many states, public service commissions establish allowable voltage tolerances for utilities. These tolerances are continually monitored, and in most instances, every reasonable precaution is taken to stay within these limits. However, some equipment is so sensitive that fluctuations within the tolerance limits can still cause problems (Figure 9-8).

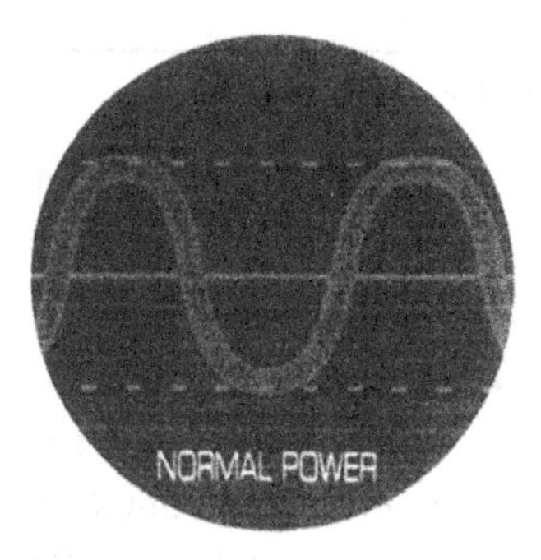

Figure 9-8 Normal power sine wave.

Voltage fluctuations usually can be detected by visible flickering of lights. High- or low-voltage conditions can result in damage to equipment, loss of data, and erroneous readings in monitoring systems (Figure 9-9).

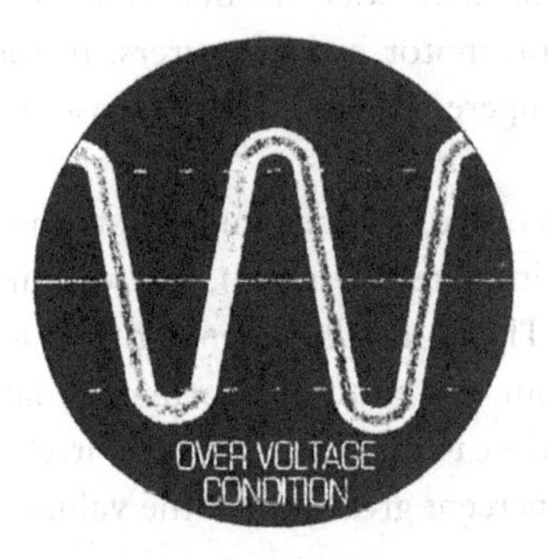

Figure 9-9 Overvoltage condition.

Undervoltage can result from overloaded power circuits. Intermittent low voltage is typically caused by starting a large, heavily loaded motor such as an air conditioner. Overvoltage conditions are less common but are more damaging and are seen frequently in facilities with rapidly varying loads (Figure 9-10).

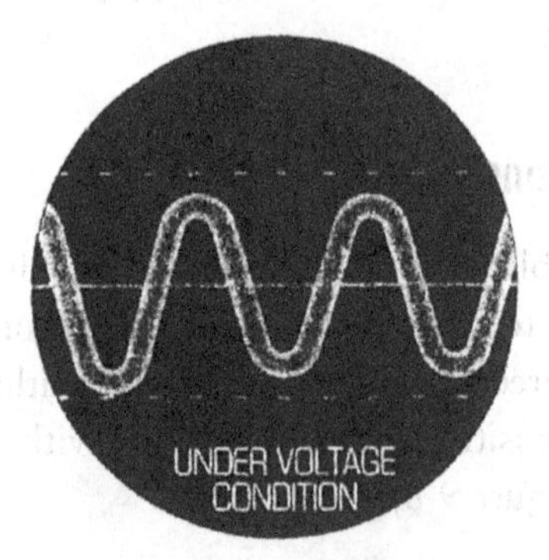

Figure 9-10 Undervoltage condition.

Transients

Voltage Spikes

Short-duration impulses in excess of the normal voltage are called *spikes* or *surges*. Although their duration is incredibly brief, a spike may exceed the normal voltage level by five- or tenfold. Spikes can wipe out data stored in memory, produce output errors, or cause extensive equipment damage. Besides the immediate damage, there are also harder-to-detect effects, particularly reduced service life. Subsequent random failures can be particularly annoying and expensive (Figure 9-11).

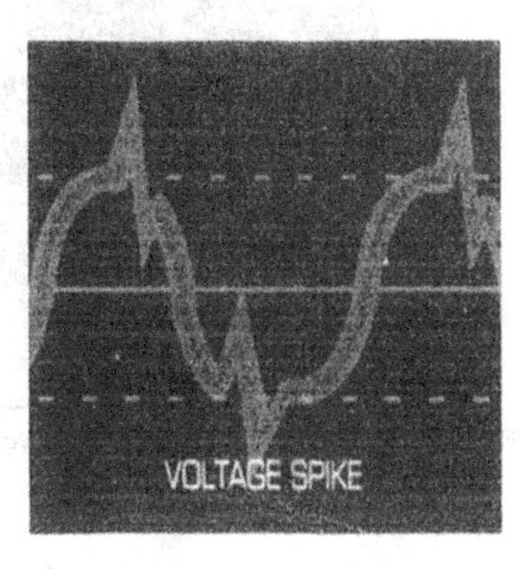

FIGURE 9-11 Voltage spike on a sine wave.

The leading day-to-day cause of small, low-energy spikes is the switching on and off of an electric motor (inductive load switching). Air conditioners, electrical power tools, furnace ignitions, electrostatic copy machines, arc welders, and elevators are particularly guilty of creating voltage spikes. The problems created by the inductive load switching are very common in industrial plants. Larger spikes are typically caused by lightning. A direct lightning hit, of course, is catastrophic but of very low probability. However, a distant lightning strike several miles away may be transmitted through utility power lines and show up as a voltage spike all along the line.

Electrical Noise

As contrasted with outright equipment damage, computer "glitches" are caused by electrical noise (Figure 9-12). The same causes of voltage spikes can (at a lower voltage magnitude) cause noise interference. Other electrical noise generators include radio transmitters, fluorescent lights, computers, business machines, and electrical devices such as light sockets, wall receptacles, plugs, and loose electrical connections. Interaction between system components may generate sufficient noise to cause errors. Although most electronic equipment has some internal noise filtering, equipment located in severe noise environments may encounter some interference. Transients are by far the most common sort of power disturbance and, fortunately, often the easiest to correct. However, they may be difficult to detect because they last such a short time.

FIGURE 9-12 Noise interference on a power-line sine wave.

Power Outages

Power outages are a total interruption of power supply. An interruption of a mere 15 milliseconds is considered a blackout to sensitive equipment. Power outages can cause problems for equipment users, most critical of which are loss of valuable data and expensive, time-consuming reprogramming.

Outages tend to be caused by a larger-scale problem than a transient. Interruptions may be caused, for example, by utility or on-site load changes, on-site equipment malfunction, or faults on the power system (Figure 9-13).

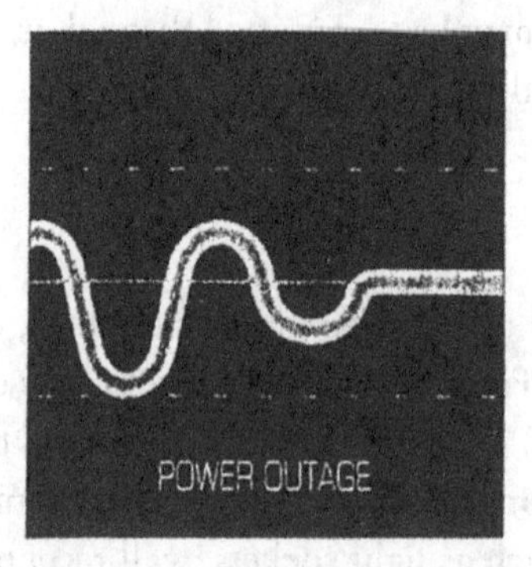

FIGURE 9-13 Power outage with sine wave diminishing to zero.

Looking for Shorts

Shorted turns in the winding of a motor behave like a shorted secondary of a transformer. A motor with a shorted winding will draw excessive current while running at no load. Measurement of the current can be made without disconnecting the lines. This means that you engage one of the lines with the split-core transformer of the tester. If the ammeter reading is much higher than the full-load ampere rating on the nameplate, the motor is probably shorted.

In a two- or three-phase motor, a partially shorted winding produces a higher current reading in the shorted phase. This becomes evident when the current in each phase is measured.

Motors with Squirrel-Cage Rotors

Loss of output torque at rated speed in an induction motor may be due to open bars in the squirrel-cage rotor. To test the rotor and determine which rotor bars are loose or open, place the rotor in a growler. Engage the split-core ammeter around the lines going to the growler, as shown in Figure 9-14. Set the switch to the highest current range. Switch on the growler, and then set the test unit to the approximate current range. Rotate the rotor in the growler, and take note of the current indication whenever the growler is energized. The bars and end rings in the rotor behave similarly to a shorted secondary of a transformer. The growler windings act as the primary. A good rotor will produce approximately the same current indications for all positions of the rotor. A defective rotor will exhibit a drop in current reading when the open bars move into the growler field.

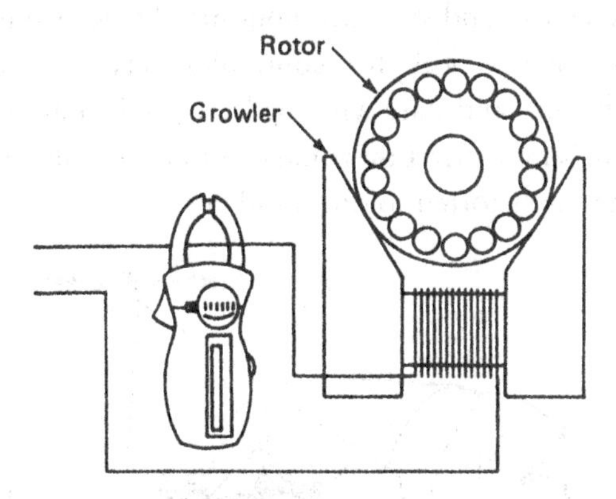

FIGURE 9-14 Using a growler to test a rotor.

Testing the Centrifugal Switch in a Single-Phase Motor

A defective centrifugal switch may not disconnect the start winding at the proper time. To determine conclusively that the start winding remains in the circuit, place the split-core ammeter around one of the start-winding leads. Set the instrument to the highest current range. Turn on the motor switch. Select the appropriate current range. Observe whether there is any current in the start-winding circuit. A current indication signifies that the centrifugal switch did not open when the motor came up to speed (Figure 9-15).

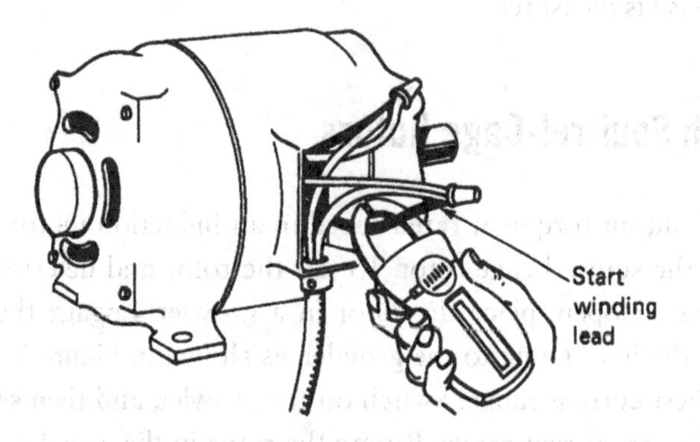

FIGURE 9-15 Checking the centrifugal switch with a clamp-on meter. (*Courtesy of Amprobe Instrument Co.*)

Testing for Short Circuits Between Run and Start Windings

A short between run and start windings may be determined by using the ammeter and line voltage to check for continuity between the two separate circuits. Disconnect the run- and start-winding leads, and connect the instrument as shown in Figure 9-16. Set the meter on voltage. A full-line voltage reading will be obtained if the windings are shorted to one another.

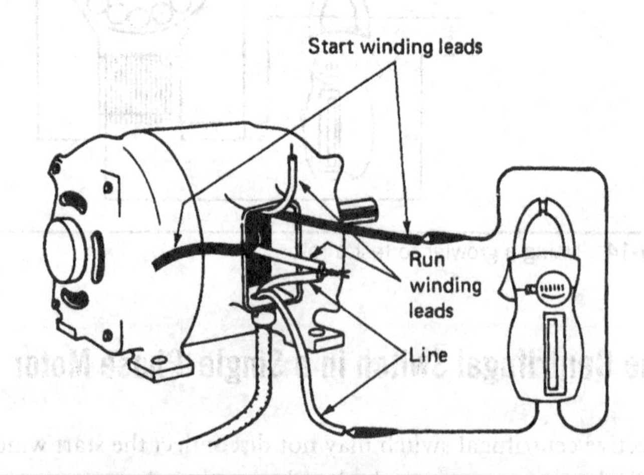

FIGURE 9-16 Finding a shorted winding using a clamp-on meter. (*Courtesy of Amprobe Instrument Co.*)

Capacitor Testing

Defective capacitors are very often the cause of trouble in capacitor-type motors. Shorts, opens, grounds, and insufficient capacity in microfarads are conditions for which capacitors should be tested to determine whether they are good. You can determine a grounded capacitor by setting the instrument on the proper voltage range and connecting it and the capacitor to the line, as shown in Figure 9-17. A full-line-voltage indication on the meter signifies that the capacitor is grounded to the can. A high-resistance ground is evident by a voltage reading that is somewhat below the line voltage. A negligible reading or a reading of no voltage indicates that the capacitor is not grounded.

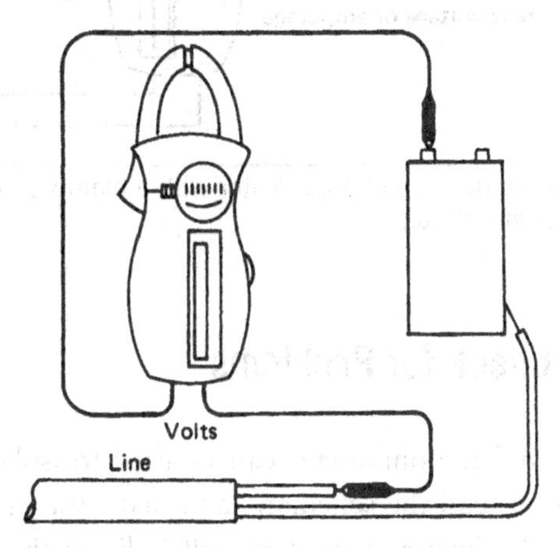

FIGURE 9-17 Finding a grounded capacitor with a clamp-on meter. (*Courtesy of Amprobe Instrument Co.*)

Measuring the Capacity of a Capacitor

To measure the capacity of a capacitor, set the test unit's switch to the proper voltage range, and read the line-voltage indication. Then set the unit to the appropriate current range, and read the capacitor current indication. During the test, keep the capacitor on the line for a very short period of time because motor-starting electrolytic capacitors are rated for intermittent duty (Figure 9-18). The capacity in microfarads is then computed by substituting the voltage and current readings in the following formula, assuming that a full 60-hertz line is used:

$$\text{Microfarads} = \frac{2{,}650 \text{ amperes}}{\text{volts}}$$

An open capacitor will be evident if there is no current indication in the test. A shorted capacitor is easily detected. It will blow the fuse when the line switch is turned on to measure the line voltage.

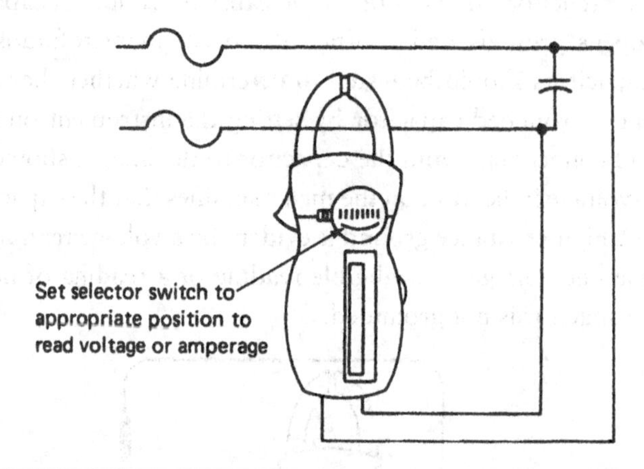
Set selector switch to
appropriate position to
read voltage or amperage

FIGURE 9-18 Finding the size of a capacitor with a clamp-on meter. (*Courtesy of Amprobe Instrument Co.*)

Using Meters to Check for Problems

The voltmeter and the ohmmeter can be used to isolate various problems. You should be able to read the schematic and make the proper voltage or resistance measurements. An incorrect reading will indicate the possibility of a problem. Troubleshooting charts will aid in isolating the problem to a given system. Once you have arrived at the proper system that may be causing the symptoms, you will then need to use the ohmmeter with the power off to isolate a section of the system. Once you have zeroed in on the problem, you can locate it by knowing what the proper reading should be. Deviation from a published reading by over 10 percent is usually indicative of a malfunction, and in most cases, the component part must be replaced to ensure proper operation and no call-backs.

Using a Volt-Ammeter for Troubleshooting Electric Motors

Most electrical equipment will work satisfactorily if the line voltage differs ±10 percent from the actual nameplate rating. In a few cases, however, a 10 percent voltage drop may result in a breakdown. Such may be the case with an induction motor that is being loaded to its fullest capacity on both start and run. A 10 percent loss in line voltage will result in a 20 percent loss in torque.

The full-load current rating on the nameplate is an approximate value based on the average unit coming off the manufacturer's production line. The actual current for any unit may vary by as much as ±10 percent of rated output. However, a motor whose load current exceeds the rated value by 20 percent or more will have a reduced life due to higher operating temperatures, and the reason for excessive current should be determined. In many cases it may simply be an overloaded motor. The percentage increase in load will not correspond with the percentage increase in load current. For example, in the case of a single-phase induction motor, a 35 percent increase in current may correspond to an 80 percent increase in torque output.

Operating conditions and behavior of electrical equipment can be analyzed only by actual measurement. A comparison of the measured terminal voltage and current will check whether the equipment is operating within electrical specifications.

A voltmeter and an ammeter are needed for the two basic measurements. To measure voltage, the test leads of the voltmeter are placed in contact with the terminals of the line under test. To measure current, the conventional ammeter must be connected in series with the line so that the current will flow through the ammeter.

To insert the ammeter, you have to shut down the equipment, break open the line, connect the ammeter, and then start up the equipment to read the meter. You have to do the same to remove the meter once it has been used. Then there are other time-consuming tests that may have to be made to locate a problem. However, all this can be eliminated by the use of a clamp-on volt-ammeter (Figure 9-19).

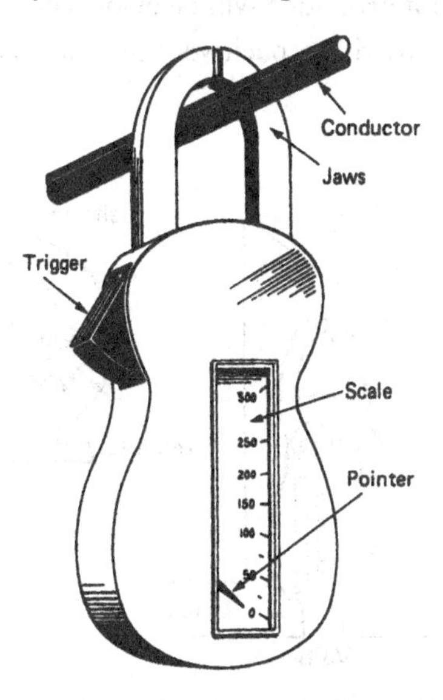

FIGURE 9-19 Clamp-on volt-ammeter. (*Courtesy of Amprobe Instrument Co.*)

Clamp-On Volt-Ammeter

The pocket-sized volt-ammeter shown in Figure 9-19 is the answer to most troubleshooting problems on the job. The line does not have to be disconnected to obtain a current reading. The meter works on the transformer principle that picks up the magnetic lines surrounding a current-carrying conductor and then presents this as a function of the entire amount flowing through the line. Remember that earlier we discussed how the magnetic field strength in the core of a transformer determines the amount of current in the secondary. Well, the same principle is used here to detect the flow of current and how much.

To get transformer action, the line to be tested is encircled with the split-type core simply by pressing the trigger button. Aside from measuring terminal voltages and load currents, the split-core ammeter-voltmeter can be used to track down electrical difficulties in electric motor repair.

Looking for Grounds

To determine whether a winding is grounded or has a very low value of insulation resistance, connect the unit and test leads as shown in Figure 9-20. Assuming that the available line voltage is approximately 120 volts, use the unit's lowest voltage range. If the winding is grounded to the frame, the test will indicate full-line voltage.

A high-resistance ground is simply a case of low insulation resistance. The indicated reading for a high-resistance ground will be a little less than line voltage. A winding that is not grounded will be evidenced by a small or negligible reading. This is due mainly to the capacitive effect between the windings and the steel lamination.

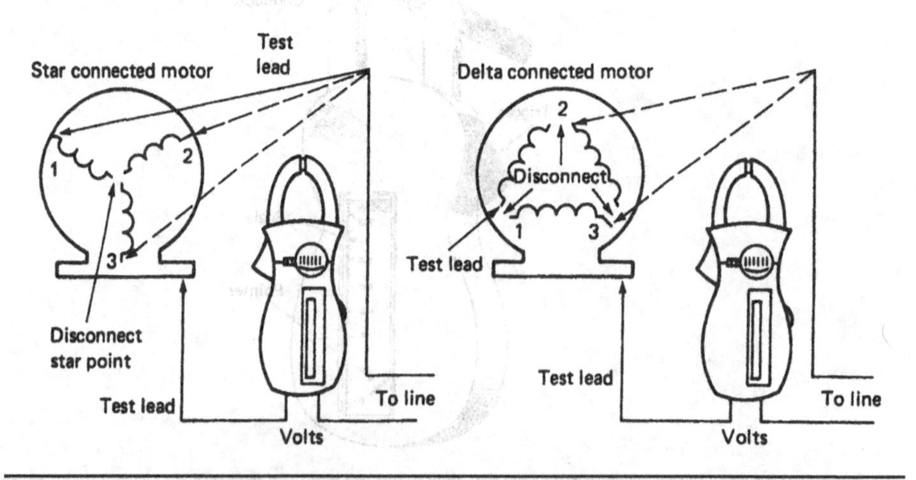

Figure 9-20 Grounded phase of a motor.

To locate the grounded portion of the windings, disconnect the necessary connection jumpers and test. Grounded sections will be detected by a full-line voltage indication.

Looking for Opens

To determine whether a winding is open, connect test leads as shown in Figures 9-21 and 9-22. If the winding is open, there will be no voltage indication. If the circuit is not open, the voltmeter indication will read full-line voltage.

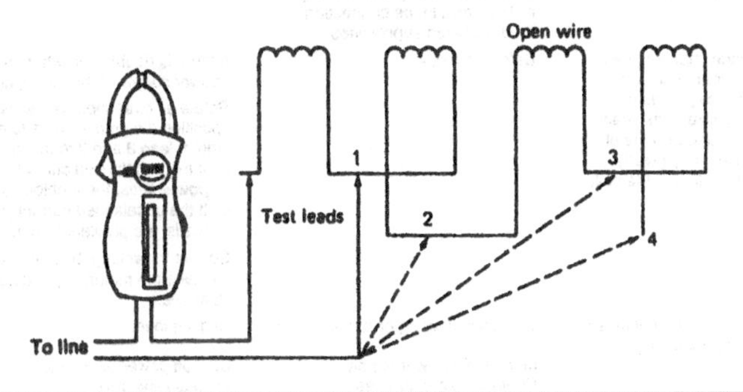

Figure 9-21 Isolating an open phase.

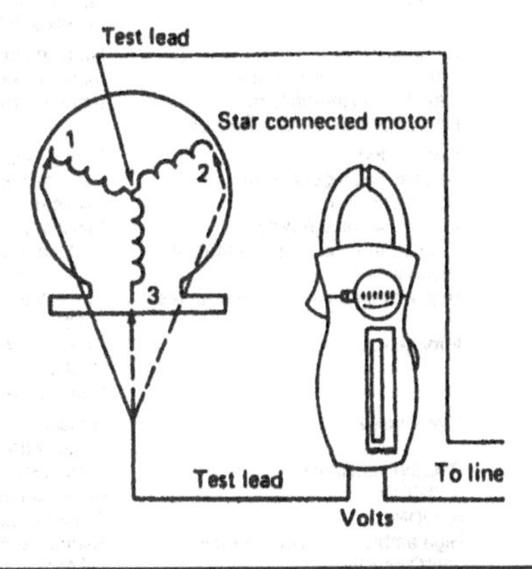

Figure 9-22 Finding an open phase.

Troubleshooting Guide

One of the quickest ways to troubleshoot is to check out symptoms and possible causes in a chart. This allows for quick isolation of the cause and suggests possible corrections. Both three-phase motors and their starters can be checked quickly this way. Tables 9-3 and 9-4 will aid in troubleshooting motors and starters.

TABLE 9-3 Three-Phase Motor Troubleshooting Guide

Symptom	Possible Causes	Correction
High input current (all three phases)	Accuracy of ammeter readings	First check accuracy of ammeter readings on all three phases.
Running idle (discounted from load)	High line voltage: 5 to 10% over nameplate	Consult power company—possibly decrease by using lower transformer tap.
Running loaded	Motor overloaded	Reduce load or use large motor.
	Motor voltage rating does not match power system voltage	Replace motor with one of correct voltage rating.
		Consult power company—possibly correct by using a different transformer tap.
Unbalanced input current (5% or more deviation from the average input current)	Unbalanced line voltage due to: a. Power supply b. Unbalanced system loading c. High-resistance connection d. Undersized supply lines	Carefully check voltage across each phase *at the motor terminals* with good, properly calibrated voltmeter.
Note: A small voltage imbalance will produce a large current imbalance. Depending on the magnitude of imbalance and the size of the load, the input current in one or more of the motor input lines may greatly exceed the current rating of the motor.	Defective motor	If there is doubt as to whether the trouble lies with the power supply or the motor, check as follows: Rotate *all three input power lines* to the motor by one position (i.e., move line 1 to motor lead 2, line 2 to motor lead 3 and 3 to motor lead 1). a. If the unbalanced current pattern follows the *input power lines*, the problem is the power supply b. If the unbalanced current pattern follows the *motor leads*, the problem is in the motor. Correct the voltage balance of the power supply or replace the motor, depending on the answer to a and b above.
Excessive voltage drop (more than 2 or 3% of nominal supply voltage)	Excessive starting or running load	Reduce load.
	Inadequate power supply	Consult power company.
	Undersized supply lines	Increase line sizes.
	High-resistance connections	Check motor leads and eliminate poor connections
	Each phase lead runs in separate conduits	All three-phase leads must be in a single conduit, according to the *National Electrical Code** (This applies only to metal conduit with magnetic properties.)
Overload relays tripping upon starting 9 see also "Slow Starting"	Slow starting (10–15 seconds or more) due to high inertia load	Reduce starting load. Increase motor size if necessary.
	Low voltage at motor terminals	Improve power supply and/or increase line size.
Running loaded	Overloaded	Reduce load or increase motor size.
	Unbalanced input current	Balance supply voltage
	Single phasing	Eliminate.
	Excessive voltage drop	Eliminate (see above).
	Too frequent starting or intermittent overloading	Reduce frequency of starts and overloading or increase motor size.
	High ambient starter temperatures	Reduce ambient temperature or provide outside source of cooler air.
	Wrong-size relays	Correct size per nameplate current of motor. Relays have built in allowances for service factor current. Refer to *National Electrical Code*®
Motor runs excessively hot	Overloaded	Reduce load or load peaks and number of starts in cycle or increase motor size
	Blocked ventilation a. TEFCs b. ODPs	Clean external ventilation system; check fan. Blow out internal ventilation passages. Eliminate external interference to motor ventilation.
	High ambient temperature over 40°C or 105°F	Reduce ambient temperature or provide outside source of cooler air.
	Unbalanced input current	Balance supply voltage. Check motor leads for tightness.
	Single phased	Eliminate.
Won't start (just hums and heats up)	Single phased	Shut power off. Eliminate single phasing Check motor leads for tightness.
	Rotor or bearings locked	Shut power off. Check shaft for freeness of rotation. Be sure proper sized overload relays are in each of the three phases of starter. Refer to *National Electric Code*®

(*Continued*)

TABLE 9-3 *(Continued)*

Symptom	Possible Causes	Correction
Runs noisy under load	Single phased	Shut power off. If motor cannot be restarted, it is single phased. Eliminate single phasing. Be sure that proper-sized overload relays are in each of the three phases of the starter. Refer to *National Electrical Code®*.
Slow starting (10 or more seconds on small motors; 15 or more seconds on large motors)		
Across the line start	Excessive voltage drop (5–10% voltage drop causes 10–20% or more drop in starting torque)	Consult power company; check system. Eliminate voltage drop.
	High-inertia load	Reduce starting load or increase motor size.
Reduced voltage start	Excessive voltage drop Loss of starting torque	Check and eliminate.
Wye-delta	Starting torque reduced to 33%	Reduce starting load or increase motor size.
PWS	Starting torque reduced to 50%	Choose starting method with higher starting torque.
Autotransformer	Starting torque reduced 25 to 64%	Reduce time delays between first and second steps on starter; get motor across the line sooner.
Load speed appreciably below nameplate speed	Overload	Reduce load or increase voltage.
	Excessively low voltage	*Note*: A reasonable overload or voltage drop of 10–15% will reduce speed only 1–2%. A report of any greater drop would be questionable.
	Wrong nameplate	If speed is off appreciably (i.e., from 1800 to 1200 rpm, check Lincoln code stamp (on top of stator) with nameplate. If codes do not agree, replace with motor of proper speed.
	Inaccurate method of measuring rpm	Check meter using another device or method
Excessive vibration (mechanical)	Out of balance	
	a. Motor mounting	Be sure motor mounting is tight and solid
	b. Load	Disconnect belt or coupling; restart motor. If vibration stops, the unbalance was in load.
	c. Sheaves or coupling	Remove sheave or coupling; securely tape 1/2 key in shaft keyway and restart motor. If vibration stops, the imbalance was in the sheave or coupling.
	d. Motor	If the vibration does not stop after checking a, b, and c above, the imbalance is in the motor; replace the motor
	e. Misalignment on close-coupled application	Check and realign motor to the driven machine.
Noisy bearings (listen to bearings)		
Smooth midrange hum	Normal fit	Bearing OK.
High whine	Internal fit of bearing too tight	Replace bearing; check fit.
Low rumble	Internal fit of bearing too loose	Replace bearing; check fit.
Rough clatter	Bearing destroyed	Replace bearing; avoid: a. Mechanical damage b. Excess greasing c. Wrong grease d. Solid contaminants e. Water running into motor f. Misalignment on close-coupled application. g. Excessive belt tension
Mechanical noise	Driven machine or motor noise?	Isolate motor from driven machine; check difference in noise level.
	Motor noise amplified by resonant mounting	Cushion motor mounting or dampen source of resonance.
	Driven machine noise transmitted to motor through drive	Reduce noise of driven machine or dampen transmission to motor.
	Misalignment on close-coupled application	Improve alignment.

Source: Courtesy of Lincoln Electric Co.

Table 9-4 Troubleshooting Motor Starters

Symptoms	Possible Causes	Correction
	Magnetic and Mechanical Parts	
Noisy magnet Humming	Misalignment or mismating of magnet pole faces	Realign or replace magnet assembly.
	Foreign matter on pole face (dirt, lint, rust, etc.)	Clean (but do not file) pole faces and realign if necessary.
	Low voltage applied to coil	Check system and coil voltage. Observe voltage variations during startup time.
Loud buzz	Broken shading coil	Replace shading coil and/or magnet assembly.
Failure to pick up and seal in	Low voltage	Check system voltage, coil voltage, and watch for voltage variations during start.
	Wrong magnet coil or wrong connection	Check wiring, coil nomenclature, etc.
	Coil open or shorted	Check with an ohmmeter and when in doubt, replace.
	Mechanical obstruction	Disconnect power and check for free movement of magnet and contact assembly.
Failure to drop out	"Gummy" substance on pole faces or magnet slides	Clean with nonvolatile solvent, degreasing fluid, possibly gasoline (with caution).
	Voltage or coil not removed	Shorted seal-in contact (exact cause found by checking coil circuit).
	Worn or rusted parts causing binding	Clean or replace worn parts.
	Residual magnetism due to lack of air gap in magnet path	Replace any worn magnet parts or accessories.
	Clean	
Contact clatter (source is probably from magnetic assembly	Broken shading coil	Replace assembly.
	Poor contact continuity in control circuit	Improve contact continuity or use holding-circuit interlock (three-wire control).
	Low voltage	Correct voltage condition. Check momentary voltage dip during start.
Welding	Abnormal inrush of Current	Use larger contactor or check for grounds, shorts, or excessive motor load current.
	Rapid jogging	Install larger device rated for jogging service or caution operator.
	Insufficient tip pressure	Replace contact springs; check contact carrier for deformation or damage.
	Low voltage preventing magnet from sealing	Correct voltage condition. Check momentary voltage dip during starting.
	Foreign matter preventing contacts from closing	Clean contacts with nonvolatile solvent. Contactors, starters, and control accessories used with very small current or low voltage should be cleaned with solvent and then with acetone to remove the solvent residue.
	Short circuit	Remove short fault and check to be sure that fuse or breaker size is correct.
Short contact life or overheating	Filing or dressing	Do not file silver contacts. Rough spots for discoloration will not harm them or impair their efficiency
	Interrupting excessively high currents.	Install larger device or check for grounds, shorts, or excessive motor currents.
	Coils	
Open circuit	Mechanical damage	Handle and store coils carefully. Replace coil.
Cooked coil (overheated)	Overvoltage or high ambient temperature	Check application and circuit. Coils will operate over a range of 85–110% rated voltage.
	Incorrect coil	Check rating and, if incorrect, replace with proper coil.
	Shorted turns caused by mechanical damage or corrosion	Replace coil.
	Undervoltage, failure of magnet to seal in	Correct system voltage.
	Dirt or rust on pole faces, increasing air gap	Clean pole faces.
	Sustained low voltage	Remedy according to local code requirements, low voltage system protection, etc.
	Overload Relays	
Nuisance tripping	Sustained overload	Check for motor or electrical equipment grounds and shorts, as well as excessive motor currents due to overload. Check motor winding resistance to ground.
	Loose connections	Clean connections and tighten. This includes load wires and heater element mounting screws.
	Incorrect heater	Check heater sizing and ambient temperature.
Failure to trip out (causing motor burn-out)	Mechanical binding, dirt, corrosion, etc.	Clean or replace.

(Continued)

TABLE 9-4 *(Continued)*

Symptoms	Possible Causes	Correction
	Incorrect heater or heaters omitted and jumper wires used instead	Recheck ratings and heater size. Correct if necessary.
	Wrong calibration adjustment	Consult factory. Calibration adjustment is not normally recommended unless factory supervised. It is customary to return units to factory for check and calibration.
	Manual Starters	
Failure to operate (mechanically)	Mechanical parts, including springs, worn or broken	Replace parts as needed.
	Welded contacts due to misapplication or other abnormal cause	Replace contacts and recheck operation.
Trips out prematurely	Motor overload, incorrect heaters, or misapplication	Check conditions and replace or adjust as needed.
	Pushbuttons	
Button inoperable Mechanical	Shaft binding due to dirt or residue	Check, clean, and clear.
	Contact board spring broken	Replace contact board.
Electrical	Contaminated contacts and corrosion	Clean.
	Excessive jogging	Install larger device or check rated for jogging or caution operator.
	Weak contact pressure	Replace contact springs; check contact carrier for deformation or damage.
	Dirt or foreign matter on contact surface	Clean contacts with nonvolatile solvent.
	Short circuits	Remove short fault and check to be sure fuse or breaker size is correct.
	Loose connection	Clean and tighten.
	Sustained overload	Install larger device or check for excessive motor load current.
	Excessive wear	Higher-than-normal voltage will cause unnecessary forces that may result in mechanical wear.
Contacts, supports, discoloring	Loose connections	Tighten hardware or replace.

Source: Courtesy of Square D.
Note: Any contact replacement should include a complete set replacement, including support springs, screws, etc.

Motor Life

The stator windings of integral-horsepower AC motors are capable of full-power operation for many years. However, winding life can be shortened by any combination of the following:

- **Mechanical damage** produces weak spots in the insulation. It can occur during maintenance of the motor or result from such operating problems as severe vibration.
- **Excessive moisture** encountered in service causes deterioration of the insulation.
- **High dielectric stress**, such as voltage surges or excess input current, can cause overheating and insulation deterioration.
- **High temperature** reduces the ability of the insulation system to withstand mechanical or electrical abuse. Overtemperature is usually a result of poor installation or misapplication of the motor.

Regardless of the reason for failure, the obvious result is thermal degradation of the insulation or burnouts. The rate of insulation degradation is increased by higher temperatures. In fact, insulation life is reduced by about half for each 10°C increase of winding temperature. Therefore, long winding life requires normal operating temperatures.

Forced ventilation is generally an inherent design feature of induction motors. Decreased cooling air volume caused by blocked air passages, blower failure, or low air density at higher altitudes leads to overheating and shortened winding life.

Ambient Temperature

The insulation system of motors is usually designed to operate at a maximum 40°C (104°F) ambient temperature. Any increase of ambient over 40°C requires derating thc motor, or its expected life will be shortened. Factors that raise input air temperature include placement of motors in discharge airstreams from other equipment and high-temperature locations.

To calculate the derating required for high ambient temperatures, multiply the ambient factor obtained from Figure 9-23 by the rated horsepower of the motor. The ambient factor from Figure 9-23 also can be used to up-rate motors used in ambient temperatures under 40°C. Whenever a motor is derated or up-rated, the starting, pull-up, and breakdown torques remain the same as the nameplate rating, but the bearings, shaft, and other components may be subjected to life that is a function of the new rating.

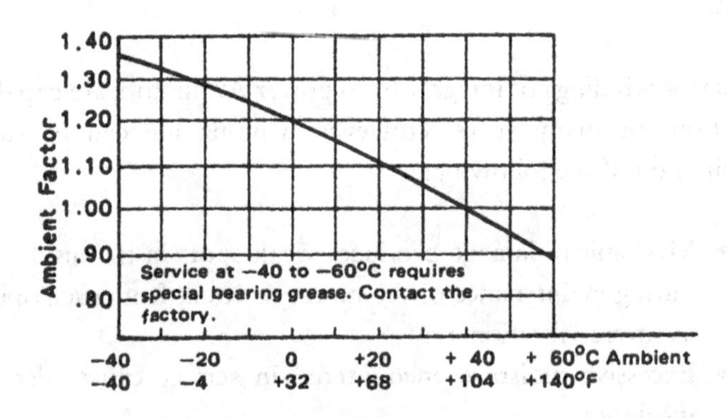

FIGURE 9-23 Derating a motor by ambient temperature. (*Courtesy of Lincoln Electric Co.*)

Performance Characteristics

Performance characteristics of motors change when they are operated at high or low voltages at rated frequency. Table 9-5 shows the characteristic and the change when operated 10 percent above rated voltage and 10 percent below rated voltage. Winding failures resulting from extreme voltage variations are identical to those of overloads because the input current is uniformly excessive.

Table 9-5 Performance Characteristic Changes When Motors Are Operated at High or Low Voltages (at Rated Frequency)

	Change, Relative to Performance at Rated Voltage	
Performance Characteristic	**When Actual Voltage Is 10% above Rated Voltage**	**When Actual Voltage Is 10% below Rated Voltage**
Starting-Pull-Up–Breakdown		
Current	Increases 10–12%	Decreases 10–12%
Torque	Increases 21–25%	Decreases 19–23%
	Idle	
Current		
1800/1200/900-rpm motors	Increases 12–39%	Decreases 10–21%
3600-rpm motors	Increases 28–60%	Decreases 21–34%
	Rated Load	
Current (1800- and 3600-rpm motors only)		
143T–182T	Varies −4 to +11%	Varies −11 to +4%
184T-256T	Decreases 1–10%	Increases 1–10%
284T-445T	Decrease 0–7%	Increases 0–7%
Current (1200- and 900-rpm motors only): These motors do not follow the pattern above. Most will respond to a 10% overvoltage with an increase (as much as 15%) in input current. A similar reduction in input current is characteristic of a 10% reduction in input voltage.		
Power factor[a]	Decreases 5–8%	Increases 2%
Efficiency[a]	Little change	Decreases 2%
Speed	Increases 1%	Decreases 1.5%
Percent slip	Decreases 1.0	Increases 1.5

[a]Note at ¾ and ½ load, these changes are approximately the same.

Voltage Unbalance

Voltage unbalance occurs when the phase voltages differ from one another. At one extreme, this phenomenon occurs as single phasing. It also can occur in a more subtle form as voltage differences among the three phases.

Single Phasing

A single-phase motor at standstill will not start, and the large inrush current overheats the windings rapidly. If the single phasing occurs when the motor is running, the motor will continue to supply torque, but the input current to the remaining two phases will increase, causing overheating.

Voltage Unbalance

By definition,

$$\text{voltage unbalance (\%)} = 100\,\frac{\text{maximum voltage deviation from average voltage}}{\text{average voltage}}$$

When the voltage varies among the three phases, the unbalanced voltage raises the current in one or two phases, causing overheating.

Voltage unbalances are caused by the following conditions:

- Unequal loading of a three-phase system
- Unequal tap settings on transformers
- Poor connections in the power supply
- Open delta transformer systems
- Improper function of capacitor banks

The winding failure pattern from either single phasing or voltage unbalance will be similar. Figure 9-24 shows two phases that are overheated in a Y-connected motor. A delta-connected motor will have one overheated phase.

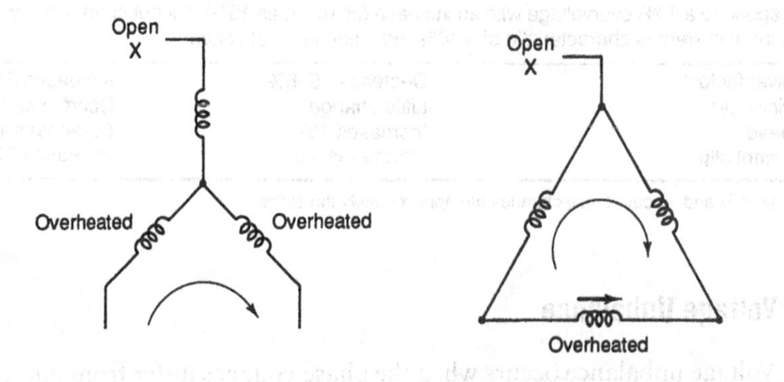

FIGURE 9-24 Heating effects on wye- or delta-connected windings when voltage is unbalanced or single phasing occurs.

Motor Protection

In each of the abnormal operating conditions listed—overloading, voltage regulation, and voltage unbalance—the risk of excessive winding temperature exists because input current is higher than nameplate current. Motor controls and protective devices must prevent input current from exceeding nameplate current for extended periods.

The *National Electrical Code* has established standards that apply to the control and protection of motors and associated circuits. For example, Figure 9-25 is a general circuit for the installation of squirrel-cage inductor motors. Consult the *National Electrical Code* for complete requirements for all installations.

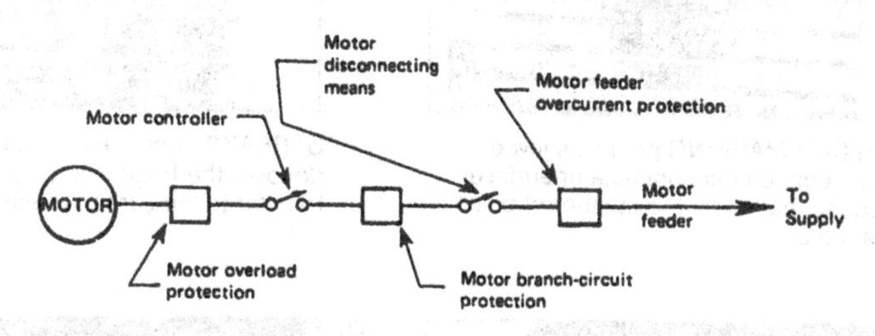

FIGURE 9-25 Diagram showing protective devices.

Motor controllers and their associated circuits will vary among different models as well as different controller manufacturers. Figure 9-26 is a functional schematic diagram of a motor controller that incorporates overload protection. Good engineering practice emphasizes using three overload relays, one in each phase line of the motor, to give protection from voltage unbalance conditions. The size of the controller and overload protection relays must be in accord with the controller manufacturer's specifications as well as the *National Electrical Code*. As a further safeguard against motor failure due to prolonged locked rotor conditions, the overload protection should trip out in 15 seconds or less at locked-rotor current.

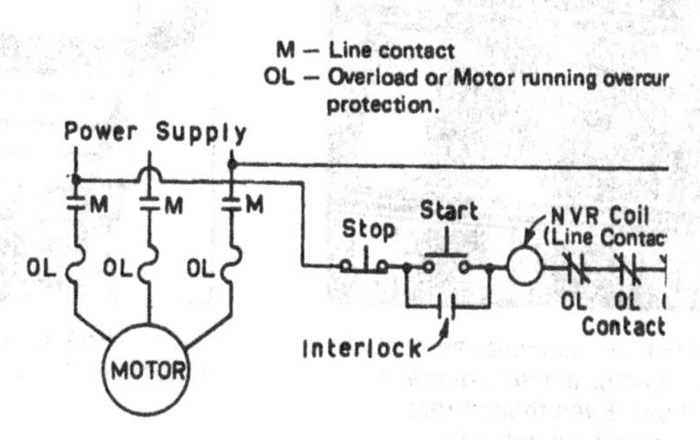

FIGURE 9-26 A motor control consists of a contactor and overload relays.

DC Motor Problems

One of the problems associated with DC motors is high maintenance costs in both time and materials. They do have a tendency to need attention at the commutator and brush level. Figure 9-27 shows some of the problems associated with worn or damaged commutators.

(A)

PITCH BAR-MARKING produces low or burned spots on the commutator surface that equals half or all the number of poles on the motor.

(B)

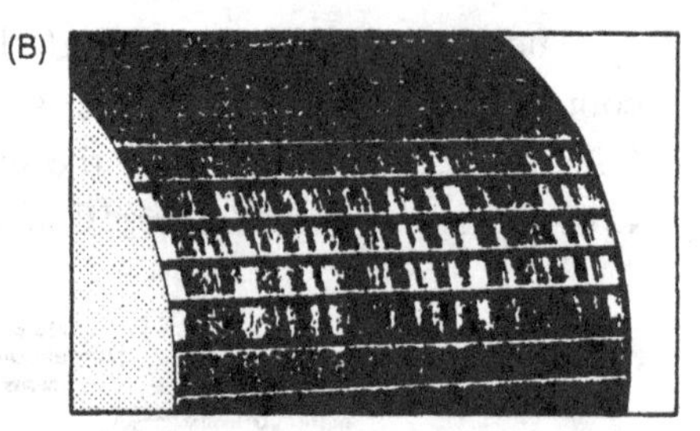

STREAKING on the commutator surface denotes the beginning of serious metal transfer to the carbon brush.

(C)

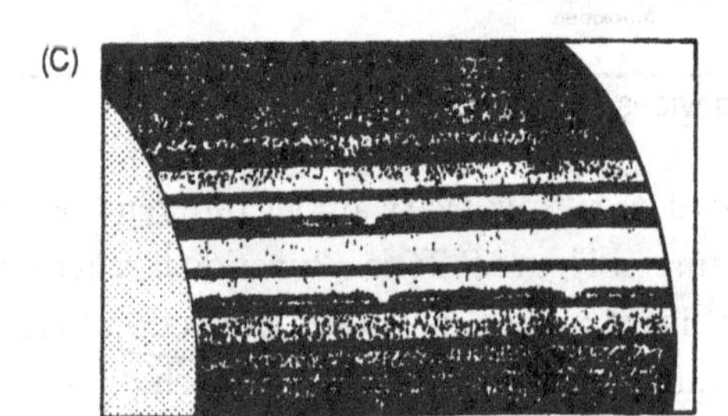

HEAVY SLOT BAR-MARKING involves etching of the trailing edge of commutator bar in relation to the numbered conductors per slot.

(D)

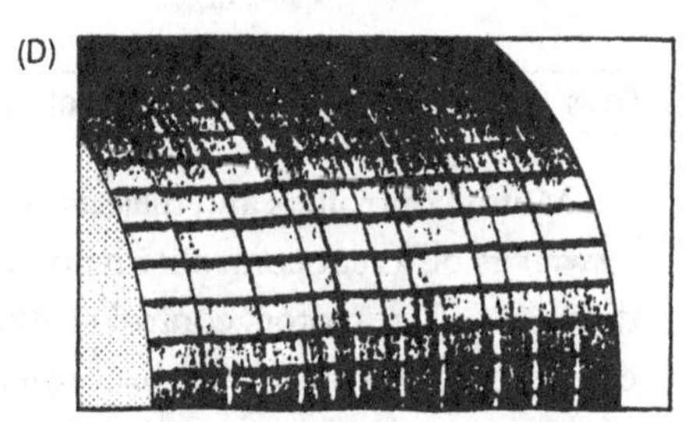

THREADING of commutator with fine lines is a result of excessive metal transfer leading to resurfacing and excessive brush wear.

(E)

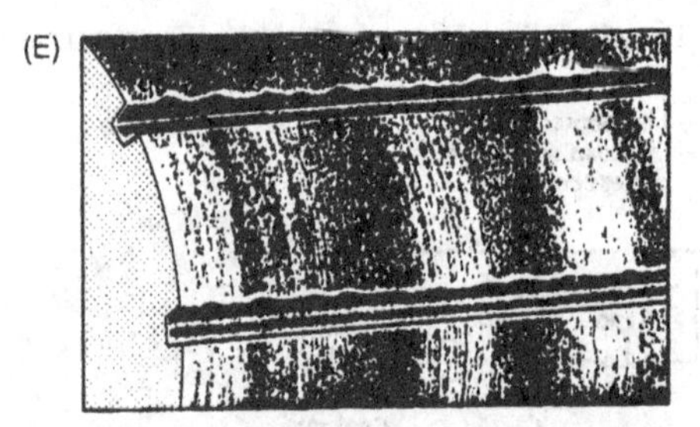

COPPER DRAG is an abnormal amount of excessive commutator material at the trailing edge of bar. Even though rare, flashover may occur if not corrected.

(F)

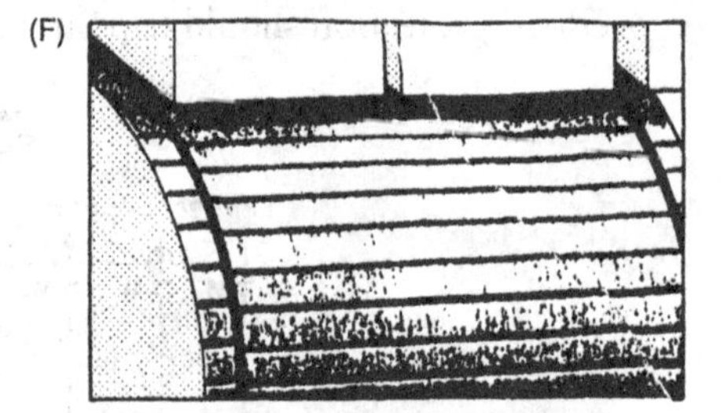

GROOVING is caused by an abrasive material in the brush or atmosphere.

FIGURE 9-27 Worn or damaged commutators. (*Courtesy of Reliance.*)

Satisfactory commutator surfaces are shown in Figure 9-28. Determine desirable and undesirable commutator surfaces by checking these with those shown in Figure 9-27. Figure 9-29 shows what can be done to correct the situation.

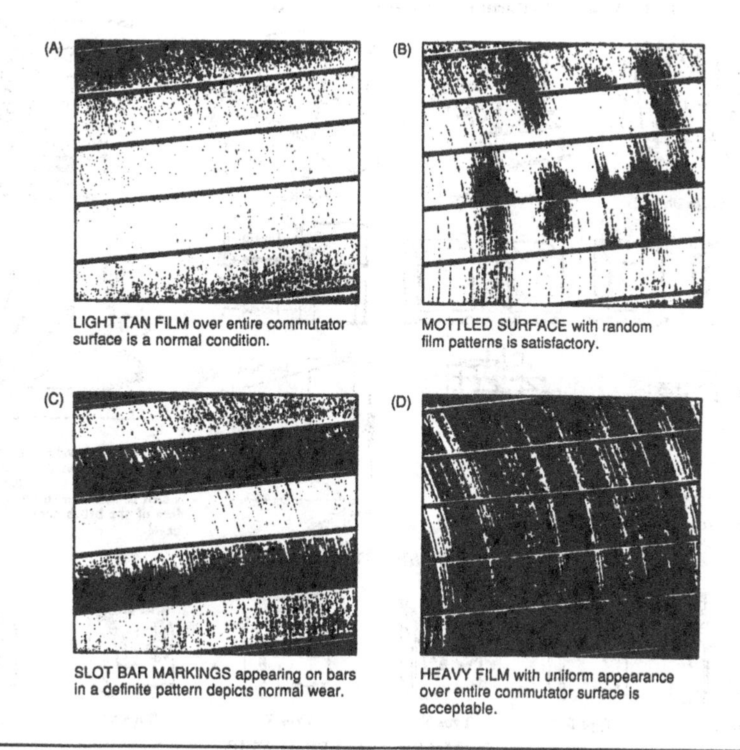

(A)
LIGHT TAN FILM over entire commutator surface is a normal condition.
(B)
MOTTLED SURFACE with random film patterns is satisfactory.
(C)
SLOT BAR MARKINGS appearing on bars in a definite pattern depicts normal wear.
(D)
HEAVY FILM with uniform appearance over entire commutator surface is acceptable.

FIGURE 9-28 Satisfactory commutator surfaces. (*Courtesy of Reliance.*)

								Contamination		Type of Brush In Use	
	Vibration	Brush Pressure (light)	Unbalanced Shunt Field	Armature Connection	Light Electrical Load	Electrical Overload	Electrical Adjustment	Gas	Abrasive Dust	Abrasive Brush	Porous Brush
Pitch bar-marking	■	■	■	■						■	
Slot bar-marking						■	■	■			
Copper Drag	■	■						■		■	
Streaking		■			■			■	■	■	■
Threading		■			■			■			■
Grooving									■	■	

FIGURE 9-29 Troubleshooting causes of worn commutators. (*Courtesy of Reliance.*)

Figure 9-30 shows 30 illustrations of brushes used in DC motors and generators.

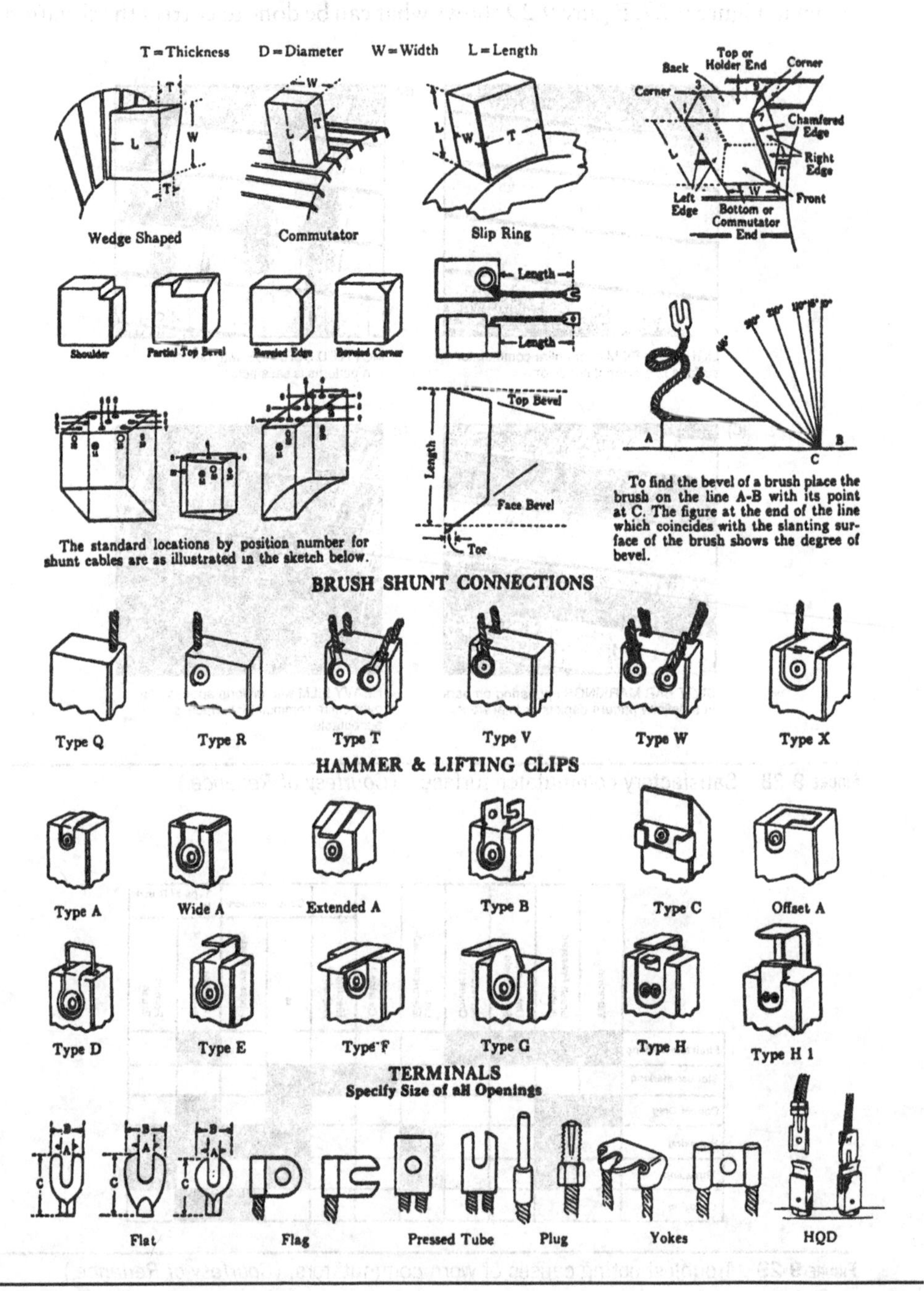

FIGURE 9-30 Brush types.

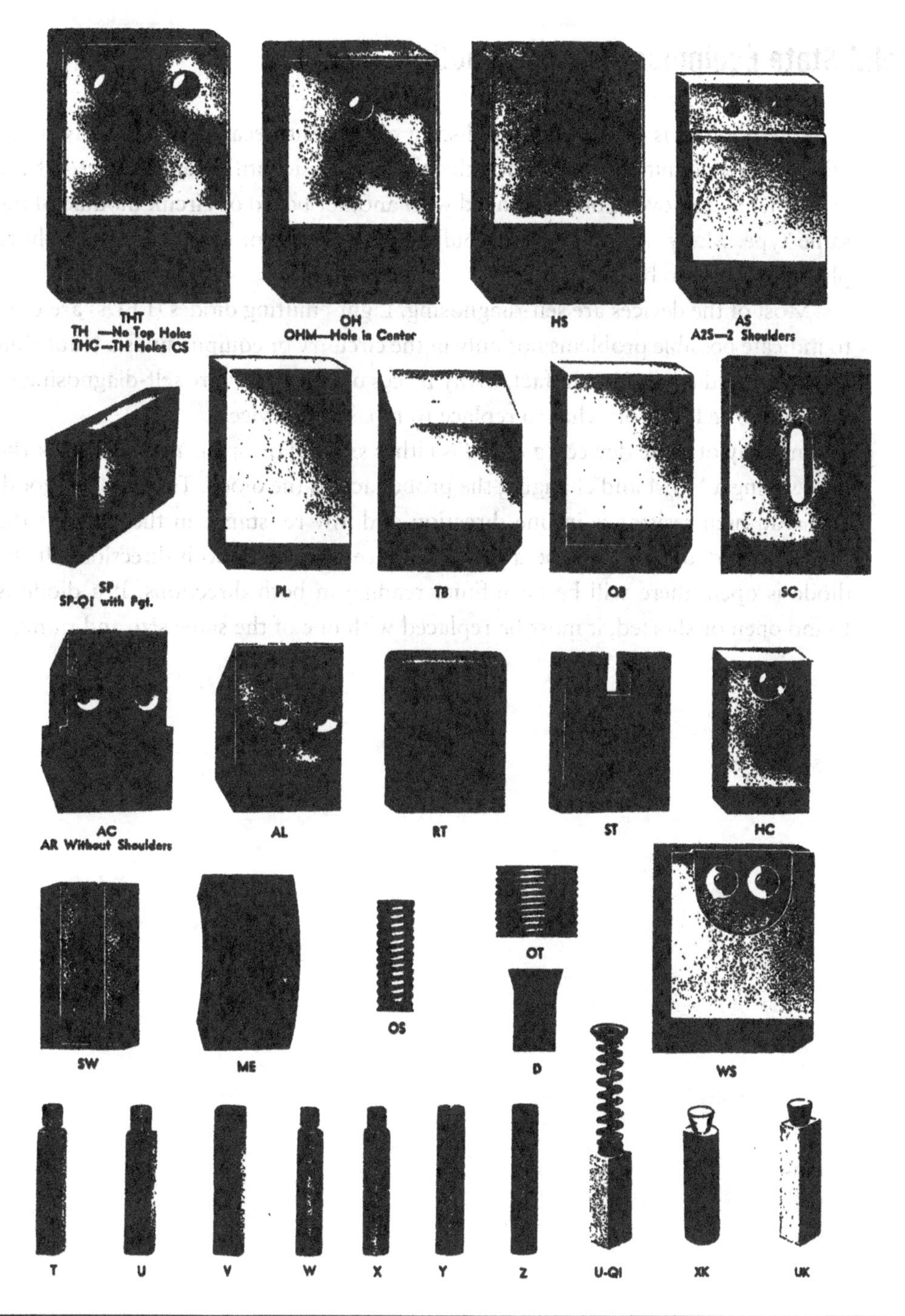

Figure 9-30 *(Continued)*

Solid-State Equipment Troubleshooting

Troubleshooting is easier with solid-state equipment because it is usually made with printed circuit boards or modules that can be identified as the cause of the problem and removed and exchanged with another board or circuit module of the same type. Many are removed by pulling the card from clips, and others have plugs that have to be disconnected.

Most of the devices are self-diagnosing. Light-emitting diodes (LEDs) are used to indicate possible problems not only in the circuitry or equipment as a whole but also in individual devices. In fact, many pieces of equipment are self-diagnosing to the point of telling you what to replace to repair the device.

In many of these devices, a diode is either shorted or open. You can check the diode using a VOM and changing the probes across the diode. The diode, if good, will read high resistance in one direction and low resistance in the other. If the diode is shorted, there will be a low-resistance reading in both directions. If the diode is open, there will be an infinite reading in both directions. If a diode is found open or shorted, it must be replaced with one of the same size and rating.

Review Questions

1. What is an indication of losing a ground in an electrical system?
2. How does a GFCI work?
3. What is the main duty of ball bearing lubrication?
4. What causes bearing failure?
5. Why is oil viscosity important to a motor?
6. How often should motors be oiled?
7. Why are commutator motors in need of more maintenance?
8. What three irregularities in a power supply affect motor power supplies?
9. Define voltage spikes. What causes them?
10. What is electrical noise? What is its source?
11. What are transients? What is their source?
12. What does a growler do?
13. How do you test for a grounded capacitor?
14. What is the main advantage of a clamp-on meter?
15. What does high temperature do to a motor?
16. What does *ambient temperature* mean?
17. What causes motor winding failures?
18. Is a light tan film over an entire commutator surface normal?
19. How is a diode checked with an ohmmeter?

CHAPTER 10

Robots and Robotics: Today and Tomorrow

Performance Objectives

After reading this chapter, you will be able to:

- Discuss the advantages of collaborative robots. You will also know what *cobot* means.
- Understand more about the history of robots and robotics.
- Know some of the "buyouts" in the robotics industry.
- Know what artificial intelligence is and how it is used.
- Know that robot use in industry is becoming more global.
- Know about advanced robotic systems.
- Know about important software and its wide use.
- Answer the review questions at the end of the chapter.

Collaborative robots, or *cobots*, are robots that can work right alongside personnel with no safety guarding based on the results of a mandatory risk assessment. One example of a collaborative robot is FANUC's CR-35iA robot (Figures 10-1 and 10- 2).

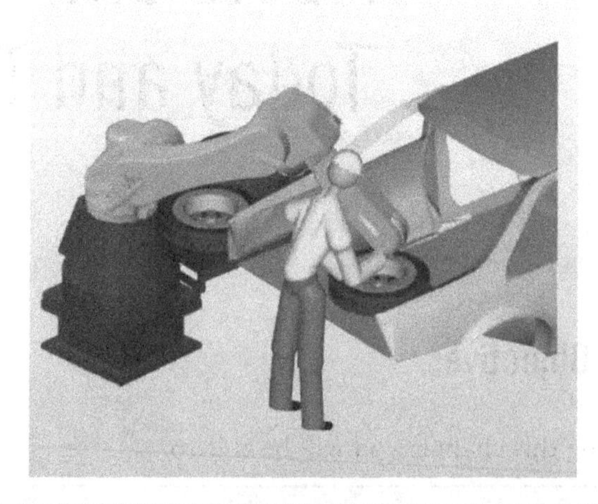

Figure 10-1 Loading spare tires into automobiles. (*Courtesy of FANUC.*)

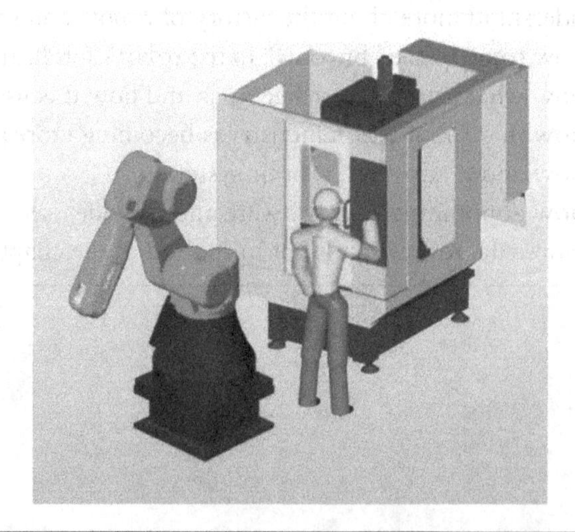

Figure 10-2 Loading and unloading machine tools. (*Courtesy of FANUC.*)

You can add value to your production by automating repetitive pick-and-place tasks with a pick-and-place robot like the one from Universal Robots (UR). You can increase accuracy and reduce shrinkage by placing a pick-and-place robot in your fulfillment center to handle pick-and-place tasks. The UR robot can run most pick-and-place applications autonomously, allowing your business to handle inventory even when employees have gone home.

The intuitive, easy-to-program interface makes the UR robot arms perfect for small-volume production. A collaborative robot arm used in repetitive pick-and-place tasks offers advantages in productivity and flexibility—letting you free up your workforce for more important tasks.

- Most pick-and-place applications can be run autonomously by the UR robot arm, making it a perfect pick-and-place robot.
- Increase productivity and flexibility with URs' pick-and-place robots. It requires superhuman abilities to repeat the same movement over and over again for many hours with exactly the same precision. This is why the URs' repeatability of ±0.1 millimeter (±0.004 inch) is perfect for automating quick-precision handling.
- Because of their small size and lightweight robot design, URs' pick-and-place robots can be easily deployed in applications in tight space conditions.
- Easy programming and a short average setup time make UR robot arms ideal even for small-volume production, where rearranging large-scale facilities would not be lucrative.
- Moving the pick-and-place robot to new processes is fast and easy, giving you the agility to automate almost any manual task, including those with small batches or fast changeovers. The robot is able to reuse programs for recurrent tasks.
- All UR robotic arms are certified IP-54. They will need protection when working in corrosive liquid environments.

History of Universal Robots

The history of Universal Robots is both interesting and informative. For instance, the company was founded in 2005 by the engineers Esben Østergaard, Kasper Støy, and Kristian Kassow. While doing joint research at the Syddansk University at Odense, they came to the conclusion that the robotics market was dominated by heavy, expensive, and unwieldy robots. So they developed their idea to make robot technology accessible to small and medium-sized enterprises.

In 2009, their first UR5 robots were available on the Danish and German markets. Since 2010, the company has expanded its activities steadily. Meanwhile, the robots are distributed globally.

In 2012, the company's second robot, UR10, was launched at the Automatica 2014 show in Munich. The company launched a totally revised version of its robots at that time. One year later, in the spring of 2015, the tabletop robot UR3 was launched globally.

In 2015, Universal Robots was purchased by Teradyne, which is headquartered in Reading, MA, for US$285 million. The robot arms are used in production by companies such as Lear Corporation, Franke (Company), Oticon, Johnson & Johnson, Clamcleats, Kunshan Dongwei, VW, and BMW.

Universal Robots' Main Products

The three main products are the compact tabletop robot UR3, the flexible robot arm UR5, and the biggest one, the UR10. All three are six-jointed robot arms with very low weights of, respectively, 11, 18, and 28 kilograms. The UR3 and UR5 have a lifting ability of 3 and 5 kilograms, respectively, and have a working radius of 500 and 850 millimeters (19.7 and 33.5 inches). In addition, the UR10 has a lifting capacity of 10 kilograms with a reach of 1,300 millimeters (51.2 inches). Each of the robots' joints can rotate through ±360 degrees and up to 180 degrees per second. Furthermore, the UR3 has an infinite rotation on the end joint. The accuracy of the robots' repetitions is ±0.1 millimeter (±0.0039 inch).

More Details on Neocortex

Neocortex makes flexible automation in unstructured environments possible for the first time. It has given industrial robots the ability to adapt and react to the world around them as well as to execute tasks in an ever-changing environment, says David Peters, CEO of Universal Robotics. "With this ability to learn, machines will be implemented in revolutionary ways across industries."

Neocortex was in development ever since 2001. It was created at Vanderbilt University with funding from the Defense Advanced Research Projects Agency (DARPA) and the *National Aeronautics and Space Administration* (NASA) and developed using Robonaut, NASA's humanoid robot. Neocortex allows a robot to learn how to complete a task. Once the task is learned, the machine observes its environment through more than 50 channels of sensor data. Drawing on what it learned from previous experiences, the Neocortex-enabled robot changes its actions as necessary in real time to complete the task.

Until now, machine automation and artificial intelligence were bound by the inability to handle unknown variables, limiting their usefulness to structured, controlled environments such as assembly lines. With its ability to allow machines to adapt and react to variables and learn from experiences, Neocortex has removed constraints to flexible automation. This leap forward unlocks the safety and productivity potential of automation for a variety of applications.

Neocortex is modeled on how we all learn. We sense our environment, act to change it, and learn from the process," said Dr. Alan Peters, UR cofounder and chief technical officer. "Neocortex allows machines to perform tasks that are beyond the capabilities of standard industrial robots. From materials handling to underwater mining, Neocortex creates robots that can operate in very dynamic environments and learn to carry out many tasks that are unsafe or impossible for people to perform."

This software will be launched in the materials-handling industry as an automated mixed-size box handler. UR partnered with Yaskawa/Motoman Robotics, the world leader in industrial robots, to provide a hardware–machine intelligence work-cell solution that features Neocortex software. Motoman's SDA Series of custom box-moving end effectors, along with a suite of sensors, includes the recently launched easy-to-use UR three-dimensional (3D) vision system called *Spaceal Vision*.

The Neocortex box-mover application replicates human object-moving capabilities in unstructured environments. This solution can be deployed to automate tasks including, but not limited to, palletization, depalletization, and floor-stacked truck off-loading, all of which involve repetitive lifting and twisting and can cause high rates of personal injury. The system not only identifies the object to be moved but also analyzes its placement, shape, orientation, and other factors to determine the best way to grasp, lift, move, and set the item down. It is able to distinguish boxes from one another, determine the fastest way to unload a pallet, and compensate for weight, orientation, and even crushed or wet containers. The system automates a difficult step in mixed-size box handling and has return on investment of 18 to 24 months. It is easy to implement in existing work cells and requires minimal modification in warehouse and distribution centers.

Some of the collaborative robots made today are shown in Figure 10-2. Universal Robots, FANUC, Elison, Cybotech, ESAB North America, Fared Robots, Binks, Cincinnati Milacron, IBM, and Comau are all still in production and are innovating as the need arises. The future looks bright for the robotics industry.

As mentioned previously, the history of robots is short. However, they are now getting a great deal of mention in a variety of magazines, such as *The Economist*, *The Financial Times*, *Time Magazine*, and the *Wall Street Journal*. It appears that interest in what robots can accomplish is just beginning to be recognized by the industrial world, which is now aware of the robot's advanced manufacturing techniques and its value in the automatic production of products.

There are many strategic acquisitions in the robotics arena, and they became more rampant in 2012 when Amazon acquired Kiva Systems to improve its warehousing methods well into the future. The $775 million transaction brought robots into consideration by a number of industries. Starting in 2013–2015, there have been big-dollar acquisitions by companies and equity firms. For instance, four of the recent acquisitions are

- **Ninebot**—a Chinese maker of mobile robots similar to Segways. Segways are used by police, vacationers, and security personnel.
- **Kuka**—a German robot manufacturer, one of the "big four" in the industry. Kuka bought Reis Robotics, which already had a factory in China. Kuka just opened a factory in Shanghai and now has two.
- **China South Rail (CSR)**—has acquired the UK-based SMD, which is a provider of underwater robotics technologies and systems. The acquisition is expected because China needed to acquire a core in deep-sea robotics technology. China now has just 30 robots per 10,000 workers employed in manufacturing industries. South Korea has 437, Japan has 323, Germany has 152, and the United States has 152. This shows how far the United States has to go to keep abreast of the demands of modern industry. The future looks bright in this area for robot expansion.
- **Teradyne**—just bought Universal Robots. Teradyne is normally an electronics manufacturer that will now have a robot manufacturing unit.

Global Robotics

Yaskawa Motoman was founded in 1989. The Motoman Robotics Division of Yaskawa America, Inc., is a leading robotics company in the Americas. Motoman has over 300,000 Motoman robots installed globally. Yaskawa provided automation products and solutions for virtually every industry and robotic application, including arc welding, assembly, coating, dispensing, materials handling, materials cutting, and materials removal.

Sales of robots in 2014 (latest year for statistics) amounted to 225,000 units, of which about 56,000 were sold in China. Chinese factories made about 16,000 of these units.

The Chinese are developing their own in-country robotics industry.

Advanced Systems

Visual-enhanced robotic systems have become the topmost or first reason for upgrading and deploying vision-enabled robots and a core reason for the steady upward growth of the robotics industry. Today, visual robotics systems are different from the old-style auto-making robots. The older system required the part to be worked on to be in a precise location at a specific time. The robot was blind and programmed to pick and process. Each step of the pick and process was hand programmed and quite detailed. Newer systems using cameras and software to identify and locate parts are more flexible and enable product movement from step to step to be less rigid and precise; as a consequence, the movement system is less costly.

Universal Robotics and its Neocortex system are now offering either vision systems that can supplement existing fixed systems or offering mobile manipulators that can find and determine how best to pick and place or handle all sorts of objects from plastic-wrapped toys to boxes, cases, and skids of materials. This cut the cost of expensive conveyor belts and movement systems. Replacing them with lower-cost mobile and bin-picking robots is rapidly gaining a big following because of the possible opportunities afforded the manufacturer and/or warehouse operator.

Artificial intelligence (AI) and various AI learning systems have been improving, especially regarding visual perception. Many new companies are now offering vision systems that can supplement existing fixed systems or offer mobile manipulators that can find and determine how best to pick and handle all sorts of objects such as plastic-wrapped toys to boxes, cases and skids, and spot welding. AI is intelligence exhibited by machines. In computer science, ideal "intelligent machines" are flexible, rational agents that perceive their environments and take actions that maximize their chance of success at some goal.

Collaborative robot arms can increase accuracy and reduce shrinkage when used properly in pick-and-place tasks. They are easy to program and are intuitive as well, and when the arms are used in repetitive pick-and-place tasks, they offer advantages in productivity and besides are flexible enough to let you free up your workforce for more important jobs. Keep in mind that these robots need protection when they are working in corrosive liquid environments.

Software for Robots

Universal Robots, Inc., has developed technology that enables mobile machines to perform tasks that are normally costly, dangerous, or impossible for humans to attempt. The company's software is known by the name *Neocortex*. This patented software allows mobile machines to learn from experiences in the physical world rather than having to be programmed to act. It enables these machines to perform highly specialized automated tasks that require them to react and adapt to their environments. They have the potential to increase productivity, profitability, and worker safety across industries worldwide.

The robotics industry has been very helpful in providing information and illustrations for this edition of this book. For this, the authors are grateful.

Review Questions

1. What does *collaborative* mean?
2. What gave robots a boost and brought about interest in robotics to work in warehousing?
3. Where are Segways used?
4. What country is the home of Kuka Robotics?
5. What is Ninebot?
6. Who bought Reis Robotics, Inc.?
7. What does CSR stand for?
8. How many robots were made in 2014?
9. What robot manufacturer is associated with Neocortex system?
10. What is artificial intelligence?
11. How can collaborative robots increase efficiency and accuracy?
12. Which robot manufacturer has technology that enables mobile machines?

APPENDIX A

Standard Units of Measurement with Conversions

The most often used measurement units have been standardized. European nations, as well as some others, use the metric system based on the meter (39.54 inches). Many of the robots available today have been made using the metric system of measurement. Those made in the United States make use of the U.S. system or the inch, foot, yard, etc. In most instances, the repair technician will encounter both standards. Therefore, a knowledge of how to convert from one to the other is necessary for the technician (see the chart on the next page). This also requires the technician to have two sets of tools available.

Weight

16 oz = 1 lb
2.2046 lb = 1 kg
2.309 ft water at 62°F = 1 psi
28.35 g = 1 oz
59.76 lb = weight of 1 cu ft of water at 212°F
0.062428 lb per cu ft = 1 kg/cu m
62.355 lb = weight of 1 cu ft water at 62°F
8⅓ (8.32675) lb = weight 1 gal water at 62°F

Power

1.3410 hp = 1 kW
2.545 Btu per hr = 1 hp
33,000 ft-lb per min = 1 hp
550-ft-lb per sec = 1 hp
745.7 W = 1 hp

Area

10.764 sq ft = 1m^2
1,273,239 circular mils = 1 sq in.
144 sq in. = 1 sq ft
645 mm^2 = 1 sq in.
9 sq ft = 1 sq yd
0.0929 m^2 = 1 sq ft

Mathematic

1.4142 = square root of 2
1.7321 = square root of 3
3.1416 = π = ratio of circumference of circle to diameter = ratio of area of a circle to square of radius
57.296 degrees = 1 rad (angle)
0.7854 × diameter squared = area of a circle

Pressure

14.223 psi = 1 kg per cm^2 = 1 "metric atmosphere"
2.0355 in. Hg at 32°F = 1 psi
2.0416 in. Hg at 62°F = 1 psi
2116.3 psf = atmospheric pressure
27.71 in. water at 62°F = 1 psi
29.921 in. Hg at 32°F = atmospheric pressure
30 in. Hg at 62°F = atmospheric pressure (approximate)
33.974 ft water at 62°F = atmospheric pressure
0.433 psi = 1 ft of water at 62°F
5196 psf = 1 in. water at 62°F
760 mm Hg = atmospheric pressure at 0°C
0.07608 lb = weight 1 cu ft air at 62°F and 14.7 psi

Volume

1728 cu in. = 1 cu ft
231 cu in = 1 gal (U.S.)
277.274 cu in = 1 gal (British)
27 cu ft = 1 cu yd
31 gal (31.5 U.S. gal) = 1 barrel
35.314 cu ft = 1 m^3
3.785 liters = 1 gal
61.023 cu in. = 1 liter
7.4805 gal = 1 cu ft

Units of Pressure

kg per cm^2 = kilograms per square centimeter
Hg = symbol for mercury
psi = pounds per square inch
psf = pounds per square foot

Units of Volume

cu in. = cubic inch
gal = gallon
cu ft = cubic feet
ml = milliliter
fl oz = fluid ounce (U.S.)

CONVERSION FACTORS

Length

To Convert:	*To:*	*Multiply by:*
meters	feet	3.281
meters	inches	39.37
inches	meters	0.0254
feet	meters	0.3048
millimeters	inches	0.0394
inches	millimeters	25.4
threads/inch	millimeter pitch	Divide into 25.4
yards	meters	0.914

Example: 10 m × 3.281 = 32.81 ft

Area

To Convert:	*To:*	*Multiply by:*
circular mil	meter^2	0.50×10^{-9}
yard^2	meter^2	0.8361

Example: 100 circular mils × 0.5 × 10^{-9} = 0.5 × 10^{-7} m^2 (0.5 × 10^{-7} circular mil = 0.00000005 m^2)

Power

To Convert:	*To:*	*Multiply by:*
watts	hp	0.00134
ft-lb/min	hp	0.0000303
hp	kW	0.746

Example: 1500 W × 0.00134 = 2.01 hp

APPENDIX B
Controllers, Teach Pendants, and Industrial Robots

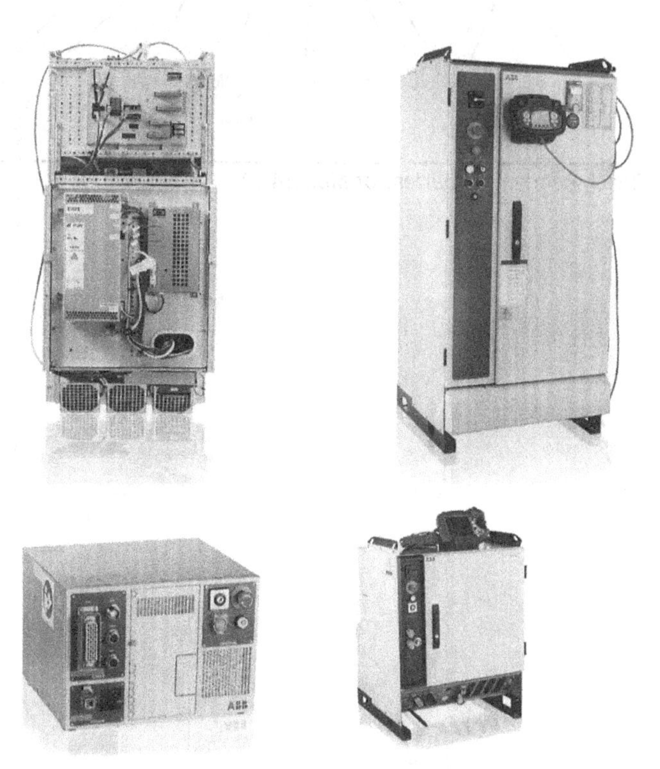

FIGURE B-1 All ABB robot systems are programmed with RAPID™, ABB's flexible, high-level programming language. On the surface RAPID's basic features and functionality are easy to use, but dig deeper and you will find that this programming language allows you to create highly sophisticated solutions. It is a truly universal language on and off the shop floor that supports structured programs and advanced features. It also incorporates powerful support for the most common robot process applications such as welding and assembly. (*Courtesy of ABB.*)

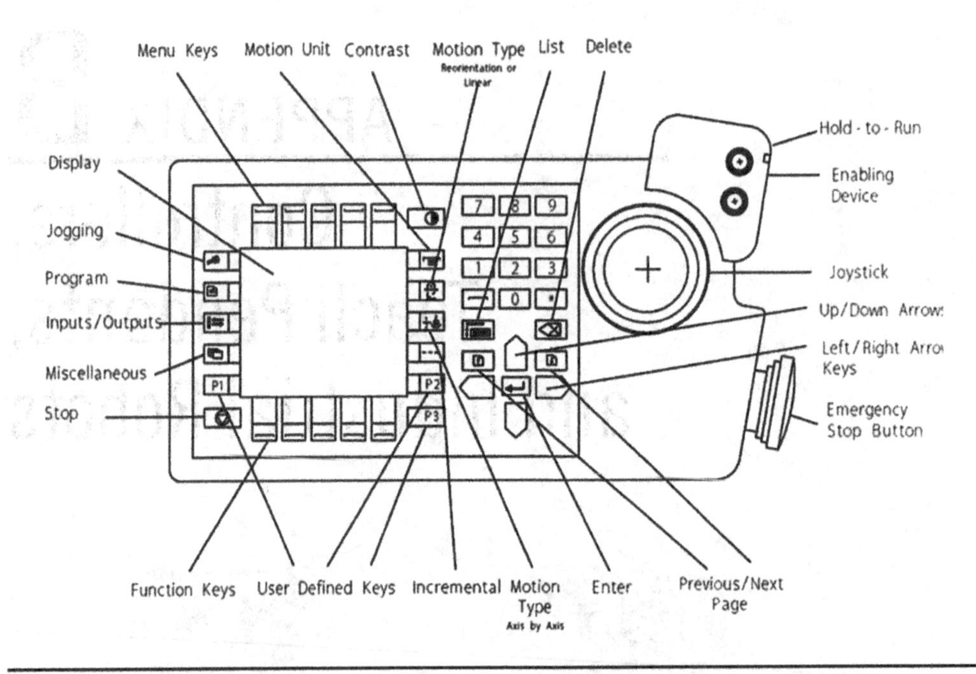

FIGURE B-2 Teach pendant. (*Courtesy of MobotArt.*)

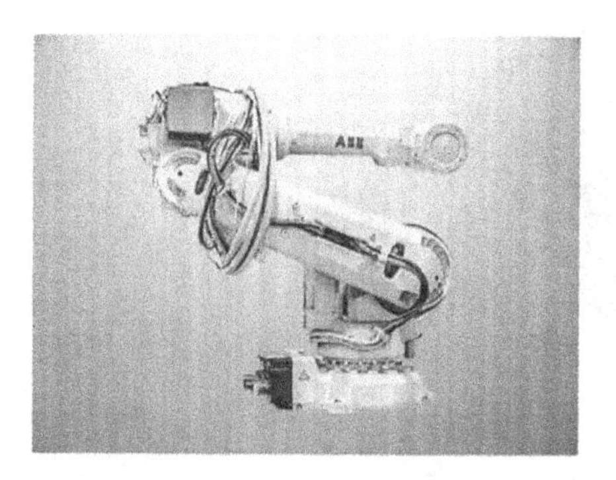

[ABB IRB 4600-60 IRC5]

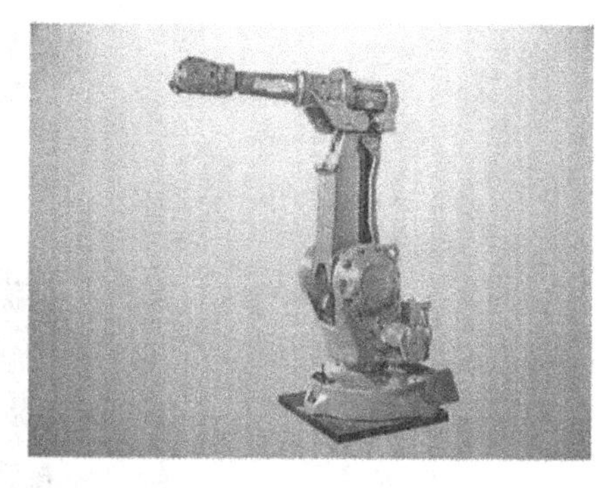

[ABB IRB 2400 IRC5]

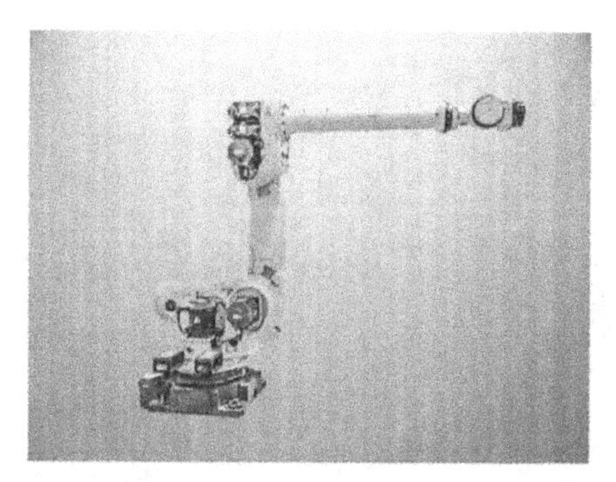

[FANUC R-2000iB/125L R30iA]

[FANUC M-900iA/400L R30iA]

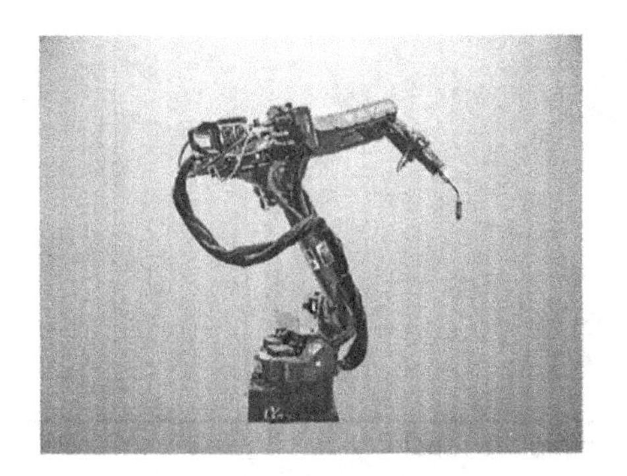

[Motoman MA1400 DX100]

[Motoman HP20 NX100]

Figure B-3 Two models of industrial robots from three different manufacturers.

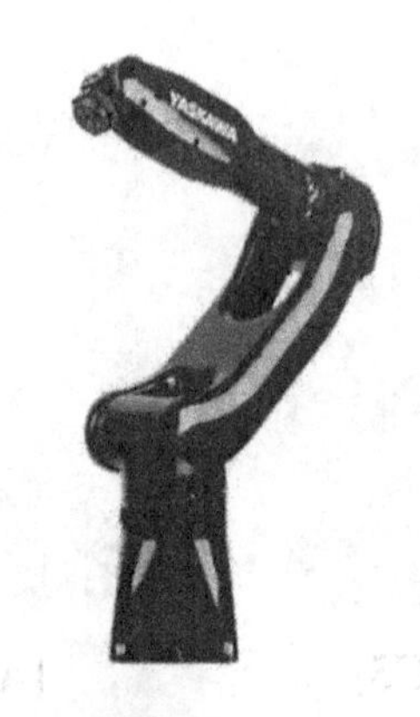

Motoman ES165D-100
DX100

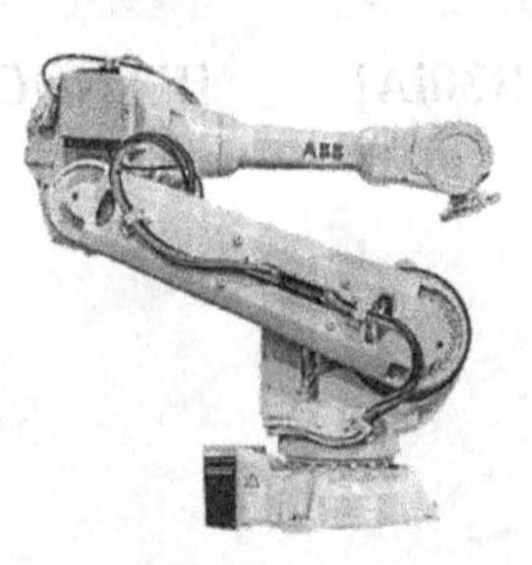

ABB IRB 4600-60
IRC5 - New!

Figure B-4 Motoman and ABB GP materials-handling robots.

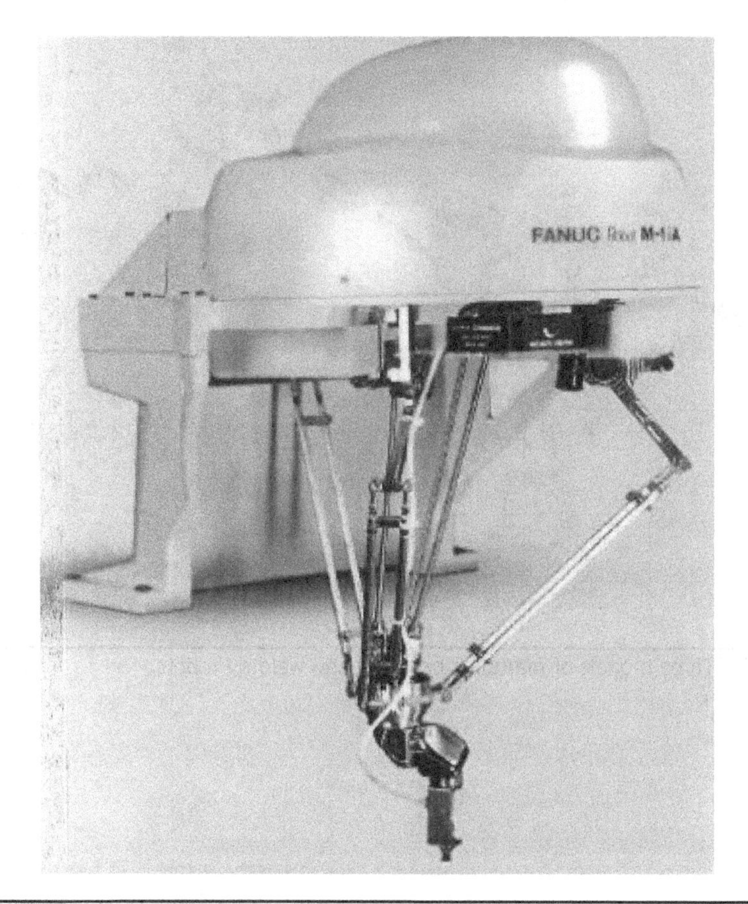

Figure B-5 FANUC's assembly line spray painter in action. (*Courtesy of FANUC.*)

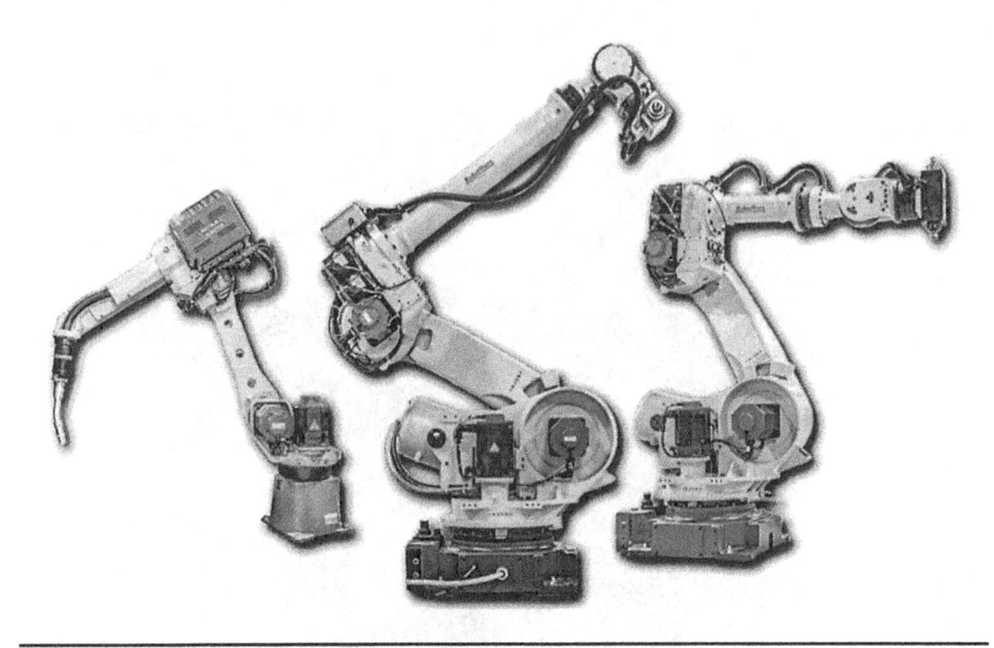

Figure B-6 Three models of materials-handling and welding robots.

APPENDIX C

Robotics Technicians and Engineers

A robotics technician must have the ability to work with robots in a number of ways. Such attention by technicians includes the building, installation, testing, and maintenance of robots. Maintaining similar automated equipment in a factory is often the responsibility of the technician as well as the whole system in which robots are used.

There are a number of titles for technical workers in the robotics industry. Some of these include

- Robot troubleshooter
- Robot engineer
- Robot repairer
- Automation technician
- Electrical and instrument technician
- Electronics technician
- Field service technician or engineer
- Instrumentation specialist
- Industrial electrician

Robots are used for a number of systems and tasks in a factory. One the most often use for industrial robots is to load and unload items from an assembly-line conveyor belt. Other duties assigned to the robot may include

- Robotic drilling
- Robotic cutting metal
- Robotic grinding
- Robotic spray painting

- Robotic polishing metal
- Robotic metal forming
- Robotic plastics handler
- Robotics soldering
- Robotics welding

Most jobs in the robotics industry require robotics engineers to have a bachelor's degree in science with a specialty in robotics engineering. A master's degree with a specialty in robotics is more valuable, especially if it comes with a certificate as a professional engineer from a four-year college or university.

A robotics engineer is involved in building robots, installing robots, and maintaining them. A technician troubleshoots robotic systems using the knowledge acquired in working with microprocessors, programmable controllers, electronic circuits, circuit analysis, mechanical sensors, or feedback systems. He or she should be able to disassemble and reassemble robots or peripheral equipment to make repairs and to replace defective circuit boards, sensors, controllers, encoders, and servo motors. A technician also should be able to perform preventive maintenance on robotic systems, whether electrically or pneumatically powered. Maintaining service records on robotics equipment is as important as with any automated production equipment.

A technician's educational requirements are somewhat different from those of a robotics engineer in that they do not usually require a four-year college degree. A large amount of emphasis is placed on the experience working with mechanical and electrical devices. Experience with pneumatically powered systems is also an added advantage.

However, most employers are more impressed with a technician who has at least a two-year college degree or certificate in this field along with on-the-job experience.

The future for robotics engineers looks great. Today's starting salary averages around $64,000. The wages or salary of an electronics engineer depends on the work requirements and the industry involved. The engineer has a four-year degree with emphasis on science and math and some experience with robots.

The robotics field is most in need of newly trained and educated individuals. Demand for both engineers and technicians in the future is expected to be high and very rewarding for those interested in a career that will be challenging.

APPENDIX D

Electronics and Fluid Power Symbols

Reproduced from *Robotics*, by James W. Masterson, Elmer C. Poe, and Stephen W. Fardo, pp. 247–254, © 1985 by Reston Publishing Co., Inc., Reston, VA with permission of Prentice-Hall, Inc., Englewood Cliffs, NJ.

ELECTRONICS SYMBOLS

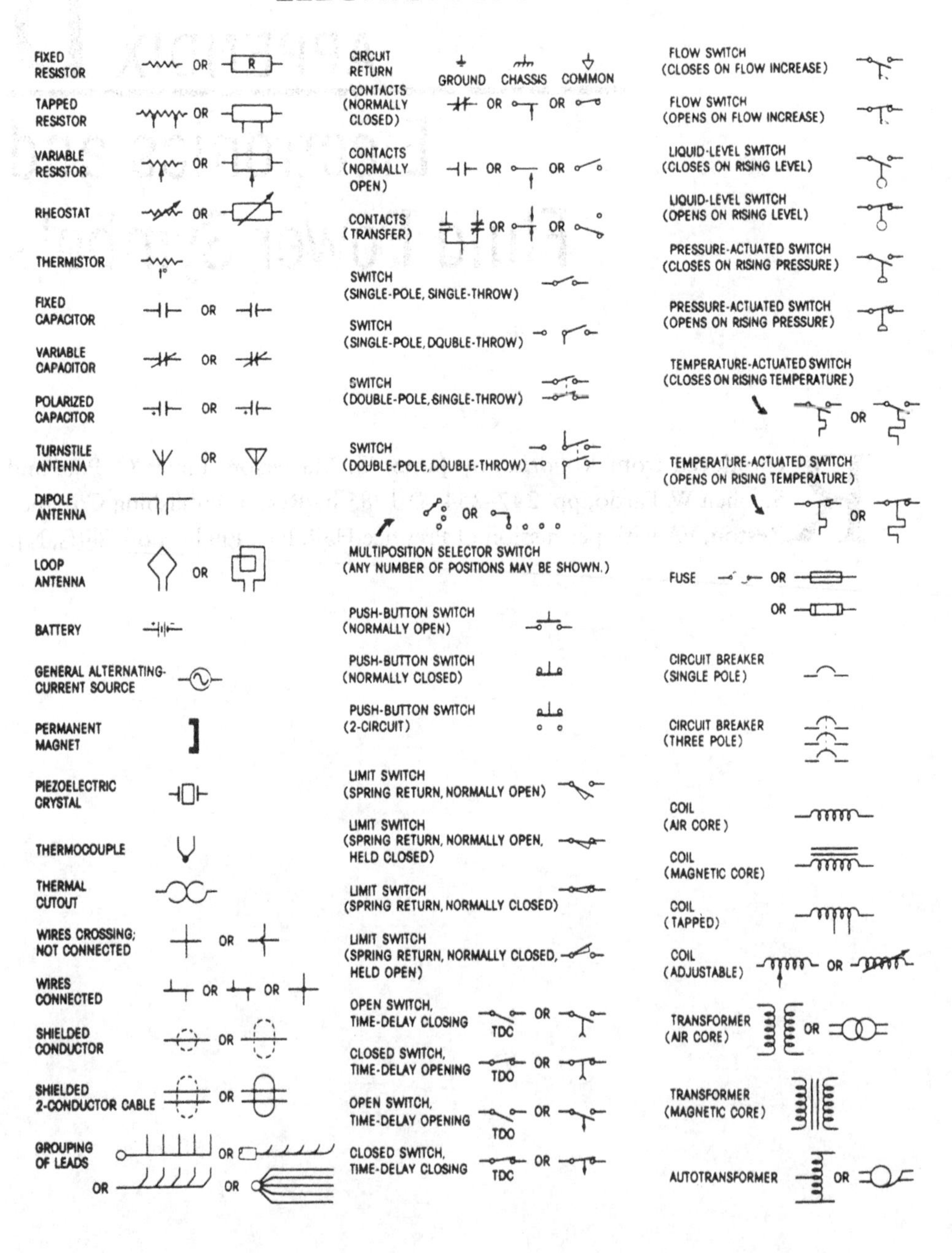

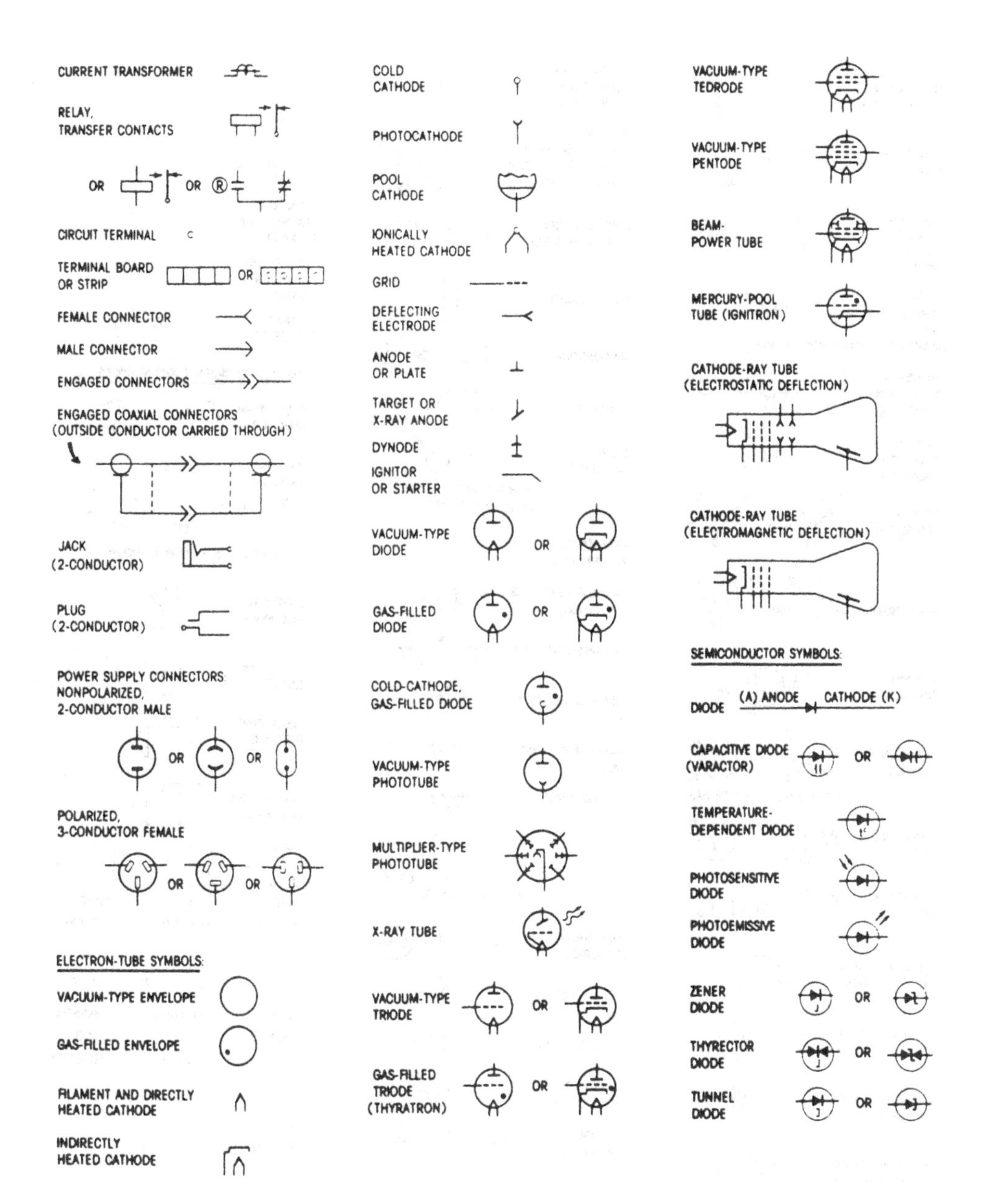
CURRENT TRANSFORMER
RELAY,
TRANSFER CONTACTS
OR
OR
CIRCUIT TERMINAL
TERMINAL BOARD
OR STRIP
OR
FEMALE CONNECTOR
MALE CONNECTOR
ENGAGED CONNECTORS
ENGAGED COAXIAL CONNECTORS
(OUTSIDE CONDUCTOR CARRIED THROUGH)
JACK
(2-CONDUCTOR)
PLUG
(2-CONDUCTOR)
POWER SUPPLY CONNECTORS:
NONPOLARIZED,
2-CONDUCTOR MALE
OR
OR
POLARIZED,
3-CONDUCTOR FEMALE
OR
OR
ELECTRON-TUBE SYMBOLS:
VACUUM-TYPE ENVELOPE
GAS-FILLED ENVELOPE
FILAMENT AND DIRECTLY
HEATED CATHODE
INDIRECTLY
HEATED CATHODE
COLD
CATHODE
PHOTOCATHODE
POOL
CATHODE
IONICALLY
HEATED CATHODE
GRID
DEFLECTING
ELECTRODE
ANODE
OR PLATE
TARGET OR
X-RAY ANODE
DYNODE
IGNITOR
OR STARTER
VACUUM-TYPE
DIODE
OR
GAS-FILLED
DIODE
OR
COLD-CATHODE,
GAS-FILLED DIODE
VACUUM-TYPE
PHOTOTUBE
MULTIPLIER-TYPE
PHOTOTUBE
X-RAY TUBE
VACUUM-TYPE
TRIODE
OR
GAS-FILLED
TRIODE
(THYRATRON)
OR
VACUUM-TYPE
TEDRODE
VACUUM-TYPE
PENTODE
BEAM-
POWER TUBE
MERCURY-POOL
TUBE (IGNITRON)
CATHODE-RAY TUBE
(ELECTROSTATIC DEFLECTION)
CATHODE-RAY TUBE
(ELECTROMAGNETIC DEFLECTION)
SEMICONDUCTOR SYMBOLS:
DIODE
(A) ANODE
CATHODE (K)
CAPACITIVE DIODE
(VARACTOR)
OR
TEMPERATURE-
DEPENDENT DIODE
PHOTOSENSITIVE
DIODE
PHOTOEMISSIVE
DIODE
ZENER
DIODE
OR
THYRECTOR
DIODE
OR
TUNNEL
DIODE
OR

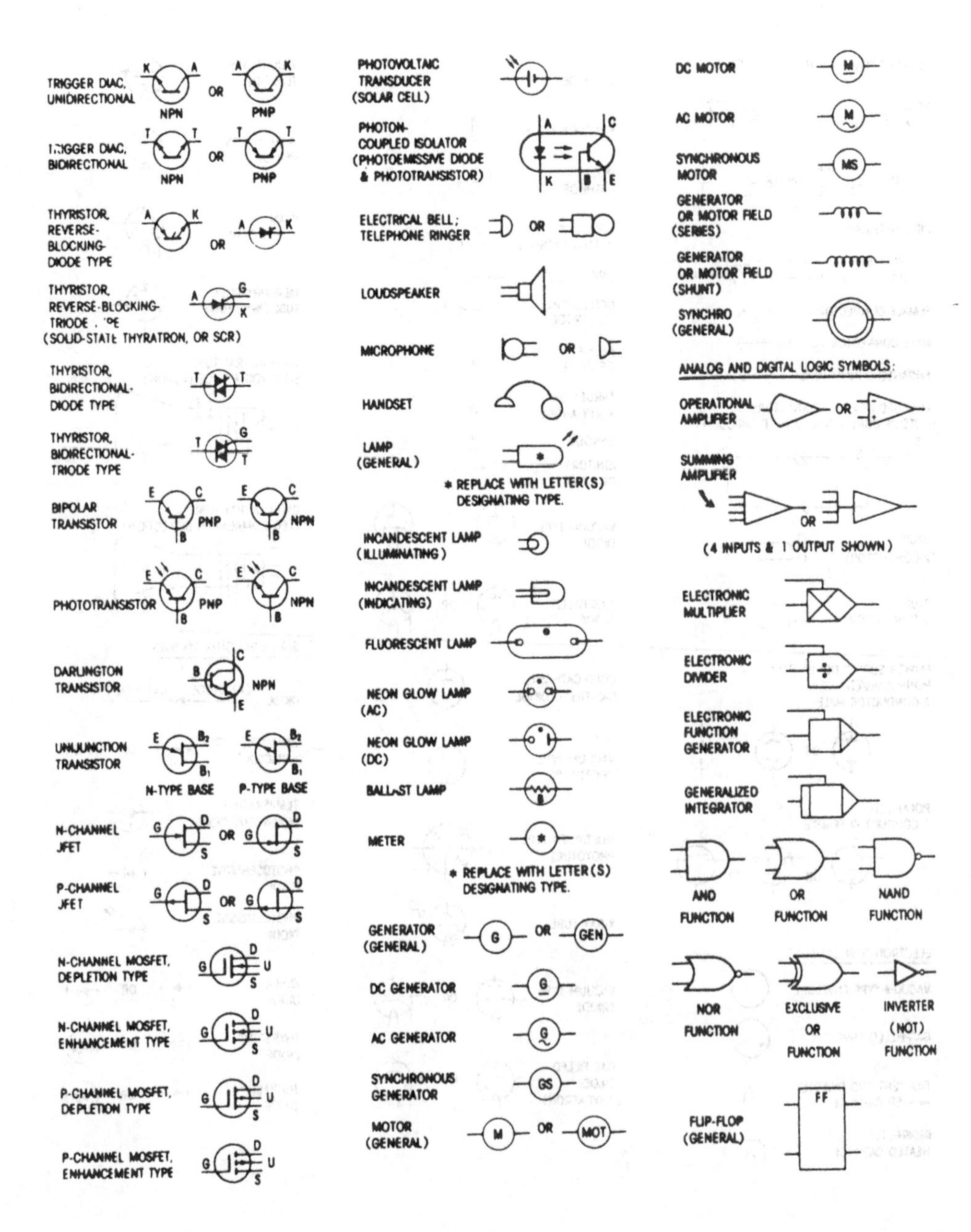

TRIGGER DIAC, UNIDIRECTIONAL
NPN
OR
PNP
TRIGGER DIAC, BIDIRECTIONAL
THYRISTOR, REVERSE-BLOCKING DIODE TYPE
THYRISTOR, REVERSE-BLOCKING TRIODE TYPE (SOLID-STATE THYRATRON, OR SCR)
THYRISTOR, BIDIRECTIONAL-DIODE TYPE
THYRISTOR, BIDIRECTIONAL-TRIODE TYPE
BIPOLAR TRANSISTOR
PHOTOTRANSISTOR
DARLINGTON TRANSISTOR
UNIJUNCTION TRANSISTOR
N-TYPE BASE
P-TYPE BASE
N-CHANNEL JFET
P-CHANNEL JFET
N-CHANNEL MOSFET, DEPLETION TYPE
N-CHANNEL MOSFET, ENHANCEMENT TYPE
P-CHANNEL MOSFET, DEPLETION TYPE
P-CHANNEL MOSFET, ENHANCEMENT TYPE
PHOTOVOLTAIC TRANSDUCER (SOLAR CELL)
PHOTON-COUPLED ISOLATOR (PHOTOEMISSIVE DIODE & PHOTOTRANSISTOR)
ELECTRICAL BELL; TELEPHONE RINGER
LOUDSPEAKER
MICROPHONE
HANDSET
LAMP (GENERAL)
* REPLACE WITH LETTER(S) DESIGNATING TYPE.
INCANDESCENT LAMP (ILLUMINATING)
INCANDESCENT LAMP (INDICATING)
FLUORESCENT LAMP
NEON GLOW LAMP (AC)
NEON GLOW LAMP (DC)
BALLAST LAMP
METER
* REPLACE WITH LETTER(S) DESIGNATING TYPE.
GENERATOR (GENERAL)
DC GENERATOR
AC GENERATOR
SYNCHRONOUS GENERATOR
MOTOR (GENERAL)
DC MOTOR
AC MOTOR
SYNCHRONOUS MOTOR
GENERATOR OR MOTOR FIELD (SERIES)
GENERATOR OR MOTOR FIELD (SHUNT)
SYNCHRO (GENERAL)
ANALOG AND DIGITAL LOGIC SYMBOLS:
OPERATIONAL AMPLIFIER
SUMMING AMPLIFIER
(4 INPUTS & 1 OUTPUT SHOWN)
ELECTRONIC MULTIPLIER
ELECTRONIC DIVIDER
ELECTRONIC FUNCTION GENERATOR
GENERALIZED INTEGRATOR
AND FUNCTION
OR FUNCTION
NAND FUNCTION
NOR FUNCTION
EXCLUSIVE OR FUNCTION
INVERTER (NOT) FUNCTION
FLIP-FLOP (GENERAL)
FF

FLUID POWER SYMBOLS

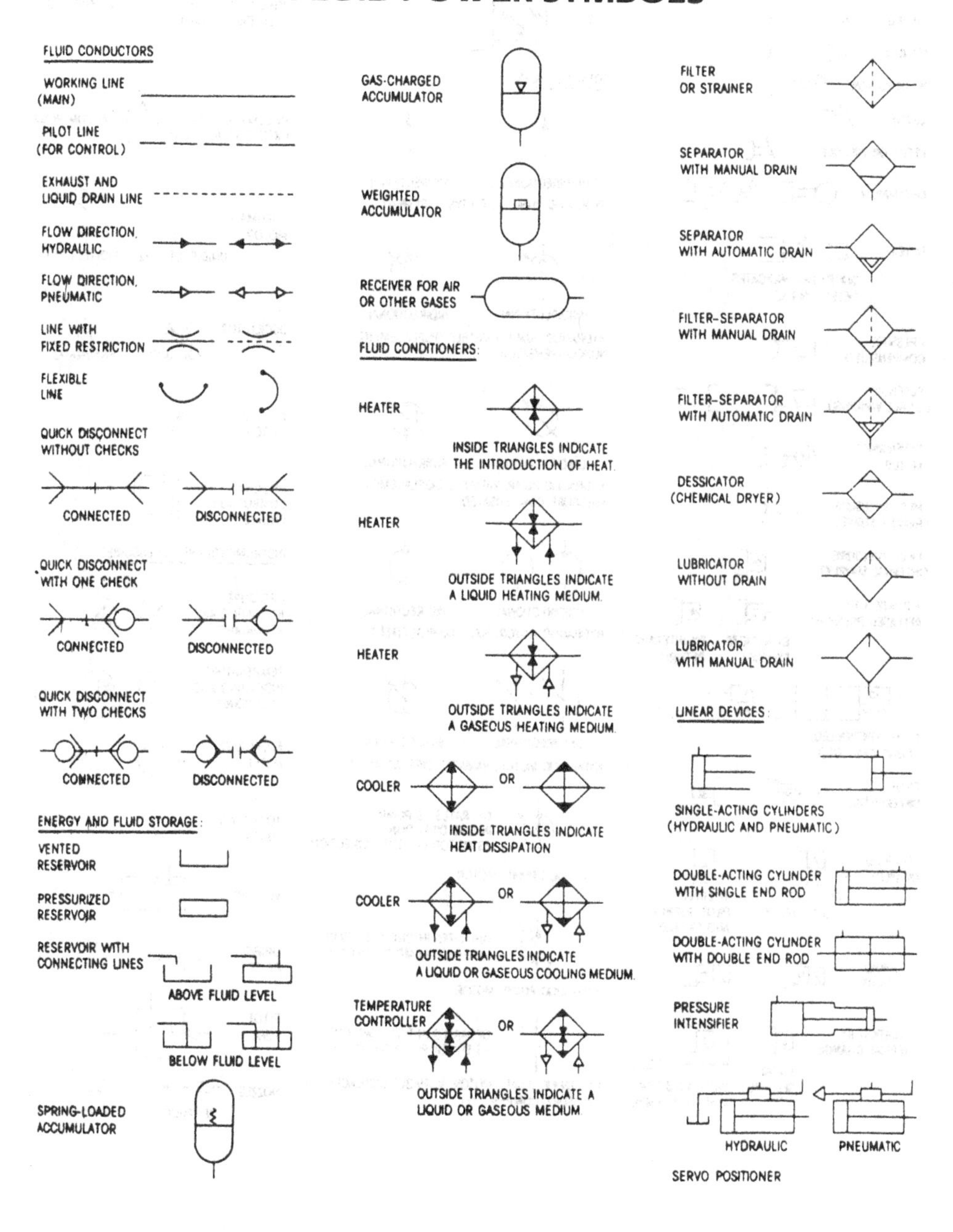

FLUID CONDUCTORS
WORKING LINE (MAIN)
PILOT LINE (FOR CONTROL)
EXHAUST AND LIQUID DRAIN LINE
FLOW DIRECTION, HYDRAULIC
FLOW DIRECTION, PNEUMATIC
LINE WITH FIXED RESTRICTION
FLEXIBLE LINE
QUICK DISCONNECT WITHOUT CHECKS
CONNECTED
DISCONNECTED
QUICK DISCONNECT WITH ONE CHECK
CONNECTED
DISCONNECTED
QUICK DISCONNECT WITH TWO CHECKS
CONNECTED
DISCONNECTED
ENERGY AND FLUID STORAGE:
VENTED RESERVOIR
PRESSURIZED RESERVOIR
RESERVOIR WITH CONNECTING LINES
ABOVE FLUID LEVEL
BELOW FLUID LEVEL
SPRING-LOADED ACCUMULATOR
GAS-CHARGED ACCUMULATOR
WEIGHTED ACCUMULATOR
RECEIVER FOR AIR OR OTHER GASES
FLUID CONDITIONERS:
HEATER
INSIDE TRIANGLES INDICATE THE INTRODUCTION OF HEAT.
HEATER
OUTSIDE TRIANGLES INDICATE A LIQUID HEATING MEDIUM.
HEATER
OUTSIDE TRIANGLES INDICATE A GASEOUS HEATING MEDIUM.
COOLER
OR
INSIDE TRIANGLES INDICATE HEAT DISSIPATION
COOLER
OR
OUTSIDE TRIANGLES INDICATE A LIQUID OR GASEOUS COOLING MEDIUM.
TEMPERATURE CONTROLLER
OR
OUTSIDE TRIANGLES INDICATE A LIQUID OR GASEOUS MEDIUM
FILTER OR STRAINER
SEPARATOR WITH MANUAL DRAIN
SEPARATOR WITH AUTOMATIC DRAIN
FILTER-SEPARATOR WITH MANUAL DRAIN
FILTER-SEPARATOR WITH AUTOMATIC DRAIN
DESSICATOR (CHEMICAL DRYER)
LUBRICATOR WITHOUT DRAIN
LUBRICATOR WITH MANUAL DRAIN
LINEAR DEVICES:
SINGLE-ACTING CYLINDERS (HYDRAULIC AND PNEUMATIC)
DOUBLE-ACTING CYLINDER WITH SINGLE END ROD
DOUBLE-ACTING CYLINDER WITH DOUBLE END ROD
PRESSURE INTENSIFIER
HYDRAULIC
PNEUMATIC
SERVO POSITIONER

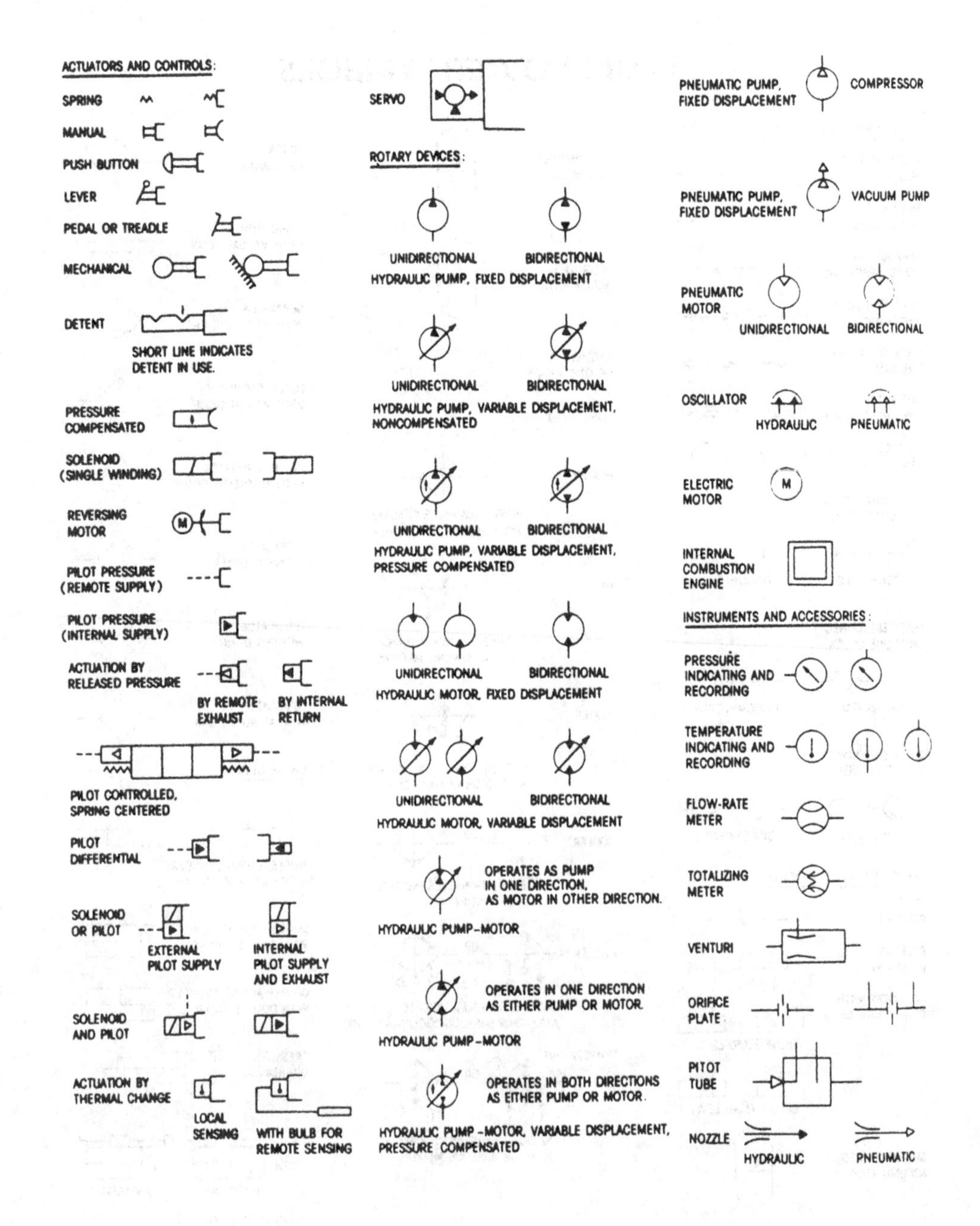
ACTUATORS AND CONTROLS:
SPRING
MANUAL
PUSH BUTTON
LEVER
PEDAL OR TREADLE
MECHANICAL
DETENT
SHORT LINE INDICATES DETENT IN USE.
PRESSURE COMPENSATED
SOLENOID (SINGLE WINDING)
REVERSING MOTOR
PILOT PRESSURE (REMOTE SUPPLY)
PILOT PRESSURE (INTERNAL SUPPLY)
ACTUATION BY RELEASED PRESSURE
BY REMOTE EXHAUST
BY INTERNAL RETURN
PILOT CONTROLLED, SPRING CENTERED
PILOT DIFFERENTIAL
SOLENOID OR PILOT
EXTERNAL PILOT SUPPLY
INTERNAL PILOT SUPPLY AND EXHAUST
SOLENOID AND PILOT
ACTUATION BY THERMAL CHANGE
LOCAL SENSING
WITH BULB FOR REMOTE SENSING
SERVO
ROTARY DEVICES:
UNIDIRECTIONAL
BIDIRECTIONAL
HYDRAULIC PUMP, FIXED DISPLACEMENT
UNIDIRECTIONAL
BIDIRECTIONAL
HYDRAULIC PUMP, VARIABLE DISPLACEMENT, NONCOMPENSATED
UNIDIRECTIONAL
BIDIRECTIONAL
HYDRAULIC PUMP, VARIABLE DISPLACEMENT, PRESSURE COMPENSATED
UNIDIRECTIONAL
BIDIRECTIONAL
HYDRAULIC MOTOR, FIXED DISPLACEMENT
UNIDIRECTIONAL
BIDIRECTIONAL
HYDRAULIC MOTOR, VARIABLE DISPLACEMENT
OPERATES AS PUMP IN ONE DIRECTION, AS MOTOR IN OTHER DIRECTION.
HYDRAULIC PUMP-MOTOR
OPERATES IN ONE DIRECTION AS EITHER PUMP OR MOTOR.
HYDRAULIC PUMP-MOTOR
OPERATES IN BOTH DIRECTIONS AS EITHER PUMP OR MOTOR.
HYDRAULIC PUMP-MOTOR, VARIABLE DISPLACEMENT, PRESSURE COMPENSATED
PNEUMATIC PUMP, FIXED DISPLACEMENT
COMPRESSOR
PNEUMATIC PUMP, FIXED DISPLACEMENT
VACUUM PUMP
PNEUMATIC MOTOR
UNIDIRECTIONAL
BIDIRECTIONAL
OSCILLATOR
HYDRAULIC
PNEUMATIC
ELECTRIC MOTOR
M
INTERNAL COMBUSTION ENGINE
INSTRUMENTS AND ACCESSORIES:
PRESSURE INDICATING AND RECORDING
TEMPERATURE INDICATING AND RECORDING
FLOW-RATE METER
TOTALIZING METER
VENTURI
ORIFICE PLATE
PITOT TUBE
NOZZLE
HYDRAULIC
PNEUMATIC

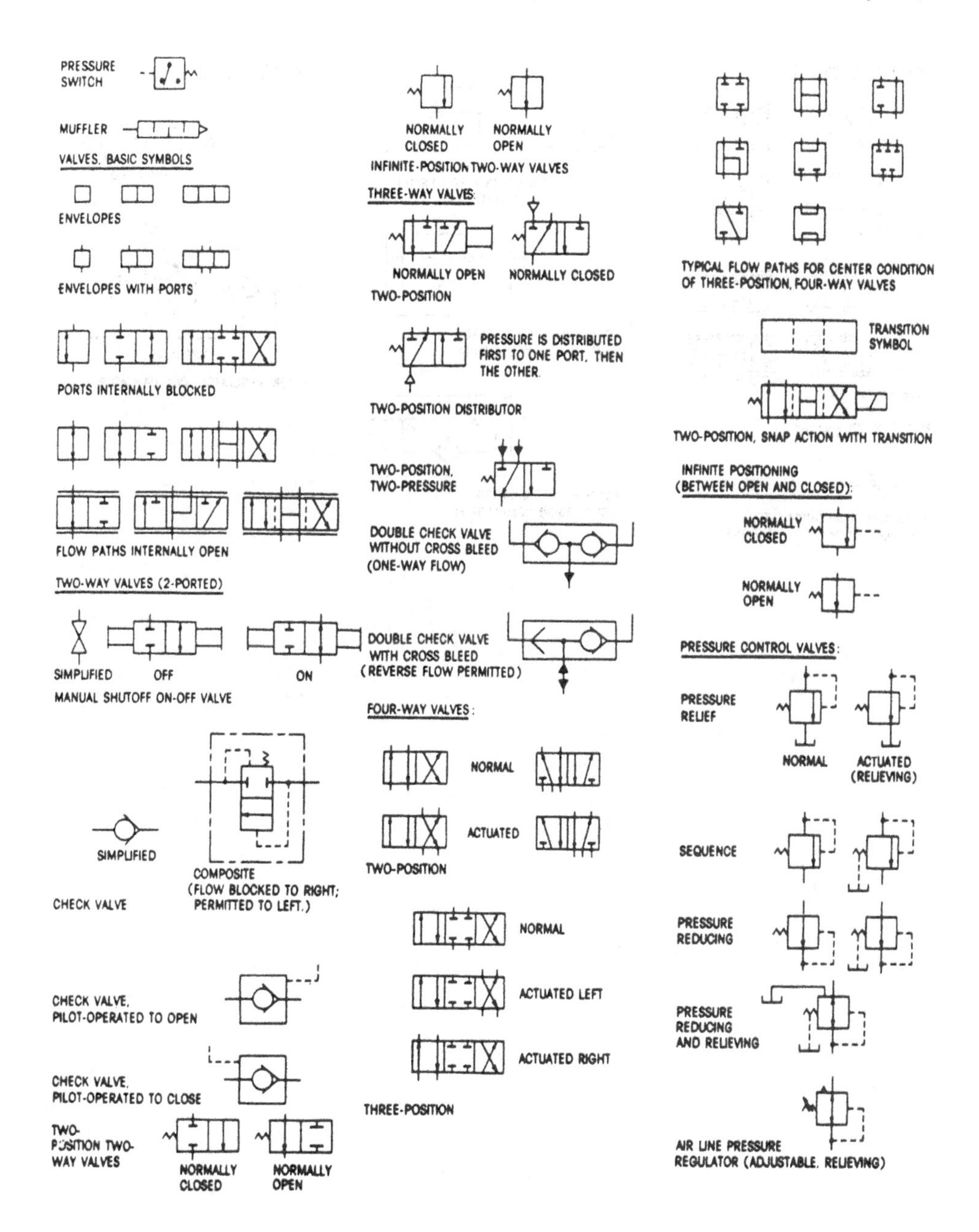
PRESSURE SWITCH
MUFFLER
VALVES, BASIC SYMBOLS
ENVELOPES
ENVELOPES WITH PORTS
PORTS INTERNALLY BLOCKED
FLOW PATHS INTERNALLY OPEN
TWO-WAY VALVES (2-PORTED)
SIMPLIFIED
OFF
ON
MANUAL SHUTOFF ON-OFF VALVE
SIMPLIFIED
COMPOSITE
(FLOW BLOCKED TO RIGHT; PERMITTED TO LEFT.)
CHECK VALVE
CHECK VALVE, PILOT-OPERATED TO OPEN
CHECK VALVE, PILOT-OPERATED TO CLOSE
TWO-POSITION TWO-WAY VALVES
NORMALLY CLOSED
NORMALLY OPEN
NORMALLY CLOSED
NORMALLY OPEN
INFINITE-POSITION TWO-WAY VALVES
THREE-WAY VALVES:
NORMALLY OPEN
NORMALLY CLOSED
TWO-POSITION
PRESSURE IS DISTRIBUTED FIRST TO ONE PORT, THEN THE OTHER.
TWO-POSITION DISTRIBUTOR
TWO-POSITION, TWO-PRESSURE
DOUBLE CHECK VALVE WITHOUT CROSS BLEED (ONE-WAY FLOW)
DOUBLE CHECK VALVE WITH CROSS BLEED (REVERSE FLOW PERMITTED)
FOUR-WAY VALVES:
NORMAL
ACTUATED
TWO-POSITION
NORMAL
ACTUATED LEFT
ACTUATED RIGHT
THREE-POSITION
TYPICAL FLOW PATHS FOR CENTER CONDITION OF THREE-POSITION, FOUR-WAY VALVES
TRANSITION SYMBOL
TWO-POSITION, SNAP ACTION WITH TRANSITION
INFINITE POSITIONING (BETWEEN OPEN AND CLOSED):
NORMALLY CLOSED
NORMALLY OPEN
PRESSURE CONTROL VALVES:
PRESSURE RELIEF
NORMAL
ACTUATED (RELIEVING)
SEQUENCE
PRESSURE REDUCING
PRESSURE REDUCING AND RELIEVING
AIR LINE PRESSURE REGULATOR (ADJUSTABLE, RELIEVING)

INFINITE POSITIONING VALVES:

THREE-WAY VALVES

FOUR-WAY VALVES

FLOW-CONTROL VALVES:

ADJUSTABLE, NONCOMPENSATED
(FLOW CONTROL IN EACH DIRECTION)

ADJUSTABLE, WITH BYPASS

FLOW CONTROLLED TO RIGHT, FLOW TO LEFT BYPASSES CONTROL

ADJUSTABLE AND PRESSURE COMPENSATED, WITH BYPASS

ADJUSTABLE, TEMPERATURE AND PRESSURE COMPENSATED

AIR LINE ACCESSORIES:

SIMPLIFIED

COMPOSITE

FILTER, REGULATOR, AND LUBRICATOR

APPENDIX E

Cross-Comparison of Robots

This appendix contains data on 166 robot models from 33 robot manufacturers. Note that the information is directional only and that some values and descriptions may vary between report sources and specifications manuals.

The table is reproduced courtesy of Patrick J. Niedbala, Schoolcraft College, Livonia, Michigan.

Manufacturer	Model	Payload lbs.	Payload Kgs.	Axes No.	Axes Action	Powered By
Accuratio Systems Clawson, MI Office (313) 288-5070	System I* Gantry	75	35	5	Rectangular coordinate	Elec.
Advanced Robotics Corporation Columbus, Ohio (614)870-7778	Cyro 750	100	45	5	Rectangular coordinate	Elec.
Advanced Robotics Corporation	Cyro 1000	22	10	5-6	Jointed-arm spherical coordinate	Elec.
Advanced Robotics Corporation	Cyro 1350	250	113	6	Rectangular coordinate	Elec.
Advanced Robotics Corporation	Cyro 2000	350	159	5	Rectangular coordinate	Elec.
American Robot Corporation Farmington Hill, MI Office (313) 477-3200	MR6200*	20	9	6	Jointed-arm spherical coordinate	Elec.
American Robot	MR6500	50	23	6	Jointed-arm spherical coordinate	Elec.
American Robot	MR6260	22	11	6	Jointed-arm spherical coordinate	Elec.
American Robot	MR6260 Overhead	22	11	6	Jointed-arm spherical coordinate	Elec.
American Robot	MR6500 Overhead	50	23	6	Jointed-arm spherical coordinate	Elec.
ASEA (Sweden) Troy, MI office (313)528-3630	ASEA* IRb 6/2	13.2	6	5	Jointed-arm spherical coordinate	Elec.

Control Systems	SERVO	NONSERVO	Sequence of Motions Taught	Programming Method	Major Applications	Repeatable Placement Accuracy
CNC micro-computer	x		Continous path	Remote pendant or off-line	Mchn load/unload, deburr, sealant, adhesive, waterjet	+/- .008"
Mini-computer w/micro-processor	x		Continous path	Remote pendant or off-line	Welding	+/- .008"
Mini-computer w/micro-processor	x		Point-to-point and continous path	Remote pendant or off-line	Arc welding	+/- .008"
Mini-computer w/micro-processor	x		Point-to-point and continous path	Remote pendant or off-line	Arc welding	+/- .008"
Mini-computer w/micro-processor	x		Continous path	Remote pendant or off-line	Welding	+/- .016"
Micro-computer w/8 micro-processors	x		Point to point continous path circular + linear interpolation	Remote pendant or off-line	Arc welding, mchn load/unload, assembly, palletizing	+/- .001"
Micro-computer w/8 micro-processors	x		Point to point continous path circular + linear interpolation	Remote pendant or off-line	Arc welding, mchn load/unload, assembly, palletizing	+/- .001"
Micro-computer w/8 micro-processors	x		Point to point continous path circular + linear interpolation	Remote pendant or off-line	Arc welding, mchn load/unload, assembly, palletizing	+/- .004"
Micro-computer w/8 micro-processors	X		Point to point continuous path circular + linear interpolation	Remote pendant or off-line	Arc welding, mchn load/unload, assembly palletizing	+/- .004"
Micro-computer w/8 micro-processors	x		Point to point continous path circular + linear interpolation	Remote pendant or off-line	Arc welding, mchn load/unload, assembly, palletizing	+/- .001"
CNC micro-computer	x		Point-to-point	Remote pendant	Mchn load/unload, pick & place, spot welding	+/- .008"

Manufacturer	Model	Payload lbs.	Payload Kgs.	Axes No.	Axes Action	Powered By
ASEA	ASEA* IRb 60/2	132	60	5	Jointed-arm spherical coordinate	Elec.
ASEA	ASEA* IRb 90S/2	198	90	5-6	Jointed-arm spherical coordinate	Elec.
ASEA	MHU Minor (Electrolux)	2.2	1	3	Cylindrical coordinate	Pnem.
ASEA	MHU Junior (Electrolux)	11	5	3	Cylindrical coordinate	Pnem.
ASEA	MHU Senior (Electrolux)	33	15	3	Cylindrical coordinate	Pnem.
Automatix Inc. Farmington Hills, MI Office (313)553-0044	AID 600	66	30	5	Rectilinear coordinate	Elec.
Automatix Inc.	AID 800	22	10	5	Jointed-arm spherical coordinate	Elec.
Automatix Inc	AID 900	33-65	15-30	6-5	Jointed-arm spherical coordinate	Elec
C. Itoh & Company Farmington, MI Office (313)473-7400	NACHI UM 5600 SP	11	5	6	Jointed-arm spherical coordinate	Hydraulic
C. Itoh & Company	NACHI UM 5601 AFI	11	5	6	Jointed-arm spherical coordinate	Hydraulic
C. Itoh & Company	NACHI UM 7500 AE (Cyro 820)	20	9	5	Jointed-arm spherical coordinate	Elec.
C. Itoh & Company	NACHI UN 8600 AK	110	50	6	Jointed-arm spherical coordinate	Elec.
Cincinnati Milacron Southfield, MI ofc (313) 557-2700	T3-566*	100	45	6	Jointed-arm spherical coordinate	Hydraulic
Cincinnati Milacron	T3-576* Extended reach	135	61	6	Jointed-arm spherical coordinate	Hydraulic

Control Systems	SERVO	NONSERVO	Sequence of Motions Taught	Programming Method	Major Applications	Repeatable Placement Accuracy
CNC micro-computer	x		Point-to-point	Remote pendant	Mchn load/ unload, pick & place, spot welding	+/- .016"
CNC micro-computer	x		Point-to-point and supplemental program	Remote pendant	Spot welding	+/- .004"
Micro-processor		x	Point-to-point	Off-line	Pick and place	+/- .004"
Micro-processor		x	Point-to-point	Remote pendant	Spot welding, deburr	+/- .004"
Micro-processor		x	Point-to-point	Manual stop set push-button sequence	Pick and place, mchn load/ unload	+/- .004"
Micro-computer	X		Point-to-point continuous	Remote pendant or off-line	Assembly & advance mtrl handling	+/- .003"
Micro-computer	x		Point-to-point continuous	Remote pendant or off-line	Arc welding & advance mtrl handling	+/- .008"
Micro-computer	x		Point-to-point continuous	Remote or pendant or off-line	Arc welding & advance mtrl handling	+/- .008"
Micro-processor floppy disc	x		Continuous path	Manual teach playback	Spray painting	+/- .080"
Micro-processor plate wire memory	x		Point-to-point	Remote pendant or manual teach playback	Spray Painting	+/- .080"
Micro-processor plate wire & cassette deck	x		Point-to-point	Remote pendant	Arc welding	+/- .010"
Micro-processor plate wire & cassette deck	x		Point-to-point	Remote pendant	Spot welding & mtrl handling	+/- .040"
Mini-computer	x		Point-to-point	Remote pendant	Mchn Load/ unload, pick & place, spot/ arc welding	+/- .050"
Mini-computer	x		Point-to-point	Remote pendant	Mchn load/ unload, pick & place, spot/ arc welding	+/- .050"

Manufacturer	Model	Payload lbs.	Payload Kgs.	Axes No.	Axes Action	Powered By
Cincinnati Milacron	T3-586*	225	102	6	Jointed-arm spherical coordinate	Hydraulic
Cincinnati Milacron	T3-595*	450	204	5	Jointed-arm spherical coordinate	Hydraulic
Cincinnati Milacron	T3-726*	14	6	6	Jointed-arm spherical coordinate	Elec.
Cincinnati Milacron	T3-746*	70	31.7	6	Jointed-arm spherical coordinate	Elec.
Cincinnati Milacron	T3-776*	150	68	6	Jointed-arm spherical coordinate	Elec.
Cincinnati Milacron	T3-786*	200	90	6	Joint-arm spherical coordinate	Elec.
Cincinnati Milacron	T3-363	110	50	3	Cylindrical coordinate	Elec.
Cincinnati Milacron	T3-364	75	34	4	Cylindrical coordinate	Elec.
Cincinnati Milacron	T3-876* Gantry	150	68	6	Rectilinear spherical coordinate	Elec.
Cincinnati Milacron	T3-886* Gantry	200	90	6	Rectilinear spherical coordinate	Elec.
Cincinnati Milacron	T3-895* Gantry	440	200	5	Rectilinear spherical coordinate	Elec.
COMAU Industriale (Italy) Troy, MI office (313)583-9900	Polar 6000	130	59	6	Spherical coordinate	Hydraulic
COMAU Industriale	Smart 6.50	132	60	6	Jointed-arm spherical coordinate	Elec.
Control Automation, Inc., Princeton, NJ (609)799-6026	Mini-sembler	5	2.3	4	Rectilinear coordinate	Elec.

Control Systems	SERVO	NONSERVO	Sequence of Motions Taught	Programming Method	Major Applications	Repeatable Placement Accuracy
Mini-computer	x		Point-to-point	Remote pendant	Mchn load/ unload, pick & place, spot/ arc welding	+/- .050"
Mini-computer	x		Point-to-point	Remote pendant	Mchn load/ unload, pick & place, spot/ arc welding	+/- .050"
Micro-processor	x		Point-to-point	Remote pendant	Mtrl handling, assembly, arc welding	+/- .004"
Micro-processor	x		Point-to-point	Remote pendant	Mtrl handling, arc welding (variable)	+/- .010"
Micro-processor	x		Point-to-point	Remote pendant	Mtrl handling, arc welding (variable)	+/- .010"
Micro-Processor	x		Point-to-point	Remote pendant	Mtrl handling, spot/arc welding (variable)	+/- .010"
Micro-processor	x		Point-to-point	Remote pendant	Mchn load/unload, pick & place	+/- .040"
Micro-processor	x		Point-to-point	Remote pendant	Mchn load/unload, pick & place	+/- .040"
Micro-processor	x		Point-to-point	Remote pendant	Mtrl handling, spot/arc welding	+/- .010"
Micro-processor	x		Point-to-point	Remote pendant	Mtrl handling, spot/ arc welding	+/- .010"
Micro-processor	x		Point-to-point	Remote pendant	Mtrl handling, spot/ arc welding	+/- .010"
Programmable controller	x		Point-to-point	Off-line	Spot welding, mchn load/unload	+/- .040"
Micro-computer	x		Point-to-point and continuous path	Remote pendant or off-line	Spot welding, assembly, mtrl handling	+/- .015"
Mini-computer	x		Continuous path	Remote pendant or off-line	Electronic assembly, mtrl handing	+/- .001"

Manufacturer	Model	Payload lbs.	Payload Kgs.	Axes No.	Axes Action	Powered By
Cybotech Novi, MI office (313)348-3335	Cybotech H80	175	80	6	Jointed-arm cylindrical coordinate	Elec.
Cybotech	Cybotech V15	33	15	6	Cylindrical coordinate	Elec.
Cybotech	Cybotech TH8	17	8	6	Cylindrical coordinate	Elec.
Cybotech	Cybotech TP15	11	4.4	6	Jointed-arm spherical coordinate	Elec.
Cybotech	Cybotech WV15	22	10	6	Jointed-arm spherical coordinate	Elec.
Cybotech	Cybotech WCX90	11	5	8	Cartesian coordinate	Elec.
Cybotech	Cybotech G80	175	80	6	Cartesian coordinate	Elec.
Durr Industries, Inc. Plymouth, MI office (313)459-6800	P-100* gantry	220	100	3-5	Cartesian coordinate	Elec.
Durr Industries, Inc.	CompArm (Hall Automation-England)	7	3	6	Spherical coordinate	Hydraulic
Gametics Roseville, MI (313)778-7220	Model 524*	50	23	3-5	Cylindrical coordinate	Hydraulic
Gametics	Model 536*	200	114	3-5	Cylindrical coordinate	Hydraulic
Gametics	Model 524E	100	57	5	Cylindrical coordinate	Elec.

Control Systems	SERVO	NONSERVO	Sequence of Motions Taught	Programming Method	Major Applications	Repeatable Placement Accuracy
Micro Processor based R6 control	x		Continuous path or point-to-point	Manual teach playback or remote pendant	Mtrl handling, spot/ arc weld welding	+/- .008"
Micro-processor based RC6 controller	x		Point-to-point	Remote pendant	Assembly, inspection/handling, arc welding	+/- .004"
Micro-processor based RC6 controller	x		Point-to-point	Remote pendant	Arc welding, mtrl handling, assembly	+/- .008"
Micro-processor based RC6 controller	x		Continuous path or point-to-point	Remote pendant	Painting, sealing, coating	+/- .008"
Micro-processor based RC6 controller	x		Continuous path or point-to-point	Remote pendant	Arc welding	+/- .008"
Micro-processor based RC6 controller	x		Continuous path or point-to-point	Remote pendant	Arc welding	+/- .008"
Micro-processor based RC6 controller	x		Continuous path or point-to-point	Remote pendant	Arc welding	+/- .008"
Micro-processor & PC	x		Point-to-point CP by interpolation	Remote pendant or off-line	Mchn load, palletizing, assembly, water cutting	+/- .008"
Micro-processor	x		Continuous path	Manual teach playback	Spray painting	+/- .040"
Micro-Processor	x		Point-to-point	Remote pendant	Mchn load/ unload	+/- .030"
Micro Processor	x		Point-to-point	Remote pendant	Mchn load/ unload	+/- .050"
Micro processor	x		Point-to-point	Remote pendant	Mchn load/ unload	+/- .030"

Manufacturer	Model	Payload		Axes		Powered By
		lbs.	Kgs.	No.	Action	
GCA Corporation Troy, MI office (313) 362-4144	DKP 200H	4.4	2	4	Horizontal joint-arm cylindrical coordinate	Elec.
GCA Corporation	DKP 200V	4.4	2	5	Jointed-arm spherical coordinate	Elec.
GCA Corporation	DKP 300H	11	5	4	Horizontal joint-arm coordinate	Elec.
GCA Corporation	DKP 300V	11	5	5	Jointed-arm spherical	Elec.
GCA Corporation	DKP 550	15	7	5	Jointed-arm cylindrical coordinate	Elec.
GCA Corporation	DKP 600	26	11.8	5	Jointed-arm spherical coordinate	Elec.
GCA Corporation	DKP 800	66	30	3	Jointed-arm spherical coordinate	Elec.
GCA Corporation	DKP 1000	66	30	6	Jointed-arm spherical coordinate	Elec.
GCA Corporation	DKP 1200	110	50	6	Jointed-arm spherical coordinate	Elec.
GCA Corporation	DKB 1440*	110	50	4-6	Cylindrical & cartesian coordinate	Elec.
GCA Corporation	DKB 3200	220	100	4-6	Cylindrical & cartesian coordinate	Elec.

Control Systems	SERVO	NONSERVO	Sequence of Motions Taught	Programming Method	Major Applications	Repeatable Placement Accuracy
Micro-processor two models	x		Continous path or point-to-point	Remote pendant or off-line	Assembly, pkging, mchn load, mtrl handling	+/- .002"
Micro-processor two models	x		Continous path or point-to-point	Remote pendant or off-line	Assembly, pkging, mchn load,, mtrl handling	+/- .002"
Micro-processor	x		Continous path or point-to-point	Remote pendant line	Assembly, pkging, mchn load, mtrl handling	+/- .004"
Micro-processor	x		Point-to-point & continuous	Remote pendant line	Assembly, pkging mchn load, mtrl handling	+/- .004"
Micro-processor two models	x		Point-to-point & continuous path	Remote pendant or off-line	Assembly arc welding, sealing, mtrl handling	+/- .008"
Micro-processor two models	x		Point-to-point & continuous path	Remote pendant or off-line	Arc welding, assembly	+/- .004"
Micro-processor two models	x		Point-to-point & continuous path	Remote pendant or off-line	Arc/spot welding, mchn load, mtrl handling	+/- .020"
Micro-processor two models	x		Point-to-point & continuous path	Remote pendant or off-line	Assembly, arc/spot welding, mtrl handling	+/- .020"
Micro-processor two models	x x		Point-to-point & continuous path	Remote pendant or off-line	Assembly, arc/ spot welding mtrl handling	+/- .002"
Micro-processor two models	x		Point-to-point & continuous path	Remote pendant or off-line	Mtrl handling, part transfer, palletizing	+/- .02"
Micro-processor	x		Point-to-point & continuous path	Remote pendant or off-line	Mtrl handling, part transfer, palletizing	+/- .04"

Manufacturer	Model	Payload lbs.	Payload Kgs.	Axes No.	Axes Action	Powered By
GCA Corporation	XR 50* series	300	136	3-6	Rectilinear coordinate	Elec.
GCA Corporation	XR 100*	2200	1000	3-6	Rectilinear coordinate	Elec.
General Electric Troy, MI (313) 362-5550	P-50	22	10	5	Jointed-arm spherical coordinate	Elec.
General Electric	GP-110	110	50	6	Jointed-arm spherical coordinate	Elec.
General Electric	GP-155	155	70	6	Jointed-arm spherical coordinate	Elec.
General Electric	GP-220	220	100	6	Jointed-arm spherical coordinate	Elec.
General Electric	A-3	4.4	2	3	Scara type	Elec.
General Electric	A-4	4.4	4	3	Scara type	Elec.
General Electric	A-40	22	10	4	Scara type	Elec.
General Electric	Allegro A-12 (DEA)	14	6.5	5 per arm	Rectilinear coordinate	Elec.
GMF Robotics Troy, MI office (313) 641-4223	A-0	22	10	3-5	Cylindrical coordinate	Elec.

Control Systems	SERVO	NONSERVO	Sequence of Motions Taught	Programming Method	Major Applications	Repeatable Placement Accuracy
Micro-processor CIMROC	x		Point-to-point & continuous path	Remote pendant or off-line	Mtrl handling, part transfer, palletizing	+/- .004"
Micro-processor CIMROC	x		Point-to-point & continuous path	Remote pendant or off-line	Heavy duty handling, arc/spot welding	+/- .008"
Micro-processor	x		Continuous path	Remote pendant	Arc welding, sealant, adhesive, grinding process	+/- .008"
Micro processor	x		Continuous path	Remote pendant	Spot welding & mtrl handling	+/- .02"
Micro processor	x		Continuous path	Remote pendant	Spot welding & mtrl handling	+/- .02"
Micro processor	x		Continuous path	Remote pendant	Spot welding & mtrl handling	+/- .04"
Micro processor	x		Point-to-point or continuous path	Remote pendant	Assembly, component insertion, mtrl handling, sealant/adhesive	+/- .002"
Micro processor	x		Point-to-point or continuous path	Remote pendant	Assembly, component insertion, mtrl handling, sealant/adhesive	+/- .002"
Micro processor	x		Point-to-point or continuous path	Remote pendant	Assembly, component insertion, mtrl handling, sealant/adhesive	+/- .004"
Micro processor	x		Point-to-point	Remote pendant or off-line	Precision assembly, pick & place	+/- .001"
Micro processor	x		Point-to-point	Remote pendant	Mtrl handling, assembly, deburr	+/- .002"

Manufacturer	Model	Payload lbs.	Payload Kgs.	Axes No.	Axes Action	Powered By
GMF Robotics	A-1	66	30	3-5	Cylindrical coordinate	Elec.
GMF Robotics	S-108R	17	8	5	Jointed-arm spherical coordinate	Elec.
GMF Robotics	S-108L (L-1000) (L-2000)	17	8	5	Jointed-arm spherical coordinate	Elec.
GMF Robotics	S360R	132	60	6	Jointed-arm spherical coordinate	Elec.
GMF Robotics	S-360L (L-1000) (L-2000)	132	60	6	Jointed-arm spherical coordinate	Elec.
GMF Robotics	M-00	44	20	5	Cylindrical coordinate	Elec. pneumatic
GMF Robotics	M-0	44	20	4	Cylindrical coordinate	Elec.
GMF Robotics	M-1	103	47	3-5	Cylindrical coordinate	Elec.
GMF Robotics	M-1A	103	47	3-5	Cylindrical coordinate	Elec.
GMF Robotics	M-2	132	60	4	Cylindrical coordinate	Elec.
GMF Robotics	M-3	264	120	3-5	Cylindrical coordinate	Elec.
GMF Robotics	S-110R	22	10	5-6	Jointed-arm spherical coordinate	Elec.
GMF Robotics	A-00	22	10	3-4	Cylindrical coordinate	Elec.

Control Systems	SERVO	NONSERVO	Sequence of Motions Taught	Programming Method	Major Applications	Repeatable Placement Accuracy
Micro processor	x		Point-to-point	Remote pendant	Mtrl handling, assembly	+/- .002"
Micro-processor	x		Point-to-point	Remote pendant	Arc welding, mtrl handling, sealing	+/- .008"
Micro-processor	x		Point-to-point	Remote pendant	Arc welding, mtrl handling, sealing	+/- .008"
Micro processor	x		Point-to-point	Remote pendant	Spot welding, mtrl handling, sealing	+/- .02"
Micro-processor	x		Point-to-point	Remote pendant	Spot welding, mtrl handling sealing	+/- .02"
Micro-processor	x		Point-to-point	Remote pendant	Mchn load/ unload	+/- .012"
Micro-processor	x		Point-to-point	Remote pendant	Mchn load/ unload	+/- .02"
Micro-processor	x		Point-to-point	Remote pendant	Mtrl handling, assembly	+/- .039"
Micro-processor	x		Point-to-point	Remote pendant	Mtrl handling, assembly	+/- .039"
Micro-processor	x		Point-to-point	Remote pendant	Mchn load/ unload, mtrl handling	+/- .039"
Micro-processor	x		Point-to-point	Remote pendant	Mtrl handling mchn load/ unload	+/- .039"
Micro-processor	x		Point-to-point	Remote pendant	Arc welding, mtrl handling & sealing	+/- .008"
Micro-processor	x		Point-to-point	Remote pendant	Assembly mtrl handling	+/- .002"

Manufacturer	Model	Payload lbs.	Payload Kgs.	Axes No.	Axes Action	Powered By
GMF Robotics	S380R	180	81	6	Jointed-arm spherical coordinate	Elec.
GMF Robotics	S380L (L-2000) (L-3000)	180	81	6	Jointed-arm spherical coordinate	Elec.
GMF Robotics	N/C	-	-	7	Rectangular coordinate	Hydraulic
Graco Robotics, Inc. Livonia, MI office (313) 523-6300	O.M. 5000	16	7.2	6	Jointed-arm spherical coordinate	Hydraulic
Graco Robotics, Inc.	O.M. 5	16	7.2	5	Jointed-arm spherical coordinate	Hydraulic
Graco Robotics, Inc.	O.M. 50	16	7.2	5	Jointed-arm spherical coordinate	Hydraulic
Graco Robotics, Inc.	O.M. 500	16	7.2	6	Jointed-arm spherical coordinate	Hydraulic
IBM Southfield, MI office (313) 827-7707	7535*	13.2	6	4	Horizontal jointed-arm cylindrical coordinate	Elec. air
IBM	7545*	22	10	4	Horizontal jointed-arm cylindrical	Elec.
IBM	7540*	55	25	4	Horizontal jointed-arm cylindrical coordinate	Elec. air

Control Systems	SERVO	NONSERVO	Sequence of Motions Taught	Programming Method	Major Applications	Repeatable Placement Accuracy
Micro-processor	x		Point-to-point	Remote pendant	Spot welding, mtrl handling, sealing	+/- .02"
Micro-processor	x		Point-to-point	Remote pendant	Spot welding, mtrl handling, sealing	+/- .02"
Micro-processor	x		Point-to-point	Remote pendant	Painting, inspection	+/- .25"
Micro-processor	x		Continuous path and point-to-point	Manual teach play-back or remote pendant	Spray coating	+/- .125
Micro-processor	x		Continuous path and point-to-point	Manual teach play-back or remote pendant	Spray coating	+/- .125
Micro-processor	x		Continuous path and point-to-point	Manual teach play-back or remote pendant	Spray coating	+/- .125
Micro-processor	x		Continuous path and point-to-point	Manual teach play-back or remote pendant	Spray coating	+/- .125
Micro-processor	2 axes	2 axes	Point-to-point	On-line with IBM PC and off-line	Small assembly and processing	+/- .002"
Micro processor	4 axes		Point-to-point	On line with IBM PC and off-line	Small parts assembly and processing	+/- .002"
Micro processor	2 axes	2 axes	Point-to-point	On line with IBM PC and off-line	Small parts assembly and processing	+/- .002"

Manufacturer	Model	Payload lbs.	Payload Kgs.	Axes No.	Axes Action	Powered By
IBM	7547	55	25	4	Horizontal jointed-arm cylindrical coordinate	Elec. air
IBM	7565	5	2.2	3-6	Rectilinear spherical coordinate	Hydraulic
Intelledex Corvallis, OR (503) 758-4700	405	12	5.4	4	Horizontal jointed-arm cylindrical coordinate	Elec.
Intelledex	605 S & T	5	2.2	6	Horizontal jointed-arm cylindrical coordinate	Elec.
Intelledex	705 S	5	2.2	7	Horizontal jointed-arm cylindrical coordinate	Elec.
KUKA (Germany) Expert Automation Sterling Heights, MI (313) 977-0100	KUKA* IR662/100	220	100	6	Jointed-arm cylindrical coordinate	Elec.
KUKA	KUKA IR 200	132	60	5-7	Horizontal joint-arm cylindrical	Elec.
KUKA	KUKA* IR 160/60	132 (1100 welding force)	60 500	4-6	Jointed-arm cylindrical coordinate	Elec.
KUKA	IR 160/15*	33	15	4-6	Jointed-arm cylindrical coordinate	Elec.
NIKO (Germany) United Technologies Dearborn, MI (313) 593-9616	NIKO 25	5	2.5	5	Jointed-arm spherical coordinate	Elec.
NIKO	NIKO 50	10	5	5	Jointed-arm spherical coordinate	Elec.
NIKO	NIKO 100	44	20	6	Horizontal Jointed-arm spherical coordinate	Elec.
NIKO	NIKO 150	33	15	5	Jointed-arm spherical coordinate	Elec.

Control Systems	SERVO	NONSERVO	Sequence of Motions Taught	Programming Method	Major Applications	Repeatable Placement Accuracy
Micro processor	4 axes		Point-to-point	On line with IBM PC and off-line	Small parts assembly and processing	+/- .002"
IBM computer	3-6		Continuous path	Remote or off-line	Small parts assembly and processing	+/- .005"
Micro-processor	x		Point-to-point	Remote pendant or off-line	Assembly	+/- .001"
Micro-processor	x		Point-to-point	Remote pendant or off-line	Assembly	+/- .001"
Micro-processor	x		Point-to-point	Remote pendant or off-line	Assembly	+/- .001"
Micro-computer	x		Point-to-point and continuous path	Remote pendant or off-line	Spot welding	+/- .04"
Micro-computer	x		Point-to-point	Remote pendant	Spot welding, mchng & handling	+/- .04"
Micro-computer	x		Point-to-point and continuous path	Remote pendant or off-line	Spot welding	+/- .04"
Micro-computer	x		Point-to-point and continuous path	Remote pendant or off-line	Assembly, mtrl handling, arc welding	+/- .04"
Micro-processor	x		Point-to-point and continuous path	Remote pendant	Light assembly, sealant	+/- .01"
Micro-processor	x		Point-to-point and continuous path	Remote pendant	Light assembly, sealant	+/- .01"
Micro-processor	x		Point-to-point and continuous path	Remote pendant	Mtrl handling, sealant applications	+/- .015"
Micro-processor	x		Point-to-point and continuous path	Remote pendant	Mtrl handling, assembly, sealant	+/- .01"

Manufacturer	Model	Payload lbs.	Payload Kgs.	Axes No.	Axes Action	Powered By
NIKO	NIKO 200	44	20	6	Cartesian coordinate	Elec.
NIKO	NIKO 450	132	60	6	Jointed-arm spherical coordinate	Elec. processor
NIKO	NIKO 451	176	80	6	Jointed-arm spherical coordinate	Elec.
NIKO	NIKO 500	176	80	6	Horizontal Jointed-arm spherical coordinate	Elec.
NIKO	NIKO 600*	176	80	6	Cartesian coordinated	Elec.
Nimak (West Germany)	Nike 600	130	59	6	Rectangular coordinate (Gantry)	Elec.
Prab Robots, Inc. Farmington Hills, MI Office (313)478-7722	Prab 4200/* 4200 HD	75 125	34 57	3-5	Spherical coordinate	Hydraulic
Prab Robots, Inc.	Prab 5800/* 5800 HD	50 100	23 46	3-5	Spherical coordinate	Hydraulic
Prab Robots, Inc.	Prab Series E	100	46	3-5	Cylindrical coordinate	Hydraulic
Prab Robots, Inc.	Prab* Series FA	250	114	3-7	Cylindrical coordinate	Hydraulic
Prab Robots, Inc.	Prab Series FB	600	272	3-7	Cylindrical coordinate	Hydraulic
Prab Robots, Inc.	Prab Series FC	2000	909	3-7	Cylindrical coordinate	Hydraulic

Control Systems	SERVO	NONSERVO	Sequence of Motions Taught	Pro-gramming Method	Major Applications	Repeatable Placement Accuracy
Micro-processor	x		Point-to-point and continuous path	Remote pendant	Light assembly, mtrl handling, arc welding, sealant	+/- .01"
Micro-	x		Point-to-point and continuous path	Remote pendant	Spot welding, mtrl handling	+/- .013"
Micro-processor	x		Point-to-point and continuous path	Remote pendant	Spot welding+ mtrl handling	/- .013"
Micro-processor	x		Point-to-point and continuous path	Remote pendant	Mtrl handling, sealant applic-ations, welding applications	+/- .015"
Micro-processor	x		Point-to-point and continuous path	Remote pendant	Spot welding, mtrl handling	+/- .013"
Mini-computer	x		Point-to-point	Remote pendant	Spot welding (will make a continuous path unit for arc welding)	+/- .02"
Pos. stops and memory drum		x	Point-to-point sequence	Set pos. stops and drum memory	Mchn load/ unload, pick & place	+/- .008"
Pos. stops and memory drum		x	Point-to-point	Set pos. stops and drum memory	Mchn load/ unload, pick and place	+/- .008"
Micro-processor	x		Point-to-point	Remote pendant	Mchn load/ unload, pick and place	+/- .03"
Micro-processor	x		Point-to-point	Remote pendant	Mchn load/ unload, pick and place	+/- .05"
Micro-processor	x		Point-to-point	Remote pendant	Mchn load/ unload, pick and place	+/- .05"
Micro-processor	x		Point-to-point	Remote pendant	Mchn load/ unload, pick and place	+/- .08"

Manufacturer	Model	Payload lbs.	Payload Kgs.	Axes No.	Axes Action	Powered By
Prab Robots, Inc.	G05	30-65	15-30	5	Cartesian coordinate	Elec.
Prab Robots, Inc.	G06	30-65	15-30	6	Cartesian coordinate	Elec.
Prab Robots, Inc.	G24	55-130	25-60	4	Cartesian coordinate	Elec.
Prab Robots, Inc.	G26	55-130	25-60	6	Cartesian coordinate	Elec.
Prab Robots, Inc.	G34	140-220	65-100	4	Cartesian coordinate	Elec.
Prab Robots, Inc.	G36	140-220	65-100	6	Cartesian coordinate	Elec.
Reis (Germany) (Represented by: E&E Engineering (313)371-2000	Model 625*	88	40	6	Cylindrical coordinate	Elec.
Reis	Model 650*	110	50	6	Cylindrical coordinate	Elec.
Reis	Model V-15	33	15	6	Jointed-arm coordinate	Elec.
Reis	Model H-15	33	15	6	Cylindrical coordinate	Elec.

Control Systems	SERVO	NONSERVO	Sequence of Motions Taught	Pro-gramming Method	Major Applications	Repeatable Placement Accuracy
Micro-processor	x		Point-to-point or continuous path	Remote pendant or off-line	Arc welding, sealing, deburring, assembly, mtrl handling	+/- .008"
Micro-processor	x		Point-to-point or continuous path	Remote pendant or off-line	Arc welding, sealing, deburring, assembly, mtrl handling	+/- .008"
Micro-processor	x		Point-to-point or continuous path	Remote pendant or off-line	Mchn load, palletizing, welding, assembly, deburring	+/- .015"
Micro-processor	x		Point-to-point or continuous path	Remote pendant or off-line	Mchn load, palletizing, welding, assembly, deburring	+/- .015"
Micro-processor	x		Point-to-point or continuous path	Remote pendant or off-line	Spot welding, palletizing, mtrl handling, mtrl process-ing	+/- .025"
Micro-processor	x		Point-to-point or continuous path	Remote pendant or off-line	Spot welding, palletizing, mtrl handling, mtrl process-ing	+/- .025"
Micro-computer	x		Point-to-point	Remote pendant	Mchn load/unload, grind-ing, stack-ing, palletiz-ing	+/- .015"
Micro-computer	x		Point-to-point	Remote pendant	Mchn load/unload, grind-ing, stack-ing, palletiz-ing	+/- .015"
Micro-computer	x		Point-to-point and continuous path	Remote pendant	Assembly gluing	+/- .004"
Micro-computer	x		Point-to-point and continous path	Remote pendant	Mchn load/unload, grind ing, stack-ing, palletiz-ing, assembly gluing	+/- .004"

Manufacturer	Model	Payload		Axes		Powered By
		Lbs.	Kgs.	No.	Action	
Rimrock Corp. Columbus, Ohio (614)471-5926 (Was Auto Place, Copperweld)	CR-10	10	4.5	4	Cylindrical coordinate	Pneu. Z-80
Rimrock Corp.	CR-50	20	9	4	Cylindrical coordinate	Pneu.
Robotek (Fraser Automation) (313)979-4500	Model FR-200-3	150	68	3	Cylindrical coordinate	Elec.
Robotek	Model FR-200-4	66	30	4	Cylindrical coordinate	Elec.
Robotek	Model FR-300-3	66	30	4	Cylindrical coordinate	Elec.
Robotek	Model FR-300-4	66	30	3	Cylindrical coordinate	Elec.
Robotek	Model FR-400-3 to 5	35	1	3-5	Jointed-arm cylindrical coordinate	Elec.
Robotek	Model FR-500	35	16	3-5	Rectilinear coordinate	Elec.
Robotics, Inc Detroit, MI office	NC 54.12* Gantry	50	23	3-5	Rectilinear coordinate	Elec.
Schrader-Bellows Clawson, MI office (313)522-4411	Motion* Mate	5	2.5	5	Cylindrical coordinate	Air
Thermwood (Represented by: Roethel Engineering Farmington Hills, MI (313) 478-7950	Paintmiser	18	8	6	Jointed-arm spherical coordinate	Hydraulic
Thermwood	Taskhandler	25	12	5	Jointed-arm spherical coordinate	Hydraulic
Thermwood	Cartesian 5 Gantry System			5	Cartesian	Elec.

Control Systems	SERVO	NONSERVO	Sequence of Motions Taught	Pro-gramming Method	Major Applications	Repeatable Placement Accuracy
Proximity switch micro-processor		x	Point-to-point	Remote pendant	Mchn load/ unload, pick and place	+/- .003"
Proximity switch micro-processor Z-80		x	Point-to-point	Remote pendant	Mchn load/ unload, pick and place	+/- .003"
Micro-processor	x		Continuous path	Remote pendant or off-line	Pick and place	+/- .016"
Micro-processor	x		Continuous path	Remote pendant or off-line	Pick and place	+/- .016"
Micro-processor	x		Continuous path	Remote pendant or off-line	Pick and place	+/- .016"
Micro-processor	x		Continuous path	Remote pendant or off-line	Pick and place	+/- .016"
Micro-processor	x		Continuous path	Remote pendant or off-line	Pick and place	+/- .003"
Micro-processor	x		Continuous path	Remote pendant or off-line	Pick and place	+/- .003"
Micro-processor	x		Continuous path	Remote pendant	Dispensing adhesives or bonding	+/- .01"
Micro-processor		x	Point-to-point	Manual teach playback	Pick and place	+/- .005"
Micro-processor hard disc option	x		Continuous path	Lead through teach edit-box	Painting, sealing, coating	+/- .125"
Micro-processor hard disc option	x		Continous path	Lead through teach edit-box	Machine load, palletizing, pick & place	+/- .125"
Micro-processor	x		Continous path	Pendant taught CAD/CAM option	Rooting drilling wet-jet adhesives	+/- .010"

Manufacturer	Model	Payload Lbs.	Payload Kgs.	Axes No.	Axes Action	Powered By
Thermwood	Cartesian 5 Electric Robot TC Series	10-150	45-68	3-4	Cartesian	Elec.
	LS Series	10-25	4.5-11.3	3-4	Cartesian	Elec.
Trallfa (Norway) US Representative: DeVilbiss Toledo, OH (419)470-2021	Trallfa 450	30	13.6	5	Jointed-arm spherical coordinate	Hydraulic
Trallfa	Trallfa 3500	30	13.6	5	Jointed-arm spherical coordinate	Hydraulic
Trallfa	Trallfa 4500	30	13.6	5	Jointed-arm spherical coordinate	Hydraulic
Unimation/Westinghouse Danbury, CT Farmington Hills, MI office (313)478-7780	Unimate* 1000A	50	22.9	3-5	Spherical coordinate	Hydraulic air
Unimation	Unimate* 2000B and 2100	300	136	3-6	Spherical coordinate	Hydraulic
Unimation	Unimate* 2571BB 2671BB	225	136	5-6	Spherical coordinate	Hydraulic
Unimation	Unimate* 4000A 4571BA 4671BA	450	205	5-6	Spherical coordinate	Hydraulic
Unimation	PUMA 260	2	0.9	6	Jointed-arm revolute	Elec.

Control Systems	SERVO	NONSERVO	Sequence of Motions Taught	Programming Method	Major Applications	Repeatable Placement Accuracy
Micro-processor	x		Continuous path	Pendant taught CAD/CAM option	Assembly, deburr load/unload	+/- .020"
Micro-	x		Point-to-point	Pendant	Pick & place	+/- .005"
Mini-computer (30 programs)	x		Continuous path	Manual teach playback	Coating, adhesives, spray painting	+/- .08"
Mini-computer (30 programs)	x		Continuous path	Manual teach playback	Coating, adhesives, spray painting	+/- .08"
Mini-computer (999 programs)	x		Continuous path	Manual teach playback	Coating, adhesives, spray painting	+/- .08"
Special purpose computer	x	x	Point-to-point	Remote pendant	Mchn load/unload, pick & place, die casting/injection molding	+/- .025"
Special purpose computer	x		Point-to-point	Remote pendant	Mchn load/unload, spot/arc welding, mtrl handling	+/- .025" +/- .04"
Mini-computer Val	x		Joint straight line/circular interpolated coordinated axis	Remote pendant or off-line	Mchn load/unload, spot/arc welding, mtrl handling	+/- .025"
Special purpose Computer/Val	x		Point-to-point or joint straight line/circular interpolated coordinated axis	Remote pendant or off-line	Mchn load/unload, spot welding	+/- .04"
Mini-Computer Val I Val II	x		Joint, straight line/circular interpolated coordinated axis	Remote pendant or off-line	Small parts handling & assembly	+/- .02"

Manufacturer	Model	Payload lbs.	Payload Kgs.	Axes No.	Axes Action	Powered By
Unimation	PUMA 550*/ 560	5.5	2.5	5-6	Jointed-arm revolute	Elect.
Unimation	PUMA 760*	22	10	6	Jointed-arm revolute	Elect.
Unimation	100	10	5	4	Jointed-arm	Elec.
Unimation	Dual Arm	100	45	9	Modified polar	Hydraulic
Unimation (Westinghouse)	6000	100-300 300	45-135	5-3	Rectilinear coordinate	Elec.
Unimation (Westinghouse)	7000	44	20	5	Rectilinear coordinate	Elec.
Yaskawa (Japan) (Represented by: Hobart Brothers)	Motoman* L-10W	22	10	5-6	Jointed-arm spherical coordinate	Elec.
Yaskawa	Motoman L-3	6.6	3	5	Jointed-arm spherical coordinate	Elec.
United States Robots Farmington Hills, MI Office (31)476-1884	Maker* 100	5	2.5	5	Jointed-arm spherical coordinate	Elec.
VSI Automation Troy, MI (313)588-1255	Charley 2*	13.2	6	4	Horizontal Jointed-arm cylindrical coordinate	Elec.
VSI Automation	Charley 4*	22	10	4	Horizontal Jointed-arm cylindrical coordinate	Elec.
VSI Automation	Charley 6*	66	30	4	Horizontal Jointed-arm cylindrical coordinate	Elec.

Control Systems	SERVO	NONSERVO	Sequence of Motions Taught	Programming Method	Major Applications	Repeatable Placement Accuracy
Mini-computer Val I Val II	x		Joint, straight line/circular interpolated coordinated axis	Remote pendant or off-line	Assembly, parts handling, arc welding, adhesive dispensing, inspection	+/- .004"
Mini-computer Val I Val II	x		Joint, straight line/circular interpolated coordinated axis	Remote pendant or off-line	Arc welding, adhesive dispensing, inspection	+/- .008"
PC	x		Point-to-point	Remote pendant	Assembly	+/- .004"
Special purpose computer	x		Point-to-point	Remote pendant	Mtrl handling	+/- .04"
Mini-computer Val II or CNC	x		Coordinated axis	Remote pendant or off-line	Arc welding, mtrl handling and burning	+/- .005"
Micro-computer	x		Continuous path	Remote pendant or off-line	Adaptive control arc welding	+/- .016"
Micro-computer	x		Point-to-point	Remote pendant & off-line	Arc welding, sealer	+/- .008"
Computer	x		Point-to-point	Manual teach playback	Arc welding and light handling	+/- .004"
Micro-processor based	x		Point-to-point continuous path	Remote pendant	Assembly, small parts and mchn load	+/- .002"
Micro-processor	x		Point-to-point	Remote pendant	Assembly	+/- .002"
Micro-processor	x		Point-to-point	Remote pendant	Assembly	+/- .002"
Micro-processor	x		Point-to-point	Remote pendant	Assembly	+/- .002"

APPENDIX F

Formulas and Conversion Factors

The data contained in this section are provided for reference only.* Many formulas are for estimating purposes only because they cannot consider all factors in every machine application. Many formulas can assist the reader by demonstrating basic physical or electrical relationships needed to understand a more abstract concept in control or motor operation. Other data, such as conversion factors, are included for your convenience to provide a more comprehensive resource when working in an international design environment.

NOTE:
The following equations for calculating horsepower are meant to be used for estimating purposes only. These equations do not include any allowance for machine friction, windage, or other factors. These factors must be considered when selecting a drive for a machine application.

HORSEPOWER FORMULAS

Rotating Objects

$$\text{hp} = \frac{T \times N}{63{,}000}$$

where T = torque (lb – in.)
N = speed(rpm)

$$\text{hp} = \frac{T \times N}{5252}$$

where T = torque (lb – ft)
N = speed(rpm)

Objects in Linear Motion

$$\text{hp} = \frac{F \times V}{33{,}000}$$

where F = force (lb)
V = velocity(ft/min)

$$\text{hp} = \frac{F \times V}{396{,}000}$$

where F = force(lb)
V = velocity(in./min)

Pumps

$$\text{hp} = \frac{\text{gpm} \times \text{head} \times \text{specific gravity}}{3960 \times \text{efficiency of pump}}$$

$$\text{hp} = \frac{\text{gpm} \times \text{psi} \times \text{specific gravity}}{1713 \times \text{efficiency of pump}}$$

where gpm = gallons per minute
head = height of water (ft)
efficiency of pump = %/100
psi = pounds per square inch

Specific gravity of water = 1.0
1 cu ft per sec = 448 gpm
1 psi = a head of 2.309 ft — for water weighing 62.36 lb per cu ft at 62°F

*This information on formulas and tables is courtesy of Allen-Bradley Co.

Fans and Blowers

$$hp = \frac{cfm \times psf}{33{,}000 \times \text{efficiency of fan}}$$

$$hp = \frac{cfm \times piw}{6356 \times \text{efficiency of fan}}$$

$$hp = \frac{cfm \times psi}{229 \times \text{efficiency of fan}}$$

where cfm = cubic feet per minute
psf = pounds per square foot
piw = inches of water gage
psi = pounds per square inch
efficiency of fan = %/100

Conveyors

$$\text{hp (vertical)} = \frac{F \times V}{33{,}000}$$

$$\text{hp (horizontal)} = \frac{F \times V \times \text{coefficient of friction}}{33{,}000}$$

where F = force (lb)
V = velocity (ft/min)

Coefficient of friction:

Ball or roller slide = 0.02
Dovetail slide = 0.20
Hydrostatic ways = 0.01
Rectangle ways with gib = 0.1–0.25

TORQUE FORMULAS

$$T = \frac{hp \times 5252}{N}$$

where T = torque (lb-ft)
hp = horsepower
N = speed (rpm)

$$T = F \times R$$

where T = torque (lb–ft)
F = force (lb)
R = radius (ft)

$$T\text{(accelerating)} = \frac{\omega k^2 \times \Delta rpm}{308 \times t}$$

where T = torque (lb–ft)
ωk^2 = inertia reflected to the motor shaft (lb – ft^2)
Δrpm = change in speed
t = time to accelerate (sec)

Note:
To change lb-ft^2 to in.-lb-sec^2: Divide by 2.68
To change in.-lb-sec^2 to lb-ft^2: Multiply by 2.68

AC MOTOR FORMULAS

$$\text{sync speed} = \frac{\text{frequency} \times 120}{\text{number of poles}}$$

where sync speed = synchronous speed (rpm)
frequency = frequency (Hz)

$$\%\text{ slip} = \frac{(\text{sync speed} - \text{FL speed}) \times 100}{\text{sync speed}}$$

where FL speed = full-load speed (rpm)
sync speed = synchronous speed (rpm)

$$\text{reflected } \omega k^2 = \frac{\omega k^2 \text{ of load}}{(\text{reduction ratio})^2}$$

where ωk^2 = inertia (lb-ft^2)

ELECTRICAL FORMULAS

Ohm's Law

$$I = \frac{E}{R} \quad R = \frac{E}{I} \quad E = I \times R$$

where I = current (amperes)
E = EMF or voltage (volts)
R = resistance (ohms)

Power in DC Circuits

$$P = I \times E \qquad hp = \frac{I \times E}{746}$$

$$kW = \frac{I \times E}{1000} \qquad kWh = \frac{I \times E \times \text{hours}}{1000}$$

where P = power (watts)
I = current (amperes)
E = EMF or voltage (volts)
kW = kilowatts
kWh = kilowatthours

Power in AC Circuits

$$\text{kVA (one-phase)} = \frac{I \times E}{1000}$$

$$\text{kVA (three-phase)} = \frac{I \times E \times 1.73}{1000}$$

where kVA = kilovolt-amperes
I = current (amperes)
E = EMF or voltage (volts)

$$\text{kW (one-phase)} = \frac{I \times E \times PF}{1000}$$

$$\text{kW (two-phase)} = \frac{I \times E \times PF \times 1.42}{1000}$$

$$\text{kW (three-phase)} = \frac{I \times E \times \text{PF} \times 1.73}{1000}$$

$$\text{PF} = \frac{\text{W}}{\text{V} \times I} = \frac{\text{kW}}{\text{kVA}}$$

where kW = kilowatts
I = current (amperes)
E = EMF or voltage (volts)
PF = power factor
W = watts
V = volts
kVA = kilovolt-amperes

Calculating Motor Amperes

$$\text{motor amperes} = \frac{\text{hp} \times 746}{E \times 1.732 \times \text{Eff} \times \text{PF}}$$

$$\text{motor amperes} = \frac{\text{kVA} \times 1000}{1.73 \times E}$$

$$\text{motor amperes} = \frac{\text{kW} \times 1000}{1.73 \times E \times \text{PF}}$$

where hp = horsepower
E = EMF or voltage (volts)
Eff = efficiency of motor (%/100)
kVA = kilovolt-amperes
kW = kilowatts
PF = power factor

Calculating AC Motor Locked-Rotor Amperes

$$\text{LRA} = \frac{\text{hp} \times \left(\frac{\text{start kVA}}{\text{hp}}\right) \times 1000}{E \times 1.73}$$

where LRA = locked-rotor amperes
hp = horsepower
kVA = kilovolt-amperes
E = voltage (volts)

$$\text{LRA at freq. } X = \frac{\text{60-Hz LRA}}{\sqrt{\frac{60}{\text{freq. } X}}}$$

where 60 Hz LRA = locked-rotor amperes
freq. X = desired frequency (Hz)

ER FORMULAS

Calculating Accelerating Force for Linear Motion

$$F\text{(acceleration)} = \frac{W \times \Delta V}{1933 \times t}$$

where F = force (lb)
W = weight (lb)
ΔV = change of velocity (fpm)
t = time to accelerate weight (sec)

Calculating Minimum Accelerating Time of a Drive

$$t = \frac{\omega k^2 \times \Delta N}{308 \times T}$$

where t = Time required to accelerate load (sec)
ωk^2 = total inertia that the motor must accelerate (lb-ft^2; includes motor rotor, gear reducer, and load
ΔN = change in speed required (rpm)
T = accelerating torque (lb-ft)

Note:
To change lb-ft^2 to in.-lb-sec^2: Divide by 2.68
To change in.-lb-sec^2 to lb-ft^2: Multiply by 2.68

$$\text{rpm} = \frac{\text{fpm}}{0.262 \times D}$$

where rpm = revolutions per minute
fpm = feet per minute
D = diameter (ft)

$$\omega k^2 \text{ reflected to motor} = \text{load } \omega k^2 \times \left(\frac{\text{load rpm}}{\text{motor rpm}}\right)^2$$

where ωk^2 = inertia (lb-ft^2)
rpm = revolutions per minute

ENGINEERING CONSTANTS

Temperature

0°C = freezing point of water
32°F = freezing point of water = 0°C
100°C = boiling point of water at atmospheric pressure
212°F = boiling point of water at atmospheric pressure
1.8°F change = 1°C
0.252 kilocalorie = 1 Btu
−270°C = absolute zero
−459.6°F = absolute zero

Length

1760 yd = 1 mile
25.4 mm = 2.54 cm = 1 in.
3 ft = 1 yd
3.2808 ft = 1 m
39.37 in. = 1 m = 100 cm = 1000 mm
5280 ft = 1 mile
0.62137 mile = 1 km

Weight

16 oz = 1 lb
2.2046 lb = 1 kg
2.309 ft water at 62°F = 1 psi
28.35 g = 1 oz
59.76 lb = weight of 1 cu ft of water at 212°F
0.062428 lb per cu ft = 1 kg/cu m
62.355 lb = weight of 1 cu ft water at 62°F
8⅓ (8.32675) lb = weight 1 gal water at 62°F

Power

1.3410 hp = 1 kW
2.545 Btu per hr = 1 hp
33,000 ft-lb per min = 1 hp
550-ft-lb per sec = 1 hp
745.7 W = 1 hp

Area

10.764 sq ft = 1m²
1,273,239 circular mils = 1 sq in.
144 sq in. = 1 sq ft
645 mm² = 1 sq in.
9 sq ft = 1 sq yd
0.0929 m² = 1 sq ft

Mathematic

1.4142 = square root of 2
1.7321 = square root of 3
3.1416 = π = ratio of circumference of circle to diameter = ratio of area of a circle to square of radius
57.296 degrees = 1 rad (angle)
0.7854 × diameter squared = area of a circle

Pressure

14.223 psi = 1 kg per cm² = 1 "metric atmosphere"
2.0355 in. Hg at 32°F = 1 psi
2.0416 in. Hg at 62°F = 1 psi
2116.3 psf = atmospheric pressure
27.71 in. water at 62°F = 1 psi
29.921 in. Hg at 32°F = atmospheric pressure
30 in. Hg at 62°F = atmospheric pressure (approximate)
33.974 ft water at 62°F = atmospheric pressure
0.433 psi = 1 ft of water at 62°F
5196 psf = 1 in. water at 62°F
760 mm Hg = atmospheric pressure at 0°C
0.07608 lb = weight 1 cu ft air at 62°F and 14.7 psi

Volume

1728 cu in. = 1 cu ft
231 cu in = 1 gal (U.S.)
277.274 cu in = 1 gal (British)
27 cu ft = 1 cu yd
31 gal (31.5 U.S. gal) = 1 barrel
35.314 cu ft = 1 m³
3.785 liters = 1 gal
61.023 cu in. = 1 liter
7.4805 gal = 1 cu ft

Units of Pressure

kg per cm² = kilograms per square centimeter
Hg = symbol for mercury
psi = pounds per square inch
psf = pounds per square foot

Units of Volume

cu in. = cubic inch
gal = gallon
cu ft = cubic feet
ml = milliliter
fl oz = fluid ounce (U.S.)

CONVERSION FACTORS

Length

To Convert:	To:	Multiply by:
meters	feet	3.281
meters	inches	39.37
inches	meters	0.0254
feet	meters	0.3048
millimeters	inches	0.0394
inches	millimeters	25.4
threads/inch	millimeter pitch	Divide into 25.4
yards	meters	0.914

Example: 10 m × 3.281 = 32.81 ft

Area

To Convert:	To:	Multiply by:
circular mil	meter²	0.50×10^{-9}
yard²	meter²	0.8361

Example: 100 circular mils × 0.5×10^{-9} = 0.5×10^{-7} m² (0.5×10^{-7} circular mil = 0.00000005 m²)

Power

To Convert:	To:	Multiply by:
watts	hp	0.00134
ft-lb/min	hp	0.0000303
hp	kW	0.746

Example: 1500 W × 0.00134 = 2.01 hp

Rotation/Rate

To Convert:	To:	Multiply by:
rpm	deg/sec	6.00
rpm	rad/sec	0.1047
deg/sec	rpm	0.1667
rad/sec	rpm	9.549
fpm	m/s	0.00508
fps	m/s	0.3048
gal/min	cm^3/s	63.09
in./sec	m/s	0.0254
km/h	m/s	0.2778
mph	m/s	0.447
mph	km/h	1.609
rpm	rad/s	0.1047
yd^3/min	m^3/s	0.01274

Example: 1800 rpm × 6.00 = 10800 deg/sec

Moment of Inertia

To Convert:	To:	Multiply by:
$newton\text{-}meters^2$	$lb\text{-}ft^2$	2.42
$oz\text{-}in.^2$	$lb\text{-}ft^2$	0.000434
$lb\text{-}in.^2$	$lb\text{-}ft^2$	0.00694
$slug\text{-}ft^2$	$lb\text{-}ft^2$	32.17
$oz\text{-}in.\text{-}sec^2$	$lb\text{-}ft^2$	0.1675
$in.\text{-}lb\text{-}sec^2$	$lb\text{-}ft^2$	2.68

Example: 25 newton-meters^2 × 2.42 = 60.5 lb-ft^2

Mass/Weight

To Convert:	To:	Multiply by:
oz	g	31.1
kg	lb	2.205
lb	kg	0.4536
newtons	lb	0.2248

Example: 50 oz × 31.1 = 1555 g

Torque

To Convert:	To:	Multiply by:
newton-meters	lb-ft	0.7376
lb-ft	newton-meters	1.3558
lb-in.	lb-ft	0.0833
lb-ft	lb-in.	12.00

Example: 30 Newton-Meters × 0.7376 = 22.13 lb-ft

Volume

To Convert:	To:	Multiply by:
cm^3 (ml)	m^3	0.000001
fl oz	cm^3	29.57
ft^3 of water (39.2°F)	kg (or liter)	28.32
cfm	m^3/s	0.000472
liters	m^3	0.001
yd^3	m^3	0.7646

Example: 250 cm^3 × .000001 = .00025 m^3

Temperature

To Convert:	To:	Use the Formula:
°F	°C	$°C = \frac{°F - 32}{1.8}$
°C	°F	°F = (°C × 1.8) + 32

Example: $68°F = \frac{68 - 32}{1.8} = 20°C$

$$20°C = (20 \times 1.8) + 32 = 68°F$$

Fractional Inch to Equivalent Millimeters and Decimals

Inch	Equivalent mm	Equivalent Decimal	Inch	Equivalent mm	Equivalent Decimal	Inch	Equivalent mm	Equivalent Decimal	Inch	Equivalent mm	Equivalent Decimal
$\frac{1}{64}$	0.3969	0.0156	$\frac{17}{64}$	6.7469	0.2656	$\frac{33}{64}$	13.0969	0.5156	$\frac{49}{64}$	19.4469	0.7656
$\frac{1}{32}$	0.7938	0.0313	$\frac{9}{32}$	7.1438	0.2813	$\frac{17}{32}$	13.4938	0.5313	$\frac{25}{32}$	19.8438	0.7813
$\frac{3}{64}$	1.1906	0.0469	$\frac{19}{64}$	7.5406	0.2969	$\frac{35}{64}$	13.8906	0.5469	$\frac{51}{64}$	20.2406	0.7969
$\frac{1}{16}$	1.5875	0.0625	$\frac{5}{16}$	7.9375	0.3125	$\frac{9}{16}$	14.2875	0.5625	$\frac{13}{16}$	20.6375	0.8125
$\frac{5}{64}$	1.9844	0.0781	$\frac{21}{64}$	8.3344	0.3181	$\frac{37}{64}$	14.6844	0.5781	$\frac{53}{64}$	21.0344	0.8281
$\frac{3}{32}$	2.3813	0.0938	$\frac{11}{32}$	8.7313	0.3438	$\frac{19}{32}$	15.0813	0.5938	$\frac{27}{32}$	21.4313	0.8438

Inch	Equivalent mm	Equivalent Decimal	Inch	Equivalent mm	Equivalent Decimal	Inch	Equivalent mm	Equivalent Decimal	Inch	Equivalent mm	Equivalent Decimal
7/64	2.7781	0.1094	23/64	9.1281	0.3594	39/64	15.4781	0.6094	55/64	21.8281	0.8594
1/8	3.1750	0.1250	3/8	9.5250	0.3750	5/8	15.8750	0.6250	7/8	22.2250	0.8750
9/64	3.5719	0.1406	25/64	9.9219	0.3906	41/64	16.2719	0.6406	57/64	22.6219	0.8906
5/32	3.9688	0.1563	13/32	10.3188	0.4063	21/32	16.6688	0.6563	29/32	23.0188	0.9063
11/64	4.3656	0.1719	27/64	10.7156	0.4219	43/64	17.0656	0.6719	59/64	23.4156	0.9219
3/16	4.7625	0.1875	7/16	11.1125	0.4375	11/16	17.4625	0.6875	15/16	23.8125	0.9375
13/64	5.1594	0.2031	29/64	11.5094	0.4531	45/64	17.8594	0.7031	61/64	24.2094	0.9531
7/32	5.5563	0.2188	15/32	11.9063	0.4688	23/32	18.2563	0.7188	31/32	24.6063	0.9688
15/64	5.9531	0.2344	31/64	12.3031	0.4844	47/64	18.6531	0.7344	63/64	25.0031	0.9844
1/4	6.3500	0.2500	1/2	12.700	0.5000	3/4	19.0500	0.7500			

Inertia of a Solid Steel Shaft (lb-ft^2 per inch of length)

Diameter (in.)	ωk^2	Diameter (in.)	ωk^2	Diameter (in.)	ωk^2	Diameter (in.)	ωk^2	Diameter (in.)	ωk^2	Diameter (in.)	ωk^2
3/4	0.00006	8	0.791	14 1/4	7.97	37	360.70	62	2844.3	87	11,028
1	0.00007	8 1/4	0.895	14 1/2	8.54	38	401.30	63	3032.3	88	11,544
1 1/4	0.0005	8 1/2	1.000	14 3/4	9.15	39	445.30	64	3229.5	89	12,077
1 1/2	0.001	8 3/4	1.130	15	9.75	40	492.78	65	3436.1	90	12,629
1 3/4	0.002	9	1.270	16	12.59	41	543.90	66	3652.5	91	13,200
2	0.003	9 1/4	1.410	17	16.04	42	598.80	67	3879.0	92	13,790
2 1/4	0.005	9 1/2	1.550	18	20.16	43	658.10	68	4115.7	93	14,399
2 1/2	0.008	9 3/4	1.750	19	25.03	44	721.40	69	4363.2	94	15,029
2 3/4	0.011	10	1.930	20	30.72	45	789.30	70	4621.7	95	15,679
3	0.016	10 1/4	2.130	21	37.35	46	861.80	71	4891.5	96	16,349
3 1/2	0.029	10 1/2	2.350	22	44.99	47	939.30	72	5172	97	17,041
3 3/4	0.038	10 3/4	2.580	23	53.74	48	1021.80	73	5466	98	17,755
4	0.049	11	2.830	24	63.71	49	1109.60	74	5774	99	18,490
4 1/4	0.063	11 1/4	3.090	25	75.02	50	1203.07	75	6090	100	19,249
4 1/2	0.079	11 1/2	3.380	26	87.76	51	1302.2	76	6422		
5	0.120	11 3/4	3.680	27	102.06	52	1407.4	77	6767	110	28,183
5 1/2	0.177	12	4.000	28	118.04	53	1518.8	78	7125	120	39,914
6	0.250	12 1/4	4.350	29	135.83	54	1636.7	79	7498	130	54,978
6 1/4	0.296	12 1/2	4.720	30	155.55	55	1761.4	80	7885	140	73,948
6 1/2	0.345	12 3/4	5.110	31	177.77	56	1893.1	81	8286	150	97,449
6 3/4	0.402	13	5.58	32	201.80	57	2031.9	82	8703	160	126,152
7	0.464	13 1/4	5.96	33	228.20	58	2178.3	83	9135	170	160,772
7 1/4	0.535	13 1/2	6.42	34	257.20	59	2332.5	84	9584	180	202,071
7 1/2	0.611	13 3/4	6.91	35	386.80	60	2494.7	85	10,048	190	250,858
7 3/4	0.699	14	7.42	36	323.20	61	2665.2	86	10,529	200	307,988

Glossary

accuracy Degree to which a robot can lace an object in a given spot repeatedly.

actuator Motor, cylinder, or other mechanism used to power a robot's ability to extend and retract, swing or rotate, and elevate its arm.

artificial intelligence (AI) Ability of a robot to perform functions that are usually possessed by humans, such as reasoning, classifying, self-correction, and improvement.

ASCII Code used to transfer information from a keyboard to a processor, MPU, or CPU.

assemble To put together.

asynchronous Not occurring at the same time.

backlash Looseness in the gears of an assembly where they mesh.

ball screw Method of using ball bearings to substitute for screw threads (the ball screw changes rotary motion to linear motion).

brushless motor DC motor that operates without brushes (an electronic circuit controls its field excitation).

Cartesian coordinates Simplest of the coordinate systems because it refers to up-and-down, back-and-forth, and in-and-out movement of a robotic arm.

chip An integrated circuit that contains many transistors, diodes, and resistors to make up various electronic circuits on a small (about 8 mm^2) piece of silicon; device that stores the computer's program and acts as a memory.

CIM Computer-integrated manufacturing; one of a number of proposed organizational methods for manufacturing products in large quantities.

collision avoidance Ability of a robot to avoid colliding with the part it is supposed to pick up.

contact sensor A sensor that detects the presence of an object by actually making contact with it.

controller Unit needed to control a robot.

coordinates Points and planes in reference to the movement capability of a robotic arm.

CPU Central processing unit.

dead zone Safety zone where a robotic arm does not move during normal operation.

degree of freedom Ability of a robot to move in six axes.

die casting Use of dies to form hot metal into desired shapes.

end effector Device mounted on the end of a manipulator or robotic arm.

fabricate To make something.

feedback Ability of a device to feed a signal back to its controller to aid in keeping track of the position of the manipulator or gripper.

flow-line transfer Robots designed to pick up two or more pieces at a time and transfer them off a machining line onto a second transfer line located parallel to the first one.

gripper Located on the end of a manipulator arm and used to pick things up.

hard automation The use of a conventional assembly-line method of producing a manufactured produce with dedicated equipment.

harmonic drive Type of drive that uses a flex spline, circular spline, and wave generator to accurately position a manipulator with no backlash and little noise.

hydraulics Use of pressure on a fluid to drive an end effector or a manipulator.

interface Proper connections between a robot and its programmed computer or microprocessor.

interfacing Matching up a device with a computer or microprocessor so that they operate as one unit.

ladder diagram Drawing of the electric circuit of the sequence of switches needed to cause a robot to perform programmed activities.

lane loader Pick-and-place robot used to adjust the feed between fast and slow or slow and fast lines.

language Method of speaking to a robot (languages used are VAL, Al, AML, Pascal, and ADA).

lead-through Leading the continuous-path robot manipulator through a sequence of motions to accomplished a task.

LED Light-emitting diode.

LERT Classification system for robots based on four basic motion capabilities.

limit switch A switch designed to be used with a moving body that should not go past a given point.

line tracking Having a robot keep up and move along with a production line, doing its work as it moves with the line.

machine vision system A system that allows a robot to recognize and verify parts.

manipulator One of the three basic parts of a robot.

microcomputer A small-scale computer.

microprocessor Electronic device consisting of integrated circuit (IC) chips that have memory and the ability to gather signals to control a robot.

MIG Type of metal-in-gas (MIG) metal welding in which an inert gas surrounds the welding spot or area while it is in the molten state.

minicomputer Midsized computer, slightly larger than a microcomputer and smaller than a mainframe.

MPU Microprocessor unit.

MVS Machine vision system; method of giving a robot the ability to see.

object recognition Ability of a robot to recognize certain shapes and choose the right one for processing.

off-line Refers to programming of a robot by using of a computer and then placing the program in the robot's controller for action.

palletizing Task in which a robot stacks parts or boxes on a pallet.

parallel port Method for connecting computers and peripheral devices so that they can share data (uses eight or more wires).

plugging Method of stopping an electric motor by reversing the polarity of its power source.

pneumatic drive Using air pressure to drive a manipulator.

positioning Ability of a robot to place an object in a desired location.

power supply Supplies power to a robot.

process flow Orderly flow of parts and materials to keep production going.

program sequence Sequence of commands instructing a robot to perform some task.

programmable robot Robot that can be programmed or taught with a teach box, keyboard, or another input device.

programmer Person who teaches a robot; person who can communicate with a robot in its language.

proximity sensor Device used to detect how close an object is.

pumps Devices, usually electrically driven, used to increase the pressure on a hydraulic fluid.

range sensor Device used to detect the precise distance between an object and a gripper or manipulator.

relay Electromechanical device that is energized and closes or opens switches (it has a coil that can be magnetized by the presence of an electric current).

repeatability Ability of a robot to place an object in the same spot repeatedly.

robot A system that simulates human activities by means of computer instruction.

roller chain Same type of chain used in bicycles; used to drive manipulators and end effectors.

RS-232C Standard A standard that uses –3 to –25 volts for logic 1 and +3 to +25 volts for logic 0.

sensor Device used to detect changes in temperature, light, pressure, sound, and other functions needed to make a robot aware of various conditions.

serial port Method for connecting computers and peripheral devices so that they can share data over distances of 50 feet or more (uses two wires).

servo motors Motors driven by signals rather than by straight power-line voltage and current; motors whose driving signal is a function of the difference between command position and/or rate and measured actual position and/or rate.

60-mA standard Uses 60 mA for logic 1 and 0 mA for logic 0.

software Information programmed on a floppy disk, hard disk, magnetic tape, or drum.

solenoid Coil of wire with a plunger that can turn on or off a fluid or air line.

spot welding The act of welding a small spot between the electrodes of a spot welder.

stepper motor DC motor whose rotor can be made to turn as little as 1.8 degrees.

strain gauge Device made of thin wire that reacts to stretching of the wire; made of semiconductor materials today.

synchronous belt Type of belt that has teeth that fit a pulley with grooves so that it does not slip.

tactile sensor Device used to detect the presence of an object by touch.

target Point to which an arm is expected to reach for picking up an object.

teach pendant Device used to teach the robot memory a new program.

thermistor Device that changes its resistance in reverse manner from normal (if temperature increases, it lowers its resistance).

thermocouple Union made by two dissimilar metals (when heated, the junction produces a small electric current).

TIG Type of welding (tungsten-in-gas).

transducer Device used to convert mechanical energy to electrical energy.

TTL Standard Transistor-to-transistor logic standard that uses a 5-volt signal for logic 1 and 0 volt for logic 0.

20-mA Standard Uses 20 mA for logic 1 and 0 mA for logic 0.

V-belt Belt shaped to fit a V pulley and used to drive a manipulator.

work envelope Space in which a robotic arm moves during its normal work cycle.

worm gear Changes linear motion to rotary motion or vise versa.

Answers to Review Questions

Chapter 1

1. A robot is a programmable multifunctional manipulator designed to move materials, parts, tools, or specialized devices. A computer is the "brain" of the robot.
2. Robots become a reality in 1921—larger ones from the 1950s to the 1970s.
3. A Czech playwright produced a play with a device called a *robota*. The word was translated into English as *robot*. The 1950s saw large robots produced and introduced to the public.
4. A true robot can be identified by the following: (a) it is programmed by a human to perform human labor, (b) it is a device or system that reacts in a preprogrammed manner, (c) it may be able to react to various conditions in terms of the five human senses, (d) it is a device that can operate without human supervision, and (e) it may make decisions based on sensors.
5. Languages used by robots include VAL, AL, AML, Pascal, and ADA.
6. A microprocessor is a device made of chips that has memory and the ability to control robots.
7. Two positive aspects of computers are that (a) they can work 24/7 without human supervision, and (b) they can improve product quality.
8. Two negative aspects of computers are that (a) robots are expensive, and (b) robots take jobs from humans.
9. Hard automation is the usual human worker and supervisor on the assembly line. Human labor must operate within the limits of local and national laws and the call of nature for a break now and then.
10. Will the work of robots produce better quality and quantity products at a lower cost?

11. Your own personal opinion goes here.
12. (a) Pick and place, (b) load, (c) unload, (d) inspect, (e) compare quality of products.

Chapter 2

1. (a) Industrial robot, (b) laboratory robot, (c) explorer robot, (d) hobbyist robot, (e) classroom robot, and (f) entertainment robot.
2. A manipulator is one of the three major parts of a robot—the arm.
3. Base.
4. Three—bending, yawing, and swiveling.
5. Located at the end of the manipulator arm is the gripper, which is used to pick up items.
6. A device mounted on the end of a robot arm.
7. Linear motion, extension motion, rotating motion, and twisting motion.
8. The motions the arm is capable of, that is, turning or twisting, usually specified in degrees.
9. The point to which the arm is expected to reach for picking up an object.
10. Cartesian coordinates are the simplest of the coordinates because they refer to up and down, back and forth, and in-and-out movements.
11. Cartesian coordinate system, cylindrical system, polar system, and articulate system.
12. The space in which a robot arm moves during a normal work cycle.
13. A teardrop-shaped work envelope uses the articulate coordinate system.
14. Yaw describes the side-to-side movement of the wrist from 9 to 270 degrees.
15. Roll axis refers to movement of end of the wrist, can be up to 360 degrees.
16. Arc welding, materials loading, materials unloading, deburring, and assembly operations.
17. Its ability to lift heavier loads.
18. It cannot lift heavy loads.
19. 50 to 300 pounds.
20. Usually 6.6 to 176 pounds.

Chapter 3

1. (a) Electric, (b) hydraulic, and (c) pneumatic.
2. Filters are needed in pneumatics to remove dirt and moisture from the air.

3. A relief valve allows air to exhaust into the atmosphere when the pressure of the compressed air becomes too high.
4. Hydrostatic and hydraulic.
5. *Pneu* means "air" in Latin.
6. Compressor, storage tank, pressure relief valve, and silencers (see Figure 3-6).
7. PM motors are made with a permanent magnet so that the armature of the rotor rotates through a magnetic field created by the permanent magnet.
8. A stepper motor is used to power educational robots.
9. DC motors work without brushes because they use transistors to reverse the current through the field coils.
10. The wound-rotor induction type and the squirrel-cage type.
11. A squirrel-cage motor is most often used to operate a robot's manipulator.
12. (a) Class A, normal torque, most popular type; (b) Class B, normal torque, low starting currents; (c) Class C, high torque, low starting currents; (d) Class D, very high slip percentages; (e) Class E, low start torque, normal starting currents; (f) Class F, low torque, low starting currents.
13. Slip is a loss of induced current between the stator poles and rotor conductors. It is used where the torque increases to meet the demands of the load.
14. Vacuum grippers are attached to the end of a manipulator, and a vacuum hose is attached to vacuum cups. They are limited by the weight of the load.
15. Magnetic grippers use electromagnets as an end effector, the object being to pick up ferromagnetic loads
16. End-of-arm tooling is also known to have a safety joint attached to the manipulator, although the tool is attached for welding, painting, or spot welding.
17. Positioning is the placing of an object in an exact place over and over.
18. Repeatability is described as the accuracy of a robot placing an object in the same spot repeatedly.
19. Robot accuracy is the degree to which a robot can place an object in a given spot repeatedly.
20. It can provide high-torque transmission and good precision.

Chapter 4

1. A transducer is a device used to convert nonelectrical energy to electrical energy.

2. A limit switch is designed to be used with a moving body that should not go past a given point.
3. Contact and noncontact.
4. A device used by a robot to sense touch, force, temperature, and/or vision. In a robot it is a limit switch that is a contact sensor.
5. A light-emitting diode.
6. An area in which a robot arm or manipulator operates or moves.
7. The ability of a robot to avoid colliding with the part it is supposed to pick up.
8. It is used to detect how close an object is.
9. (a) RC circuits, (b) pulsed infrared photoelectric control, and (c) eddy current detectors.
10. (a) Touch and (b) stress.
11. Pick-and-place robots are the most often used in industry—assembly of cars, pickup and delivery of parts for the assembly line. In other words, it is the most often used type of robot.
12. ACS as a touch sensor is possible because it is electrically conductive only along one axis in the plane of the sheet.
13. (a) Thermocouple and (b) thermistor.
14. LVDT, RVDT encoders, pots, resolvers. Small changes in resistance, capacitance, and inductance are amplified and metered.
15. A simple stain gauge can be made by using a piece of wire and stretching it to check its then elongated resistance.
16. Speed sensing can be used when a tachometer or photo cell is used in a circuit.
17. Torque is measured in inch-pounds, inches per ounce, or newton-meters.
18. MVSs are used for inspection, sorting of parts, and verification of parts. They are needed to improve the quality of the finished job. They can start by verifying and identifying parts. They can be used for inspection, sorting of parts, and making noncontact measurements.

Chapter 5

1. Single- and three-phase electricity for robots in industrial applications.
2. (a) Split phase, (b) capacitor start, and (c) shaded pole.
3. Capacitor start.
4. (a) Electric, (b) pneumatic, and (c) hydraulic.
5. Servo control can move up and down and back and forth.

6. A servo-controlled robot can do more things than a non-servo-controlled robot. Basic difference is feedback.
7. An actuator is a motor or cylinder or some kind of mechanism used to power a robot. Actuators convert one type of energy to another. They may be stepper motors, DC servo motors, or pancake motors.
8. Electrically.
9. (a) Rotating drum, (b) air logic, (c) relay logic, (d) programmable, (e) microprocessor-based, and (f) minicomputer.
10. Because of the rapid development of an inexpensive electronic controller.
11. An electric circuit drawing of the switches needed to cause a robot to perform a programmed activity.
12. Because they are easily programmed.
13. The programmed controller uses electronics for timing and sequencing. The keyboard is used to enter the proper orders or the switch closings.
14. A building-block device with a memory and keyboard. Sometimes it uses a cathode-ray tube to display the contents of memory.
15. A device used to "teach" a robot its program and how to follow it. It is a handheld device with axis indicator, keys, and display.
16. By the lead-through method.
17. Use a computer program to train the robot.
18. Lead-through programming is done by a human doing it first. The robot's computer memorizes the program and its every motion or movement.
19. An integrated circuit is an outgrowth of the development of the transistor back in the 1960s.
20. A proper connections between a computer and a robot or microprocessor.

Chapter 6

1. A controller has input parts for interfacing with various computer controls.
2. Microprocessor unit.
3. The MPU is located in the computer or is the central procession unit's other name.
4. (a) VAL, (b) HELP, (c) AMI, (d) MCL, (e) RPL, and (f) RAIL.
5. Software is information programmed on discs, magnetic tape, or drums.
6. Interfacing is matching up of a device with a computer or microprocessor so that they can operate as one unit.
7. ASCII code, by means of a keyboard, can be used to communicate with a robot's computer or microprocessor. Parallel ports are used for separations of less than 50 feet. A serial port is used for longer distances between units.

8. The TTL standard is compatible with transistor-transistor logic and interfaces directly. RS232C voltage used is –3 and –25 volts. TTL uses 60 and 20 milliamperes for standard currents and relies on voltage signals as a result of converting. TTL relies on logic 0 and 1 conversions.
9. The two standards, 60 and 20 milliamperes, both have a separation of 40 milliamperes. Basically, both standards are about the same.
10. An older term—it was used to describe the synchronizing of code to mode "C" in the responder.
11. A parallel port is used for short separations of less than 50 feet; a serial port is good for longer distances.
12. A serial port uses two wires; a parallel port uses eight wires.
13. A service request deals with the interfacing operation for peripheral components and provides program control for peripherals when the data are transmitted and received at the input ports. A robot request requires signals for operation, input-output signals that come from either the MPU or the controller.
14. A good reason for vision robots is to improve quality in the production of various products. It provides a closer inspection by scanning grayscale signals and converting them to digital form for computer analysis.
15. Grayscale signals from video camera are analyzed and checked against stored templates and other information in the computer.

Chapter 7

1. Work handling must be fast, accurate, smooth, and dependable.
2. Another application of a pick-and-place robot. A robot that can load and unload multiple parts fixtures and balance flow rates between slow and fast.
3. Flow-line transfer enables robots to be programmed to pick up or move two or more pieces at a time and transfer them from a machining line onto a second transfer line located parallel to the first one.
4. Numerically controlled.
5. A conveyor is a device that transfers or moves something along a line or beltway.
6. The value of an object is increased as a result of work done on it.
7. Die casting involves working with hot metal. It can be dangerous for human health.
8. A lot of pressure is used to close a two-piece mold; temperature of the metal is very hot.

9. Act of using a robot to place parts of boxes on a pallet.
10. They are not affected by the paint fumes, sparks, and the brilliance of the arc.
11. (a) Spot, (b) MIG, and (c) TIG.
12. (a) Obtain an economic order quantity approaching one, (b) approach a setup time of zero, (c) obtain family-of-parts programming and production, (d) integrate design and manufacturing, (e) establish inventory integrity and just-in-time parts delivery, (f) establish absolute control of the total process, and (g) maximize efficient use of available workspace.

Chapter 8

1. It is usually mounted on the robot.
2. By a control station or by a handheld teach pendant.
3. A safety fence is used to protect people from being hit by a moving robot arm.
4. The 88-800 robot is make by Binks Manufacturing Company in Franklin Park, IL.
5. A SMART robot is one electrically run that has six axes.
6. SMART robots are used in industrial applications especially in the manufacture of automobiles.
7. The work envelope is the space in which a robot arm moves during its normal work cycle.
8. (a) Push, (b) pull, (c) grip, and (d) grasp and release.
9. A teach pendant is a handheld device used to teach a robot's memory a new program.
10. A spec sheet is a listing of the capabilities and working parts of a particular robot.

Chapter 9

1. Open ground on a three-phase motor causes it to slow down. If an open phase is present in a three-phase system that furnishes light for a building, some lights will get very bright and some will glow very dimly.
2. A ground-fault circuit interrupter can trip if a person is completing the circuit to ground. It detects current in milliamperes and trips the main circuit breaker.

3. Ball bearings on a motor are lubricated to dissipate heat, protect against rust and corrosion, and keep out foreign materials.
4. Motor bearings fail because of certain conditions, such as dirty grease, lack of grease, and foreign-matter contamination.
5. Oil viscosity is important to a motor to ensure lubrication when needed. Viscosity means a resistance to a liquid's flow.
6. Motors should be oiled at least every 6 months. Of course, sealed motors do not need to be oiled.
7. Because of their commutators and brushes. The commutator structure causes the brushes to wear down quickly.
8. Voltage fluctuations, transients, and power outages.
9. Voltage spikes are short-duration voltage impulses in excess of normal voltage. They can be caused by an on-off action of switches, motors, and other devices.
10. Electrical noise is a short-lived increase in voltage. Electrical noise can be produced by generators, radiofrequency transmitters, fluorescent lights, computers, business machines, and electrical devices.
11. Transients are produced by inductive loads. They are short-duration impulses.
12. A growler is used to check for shorts and opens in a squirrel-cage motor.
13. By setting the test instrument on the proper voltage range and connecting it and the capacitor to the live full-line voltage. The meter will indicate whether the capacitor is grounded to the can.
14. No connections have to be made or wiring interrupted. Also, it is safer for the operator.
15. High temperatures cause shorts in the windings due to insulation breakdown.
16. Ambient temperature is that which surrounds a motor in its operating location.
17. Overheating and misuse of motor load capabilities.
18. Yes.
19. Take the meter probes and check the leads on the diode. Then reverse the probes and check the diode. What you will find is a path for current flow in one direction but not in the other. This indicates normal operation. However, if you do not get a reading on the meter in either direction, or if you get a reading in both directions, this means a problem with the diode. Then reverse the probes and check the diode. What you will find is a path for current to flow in one direction but not the other. This indicates normal operation. However, if you get a reading on the meter in either direction or

if you get a reading in both directions, this means a problem with the diode. Reading in one direction of the probes—diode is okay. Reading in both directions—diode is shorted. No reading in either direction—diode is open.

Chapter 10

1. Collaborative robots can work safely with humans.
2. When Amazon bought out Kiva Systems and installed it in their warehouses.
3. Segways are used in police work, vacations, and security areas.
4. Germany.
5. Ninebot is a Chinese maker of robots similar to Segways.
6. Reis Robots were bought by Kuka.
7. Chinese South Rail.
8. 225,000.
9. Universal Robots (Neocortex is a software used in robots).
10. Artificial intelligence is intelligence exhibited by machines (robots). In computer science, ideal intelligent machines are flexible, rational agents that perceive their environment and take actions that maximize their chance of success at some goal.
11. Use Neocortex software that learns from experiences in the physical world instead of relying on programmed instructions.
12. Universal Robots, Inc.

Index